Pollution and the Environment
in Ancient Life and Thought
Edited by Orietta Dora Cordovana
and Gian Franco Chiai

T0139661

GEOGRAPHICA HISTORICA

Begründet von Ernst Kirsten,

herausgegeben von Eckart Olshausen und Vera Sauer

Band 36

Pollution and the Environment in Ancient Life and Thought

Edited by Orietta Dora Cordovana
and Gian Franco Chiai

 Franz Steiner Verlag

Gedruckt mit freundlicher Unterstützung der Fritz Thyssen Stiftung

Bibliografische Information der Deutschen Nationalbibliothek:
Die Deutsche Nationalbibliothek verzeichnet diese Publikation in der Deutschen
Nationalbibliografie; detaillierte bibliografische Daten sind im Internet über
<http://dnb.d-nb.de> abrufbar.

© Franz Steiner Verlag, Stuttgart 2017
Druck: Hubert & Co., Göttingen
Gedruckt auf säurefreiem, alterungsbeständigem Papier.
Printed in Germany.
ISBN 978-3-515-11667-1 (Print)
ISBN 978-3-515-11670-1 (E-Book)

Zum Geleit

Nicht wenige antike Zeugnisse berühren, ja thematisieren geradezu, was man modern als Umweltproblem bezeichnet, so etwa Wasser- und Luftverschmutzung oder Landschaftszerstörung. Ob freilich ein Umweltbewußtsein in dem Sinn vorhanden war, daß große Teile der Bevölkerung ein gemeinsames Verständnis von ‚Umwelt' und ‚Umweltproblematik' hatten, ist nicht offensichtlich. Infolgedessen versuchen die Autoren dieses Bandes, aus verschiedenen Perspektiven antike Einstellungen zu Umweltverschmutzung und Ressourcenausbeutung zu analysieren. Dabei stellen sie sich der Herausforderung, mit dem modernen ökologischen Konzept ‚Umwelt' in methodisch angemessener Weise zu operieren.

Wir freuen uns sehr darüber, daß mit diesem Band die Ökogeschichte, das heißt die Untersuchung der Wechselwirkungen zwischen Mensch und Umwelt in historischer Perspektive, abermals in der Reihe *Geographica Historica* vertreten ist.

Eckart Olshausen und Vera Sauer

Contents

Environment, pollution, and diseases

Pollution and the Environment in Ancient life: Material Evidence

Preface

The research project at the core of this volume was developed during a Conference supported by the Fritz Thyssen Stiftung in Köln and the Excellence Cluster 'TOPOI' and held in Berlin from 16th to 18th October 2014. The volume collects many of the papers presented on that occasion and the publication has been generously funded by the Fritz Thyssen Stiftung.

The editors are extremely grateful to all the participants, who enthusiastically accepted to develop different topics related to this new research theme. A special thank is to Eckart Olshausen and Vera Sauer, who carefully followed the peer review process and all the stages of the publication.

This book is dedicated to the lovely memory of Isabella Andorlini.

December 2016   O. D. C. – G. F. C.

Orietta Dora Cordovana / Gian Franco Chiai

Introduction. The Griffin and the Hunting

1. Anthropocene

The proposition of determining the current geological era 'Anthropocene' is a debated one. Soviet biologists apparently introduced the word for the first time in the 1960s, but the modern scientific community has not yet unanimously accepted it. In common meaning and usage, Anthropocene is characterized by the massive impact of human activities on nature, environment, climate, and ecosystems. The high number of technological conquests constantly affects the planet's natural resources and it is undeniable that alteration of the natural balance is a constant threat. To scientists, it is still difficult to define the beginning of this new geological era and distinguish it from the previous Holocene. Some scholars would fix the start of the Anthropocene after the industrial revolution in the 19th century; others think that the main turning point is marked by the massive industrial production of 'technofossils' (e. g. plastic, aluminiums, concrete, and any inorganic material) arising especially since the mid-20th century. The investigation of the specific climatic, biological, and geochemical marks of the technofossils, and their record in earth's sediments and in ice cores is still work in progress.[1] Nevertheless, we are already submerged in the Anthropocene and the dramatic global climate change of the planet is one of the most tangible pieces of evidence.

Whatever the chemical and physical parameters, as well as the geological features of the Anthropocene might be, there is a major contribution to be made by thinking about this new era. As multifaceted phenomenon of human and scientific relevance, it is part of a process of transformation and development. In other words, it undoubtedly forces us to question our past. It is crucial for us to understand the historical process, which has led *homo sapiens* in his path toward the Anthropocene and how this has been shaped. This is a difficult task, and scholars are aware of being only at the beginning of it. To geologists and natural scientists, the study of the Anthropocene mainly consists of the analysis of internal material structures and elemental evidence of earth's sediments. Specialists in the humanities look at the other side of the coin. It is often a matter of following transformations and discerning ancestral roots of what was originally different, especially in terms of past societies' habits and life-

---

* Paragraphs 1, 3, and 4 of this Introduction are by Orietta D. Cordovana; paragraph 2, 5, and the bibliographical note are by Gian Franco Chiai.

[1] Seminal studies have been conducted by Nobel Prize PAUL CRUTZEN and his research group: ZALASIEWICZ, WILLIAMS, STEFFEN, and CRUTZEN, 2010: 2228–31, DOI: 10.1021/es903118j; STEFFEN, GRINEVALD, CRUTZEN and J. MCNEILL, 2011: 842–67, DOI:10.1098/rsta.2010.0327. See also JAMES 2014: 1–6. The development of the scientific debate can also be followed in *Science* 8 Jan 2016, v. 351, iss. 6269, DOI: 10.1126/science.aad2622 and updates.

styles. The practice of 'recycling', for instance, was very common among ancient societies; it was normal to reuse glass bottles, clay jars, wineskins, and metal containers to preserve liquids and solid food. However, we cannot infer that this habit underlined a conscious idea of environmental protection in the daily life of ordinary people.

2. Common-sense environment

The question whether ancient societies had awareness of environmental problems, such as the pollution of rivers and deforestation, has been heavily disputed. Most of the studies, however, focused on the relation between nature (considered as *kosmos*), religion, and man, as well as on landscape archaeology, in order to reconstruct how man manipulated the environment, for example through urbanization and agriculture. The literature on this topic is scattered; the literary and documentary evidence (e. g. inscriptions), as well as archaeological remains have not yet been systematically collected. One of the main problems consists of projecting our modern understanding of ecological problems to the ancient world. If such an ecological awareness really existed, this probably was not widely diffused, but restricted to the intellectual and upper class within Greek and Roman society. Among many questions one is whether a common-sense understanding of the environment could have really existed in ancient times. Under common-sense environment we mean the shared common knowledge and perception of the environment by ordinary people. This concept also recalls a recent book, which deals with the idea of common-sense geography, which is considered as lower geographical knowledge and is distinguished from professional or higher geography.[2]

In the framework of the literary and epigraphic evidence, we find an impressive number of references concerning the importance of a clean environment. Vitruvius, for example, highlights the central role of clean water and a good ventilated place for human life in the cities. Moreover, the numerous admonitions against pollution-acts concerning public fountains, rivers, and wells suggest the presence of a shared sensibility for a clean environment, which was one of the targets of the authorities. Indeed, in the ancient world the common man knew the importance of a clean environment for a good life. Nobody will of course live in a stinking and dirty place, polluted by rubbish and intoxicated water; many inscriptions against *cacatores*, for example, found in context of private houses in Pompei, are good evidence of this sensitivity. The numerous literary eulogies of a bucolic landscape, which belongs to the world of literary fiction, could also be considered as a source to detect awareness for a clean environment. That many Roman villas have been constructed outside the cities, in the countryside and near the coasts in good ventilated places, is not an accident. This resembles a diffused and shared knowledge (at least within the upper classes) that clean places are good for the health and can help to relax body and soul. The Roman legislation also is rich with norms concerning the prohibition to pollute public streets and generally public spaces. These laws reveal a sensitivity for problems concerning the environmental pollution in order to avoid the rise of diseases among the population.

A critical selection of documentary sources, considered together with the archaeological evidence, can show the presence of a common-sense environment in the Greek and Roman world, whereas the literary evidence reflects rather the thought of the intellectual

---

[2] See GEUS and THIERING 2014.

class about destruction and pollution of the nature, as well as about the vantages of living in a clean environment.

3. The griffin and the hunting

To question ancient societies and cultures about current problems and present concerns, such as environmental protection and impact, is, therefore, part of our continuous interaction with and understanding of 'Classics' (broadly meant); any historian and classicist is very familiar with these matters, the implications of which are both conceptual and methodological.[3] Precisely in terms of methodology and semantic concepts, the genesis of this research and this volume can be traced in a specific way. A practical example may illustrate the range of nuanced interpretations that usually we face in the reconstruction of the past.

A well-known Sicilian mosaic shows the remarkable iconography of a griffin, which grasps a cage with a man inside (fig.1). Aside from this mosaic, the picture can be compared only with a similar figure on a silver casket with hunting scenes discovered in the Mithraeum of London.[4] The subject is apparently a classical one, but it can provide important insights from the past for modern questions. It is puzzling to identify the precise meaning of this picture; nonetheless, it entices us to enquire about the ancients' sensitivity for problems related to environmental impact and depletion of natural resources. The mosaic of the griffin denotes one of the multiple paradigms that allow different readings and highlight the coexistence of different symbols and never static meanings. More importantly, it shows the possible insights that we can hope to understand which are related to environmental consciousness in past societies, especially if we remain open to manifold approaches of interpretation and semantic decoding.

The Roman Villa of Casale near Piazza Armerina in Sicily represents an authentic microcosm of the aristocratic society's values and status symbols during Late Antiquity. It would be beyond the present purposes to provide full details of the architectural structures and mosaic decorations which make the building one of the most impressive examples of senatorial luxury in Roman imperial society during the Tetrarchy.[5] Symbols, images, and details of the classical myth in the villa's mosaic floors are frequent and overwhelming. The figurative decoration is complex and varyingly arranged according to the variety of mythical themes and genre scenes, which fit specifically in each room. Yet, within the 3500 square metres of mosaics, precisely between the peristyle and the basilica, the imposing passageway of the 'Great Hunt' consists entirely of a magnificent mosaic strip which depicts lively and colourful chase episodes and the transport of wild animals for exhibitions in amphitheatres and circuses. Two apses delimit the corridor on both sides, each showing two goddesses that scholars identify as Africa/Ethiopia and Asia/India.[6] The deities allegedly provide a general indication of undefined western and eastern locations where the animal traffic is supposed to take place. The visual perspective ranges from African western- (on the left) to Asiatic eastern-countries (on

---

[3] The problem is tackled in incisive way and clearly explained by BEARD and HENDERSON 2000.

[4] TOYNBEE 1963: 5, 10–2.

[5] It is more useful to refer to the most recent literature on the topic for any further investigation. See esp. CARANDINI, RICCI and DE VOS 1982; PENSABENE 2009: 87–116; SFAMENI 2013: 159–79.

[6] See fig. 3.1 in NELIS-CLÉMENT's paper in this volume.

Fig. 1: Roman Villa of Casale (Piazza Armerina, Sicily), mosaic of the griffin in the 'Great Hunt' Corridor, 4th cent. AD. (Photo courtesy of the Museo Regionale della Villa Romana del Casale di Piazza Armerina).

the right), but the observer's main standpoint seems to be kept to the central north-south axis of the Mediterranean basin. Different capture techniques are distinguishable in these *venationes* (hunting), which are represented with very lifelike detail. The decidedly simple scene of a griffin and a man inside a cage captures the attention of any observer who walks along the 'Great Hunt' corridor. The insertion of such a mythical beast, the griffin, at the right corner of the corridor and close to the apse of Asia/India determines a rupture in the general realistic sequence of the animals' capture. The scene is unique in the context and disengaged from the surrounding mosaic, since it moves the observer into a fantastic dimension, somewhere in the East, and centres on the fictional contest between the mythical bird and the man inside the wooden cage held tightly by the griffin. *Pace* CARANDINI and

SETTIS, there is no sign, indeed, in the nearby scenes that 'the griffin is being lured into captivity' (WITTS).[7]

The disturbing fascination of this picture inevitably stimulates questions. What does such image represent to viewers? Which kind of meanings and inferences can we detect? Is it a specific semantic symbol related to a definite cultural context? Two opposite theses have divided scholars. On the one hand, in the general debate we can distinguish a thesis based on a (pagan) principle of 'contrapasso', in terms of religious and mystical symbolism (MANGANARO). The fantastic animal would represent the goddess Nemesis' revenge for human violence against animals (*ineffugibilis necessitas ultionis*).[8] On the other, the iconography of the griffin and the man in the cage has been more simply connected to a late antique text (5th–6th c. AD), which describes the capture technique of the griffin. The animal is lured by oxen that are yoked to a very heavy wagon, on which it remains entrapped by its own claws:

περὶ τίγρεως ἐν ταὐτῷ καὶ γρυπός. […] ἀνὴρ ὑπὸ τὴν τοιαύτην ἅμαξαν κρύπτεται, καὶ ὅτε ἐνσχεθῇ, ἐπιπηδῶν καίει αὐτοῦ τὰς πτέρυγα.

About the tiger and in the same chapter the griffin. […] a man is hidden under such a wagon, and when it is entangled, jumping on it he burns its wings.[9]

This is the evidence of Timotheus of Gaza, whose *bestiarium* pertained to the tradition of *cynegetica* and to the *mirabilia* literary genre. The text was transmitted *per excerpta* in a Byzantine codex of the 11th century and it is the unique source about this very odd and fantastic hunting technique. By contrast, other authors (and Timotheus himself) report the description about the capture of tigers by the stratagem of a glass ball. This scene is also depicted beside the griffin's mosaic at Piazza Armerina. A hunter on a horse, who has stolen tiger cubs, is visible while escaping and stopping the tiger's pursuit by means of a glass ball. The tiger is deceived by her own image on the glass, since she believes it to be one of her own cubs, and in trying to recover it, unwittingly lets the horseman flee.[10]

This juxtaposition of the tiger and the griffin's capture, both in Timotheus' account and in the mosaic floor, has reinforced the thesis that the image of the griffin would not convey any religious and mystic symbolism. The mythical beast and the man in the cage, by contrast, would illustrate a hunting episode amongst the several examples in the sequence of the corridor. The main message, therefore, was aimed to celebrate the imperial power able to subjugate any exotic animal and even mythical creatures. This interpretation has been mainly defended by SETTIS and SETTIS FRUGONI, who also thought of the villa's owner as

---

[7] CARANDINI, RICCI and DE VOS 1982: 228; SETTIS 1975: 949. WITTS 1994: 112–3.

[8] See especially: MANGANARO 1959, 1960; FOUCHER 1969: 232–8; DUNBABIN 1978: 203; FERNÁNDEZ GALIANO 1995: 45–67.

[9] Timotheus Gazaeus, *Excerpta ex libris de animalibus (e cod. Paris. gr. 2422)*, frg. 9. The connection has been highlighted by SETTIS FRUGONI 1975: 21–32, also followed by MARROU 1978: 281–3; CARANDINI, RICCI and DE VOS 1982: 102–3, 228–30; PENSABENE 2009: 71. Without taking a defined position on the different views, a synthesis of the status quaestionis is in BLÁZQUEZ 1997: 155–63.

[10] Two versions differ in classical and late antique authors. The hunter releases one of the cubs to stop the tiger: Plin., *nat. hist.* 8.66; Pomp. Mela, 3.43; Solin, 18.6–7. The hunter throws the glass ball: Ambr., *exam.* 6.4 (Migne *P.L.* 14.265); Claud., *de rap. Pros.* 3.265. See CARANDINI, RICCI and DE VOS 1982: Foglio 31, sc. VII A.

one of the Tetrarchs, allegedly Emperor Maximian Herculius. Nevertheless, this hypothesis concerning the imperial ownership of the villa revealed inconsistencies and, subsequently, the thesis committed to senatorial/aristocratic ownership prevailed.[11] However it might be, the coexistence of imperial and senatorial aristocratic values are non-conflicting, either in the figurative apparatus of the 'Great Hunt' corridor or in the villa as a whole.[12] The rule of imperial power over the animals' world is a topic compared with that of Roman soldiers' discipline under the supervision of generals of senatorial rank. Hunting was both one of the most frequent aristocratic occupations and of military training during breaks in warfare; a good soldier also was a good hunter and *venationes* improved cohesion among comrades, as well as tested the ability and precision of an army's field manoeuvres.[13]

The parallel between Timotheus' description and the iconography of the mosaic is captivating; it has to be admitted, however, that it presents some dubious elements and shortcomings. Timotheus mainly refers to the capture technique performed by a (free-moving) hunter, who hides under a wagon (ἅμαξα) and is able to catch the bird by burning its wings. By contrast, the mosaic shows a captive man inside an animal cage, who is unable to come out and chase the beast, nor is there any indication of the griffin's capture.[14] More importantly, the main visual focus of this scene diverges immediately from the griffin to the human face inside the cage; the main emphasis is not on the griffin's capture, but on the man's captivity. The precise semantic meaning of the scene remains cryptic – perhaps deliberately; nevertheless, it subverts the predator-prey order, which is the guideline of the whole 'Great Hunt' mosaic. Although the nexus with Timotheus' *bestiarium* is indicative, we cannot dismiss the massive coeval evidence, both in the literature and in the material culture, of the griffin's role and its connection with Nemesis.[15] This is a matter of fact in fight contexts – specifically in hunting and public exhibitions of theatres, amphitheatres, and circuses. In ancient myth and culture, the griffin is a 'totemic' animal of Apollo, Dionysus and, above all, Nemesis, goddess of justice and right balance in a subverted and unfair collision among unequal forces. Indeed, the divine bird appears a steward for fair competition between opponents.[16] The evidence of material culture, inscriptions, and the literary sources are impressive and offer a clear idea of the cultural components in the background of these artistic and figurative productions.[17]

For the general timeframe of this period, these elements cannot be neglected. It follows that questions arise regarding the existence of certain environmental awareness, and to investigate in what sense and extent ancient artists, intellectuals, and societies perceived the environment and the exploitation of natural resources. To take only a few examples from select authors and passages, it seems that the capture and killing of animals posed both

---

[11] Settis 1975 and Settis Frugoni 1975. It is more probable that a member of high rank aristocracy was the owner of the villa and the estate: Mazzarino 1953: 417–21; Dunbabin 1978: 204–12; Marrou 1978: 254–8; Cracco Ruggini 1980: 3–96; Carandini, Ricci and De Vos 1982: 28–46.

[12] See Marrou 1978: 253–95; Pensabene 2009: 87–116; Sfameni 2013: 159–79.

[13] Evidence in ancient literature is reported by Marrou 1978: 272–8.

[14] Similar doubts and observations are in Marrou 1978: 282–3 and Witts 1994: 112–3.

[15] Especially Paus. 7.5.1–3; Amm. 14.11.25–6; Macr., *sat.* 1.18.17; Nonn., *dion.* 48.378–88.

[16] See evidence and literature in Simon 1962: 749–80; Cordovana 2007: 395–8.

[17] Manganaro 1960; Delplace 1980: 284–397; Karanastassi 1992: 733–62; Rausa 1992: 762–70; Linant De Bellefonds 1992: 770–3.

ethical problems and concerns related to their extinction, as well as the dangers of their over-exploitation. Cicero clearly complains about these matters:

> But what pleasure can a cultivated man get out of seeing a weak human being torn to pieces by a powerful animal or a splendid animal transfixed by a hunting spear? (…) The last day was for the elephants. The groundlings showed much astonishment thereat, but no enjoyment. There was even an impulse of compassion, a feeling that the monsters (i. e. the elephants) had something human about them.[18]

Cassius Dio confirms such feelings among the common people in Pompey's time, and reports that some elephants,

> contrary to Pompey's wish, were pitied by the people when, after being wounded and ceasing to fight, they walked about with their trunks raised toward heaven.[19]

Seneca even wonders about whether the value of a *communis ius animalium* may be admissible.[20] A clear concern for the extinction of some species is evident in Ammianus and Themistius, who complain of the diminishing number of hippopotami, lions, and elephants.[21] In terms of a honourable fight, Herodian strongly criticized the unfair hunt of the 'gladiator' Commodus, when

> At last the day of the show came and the amphitheatre was packed. A special raised enclosure was put up for Commodus' benefit so that h e c o u l d s p e a r t h e a n i m a l s s a f e l y f r o m a b o v e without endangering himself from close quarters, a d e m o n s t r a t i o n o f h i s s k i l l b u t n o t o f h i s c o u r a g e .[22]

Together with the dramatic numbers of animals killed by Commodus reported in the same chapter, Herodian underlines a basic principle of the Roman warrior culture in this episode: courage pertains to a fair fight between equal forces. These conditions of equanimity had to be respected. Nemesis' presence (also via her symbols, i. e. the griffin) had specific role and

---

[18] Cic., *ad fam.* 7.1.3: *sed quae potest homini esse polito delectatio, cum aut homo imbecillus a valentissima bestia laniatur aut praeclara bestia venabulo transverberatur? (…) extremus elephantorum dies fuit. in quo admiratio magna vulgi atque turbae, delectatio nulla exstitit; quin etiam misericordia quaedam consecuta est atque opinio eius modi, esse quandam illi beluae cum genere humano societatem.* (Engl. transl. SHACKLETON BAILEY 2001).

[19] Cass. Dio 39.38.2–3: ἠλεήθησαν γάρ τινες ὑπὸ τοῦ δήμου παρὰ τὴν τοῦ Πομπηίου γνώμην, ἐπειδὴ τραυματισθέντες τῆς μάχης ἐπαύσαντο, καὶ περιιόντες τάς τε προβοσκίδας ἐς τὸν οὐρανὸν ἀνέτεινον. (Engl. transl. CARY and FOSTER 1914). On the same episode also Plin., *nat. hist.* 8.20–1. In this volume NELIS-CLÉMENT offers an exhaustive collection of sources referring to animals' exploitation and trade for spectacles.

[20] Stoic ethics is familiar with these concepts: Sen., *de brev. vitae* 13.6–7; *nat. quaest.* 3.17–8. On the topic see NEWMYER 2006: 19–22; TUTRONE 2012; 2012–2013: 511–50.

[21] Amm. 22.15.24; Them., *or.* 10.140.

[22] Hdn. 1.15.2: ἐπεὶ δὲ κατέλαβον αἱ τῆς θέας ἡμέραι, τὸ μὲν ἀμφιθέατρον πεπλήρωτο, τῷ δὲ Κομόδῳ περίδρομος κύκλῳ κατεσκεύαστο, ὡς μὴ συστάδην τοῖς θηρίοις μαχόμενος κινδυνεύοι, ἄνωθεν δὲ καὶ ἐξ ἀσφαλοῦς ἀκοντίζων εὐστοχίας μᾶλλον ἢ ἀνδρείας παρέχοιτο. (Engl. transl. WHITTAKER 1969).

function to this purpose and usually demanded the fulfilment of these conditions that she was supposed to supervise and correct.

In this general context therefore, we cannot exclude that the mosaic of the griffin at Piazza Armerina might represent a symbol of warranty for the equal development of the fight between man and wild beasts. Bearing in mind that since, even to ancient societies, prey and predator are not absolute concepts based upon natural order, and they might be subverted, environmental inferences cannot be totally excluded. Because of his own uncontrolled exploitation of natural resources, man can become himself a victim. Undoubtedly the scene provides a clear example of the multilayered interpretations and even subliminal meanings that we can infer from ancient evidence. The exegesis of this composition is not simple at all, and in some way has represented a meaningful starting point for the questions developed in this volume.

In what sense and to which extent does the ancient evidence allow us to identify the roots of modern environmental sensitivity?

4. Pollution, environmental awareness, and the ancient evidence

In the ancient world, the nexus of man and environment appears problematic and quite ambiguous. In recent years several scholars have focused on this topic, which has become more and more challenging and contemporary. In the latest studies it is evident that understanding nature involved a twofold aspect in the ancient world. Both a religious and a rational-philosophic level coexisted in the approach of ancient societies toward nature, as well as in their descriptions of the intertwined dynamics between human beings and environment.[23] It is possible to distinguish, on the one hand, divine and supernatural powers that bounded and dominated nature. To humankind the only possibility for a partial control of those powers and deities consisted in specific systems of rituals and cults. On the other, especially during the classical age, the Greeks developed a rational approach, which, rooted in philosophy and physics, aimed to search for suitable answers to understand natural phenomena, as well as to bend nature herself and her resources to human needs by technological conquests. In his important work, LUKAS THOMMEN emphasized that ancient sources concerning environmental history are very limited and often return a unilateral perspective on the problem. Detailed environmental descriptions and environmental concerns did not exist. Rather, destruction of natural habitats, pollution and depletion of resources were noticed and criticized, but these observations did not receive any analysis of data.

As legitimate and true though THOMMEN's critique might be, it implies a certain value to quantitative data and statistics. Such an approach is more structured in the modern socio-economic culture than consistent with the habits of ancient societies, which were not used to statistics and to the management of quantified data. Nevertheless, from different sources, especially in epigraphy and in legal literature, it clearly emerges that both individuals and social groups were specifically responsible for certain specific measures of environmental protection. Detrimental behaviours against *publica salubritas* were subjected to

---

[23] This is especially evident in THOMMEN 2012.

sanctions, especially for that which concerned, for instance, pollution of waters and waste disposal.[24]

On the basis of this premise, two main theses fueled the modern debate. Some scholars were positively persuaded of the existence of a certain and well-structured environmental legislation.[25] Others, more pessimistically, firmly denied the existence of a general system of values related to protection of nature and natural resources, but individual regulatory actions were created to protect specific natural elements, mainly waters. This is particularly evident in the epigraphic and legal sources, which often detail the prosecution for harmful actions toward the *publica salubritas*.[26]

Ecology and environment, exploitation of natural resources and environmental impact caused by pollution-factors are all elements of the modern contemporary awareness, and the product of modern culture in post-industrial societies. Nevertheless, ancient law includes several norms regarding environmental protection.

It would be extremely hard to focus on and identify 'Environmental Law' in the whole Roman legal system; the same happens, for example, in terms of 'Constitutional', 'Public', 'Commercial' and 'Tax' Roman law, which are, as such, not *stricto sensu* distinguishable. Nonetheless, it is clear that ancient cultures had to regulate 'commercial' and 'fiscal' transactions, and 'public' properties and institutions, despite of the lack of fully evolved volumes of procedures, as well as specific *corpora* and *codices* of law pertaining to these definite fields. In any event, this does not mean that ancient societies had no notion of the importance of natural resources and environmental protection in specific contexts. Apart from the fundamental problem of focusing on the specific elements of the ancient Roman 'environmental law' (which did not exist as such), the corpus of evidence on environmental awareness and concerns is unexpectedly impressive and ranges across a variety of different fields. It is very easy to infer, therefore, that especially in the Roman imperial period the governmental regulation on this subject might have been a starting point only and, especially, it is not the unique body of evidence on the topic.[27]

All the aspects that we developed in this volume have only sketched the problem in the aim of fixing and limiting meanings and systems of values for that which cannot be fully assimilated to contemporary circumstances and must be kept distinct from modern sensitivities. This was the big initial challenge for the authors, who enthusiastically joined this project. All the contributions approach, by different perspectives, the basic questions to fix meanings and limits on the ancient attitude toward environmental pollution and the exploitation of resources. The picture that we framed is not a defined one and, above all, is far from being exhaustive.

---

[24] See e. g. CAPOGROSSI 1996: 1–69; WACKE 2002: 1–14; LABRUNA 2008: 277–80; BRADLEY and STOW 2012; CAPOGROSSI 2014: 75–91; CAPOGROSSI's paper in this volume.

[25] See the debate especially in DI PORTO 2015 and MARCONE's contribution to this volume.

[26] See especially WACKE 2002: 1–14.

[27] This is evident in SIMONETTI's paper, which analyses Babilonian legal rules concerning water management.

5. The structure of this volume

The choice of approaching and comparing ancient evidence in four distinct sections is characterized by an interdisciplinary perspective. Ancient legal evidence forms the nucleus of the first section ('Environment, pollution, and legal sources'), which aims to provide insights on different legal aspects of environmental protection and sensitivity, as well as their extents and limitations, in the ancient Graeco-Roman cultures and societies. Via an analysis of the situation in Mesopotamia, CRISTINA SIMONETTI provides an insight into the importance of water and water-regulation within this arid country. She shows how relevant a clean environment was in the ancient Mesopotamian regulation, because water represented a central element for the local economy. LUIGI CAPOGROSSI COLOGNESI examines the development of specific laws for environmental safeguard from the archaic period until the Republican age. It is evident that over the time a large amount of legal tools was given to the city inhabitants to defend themselves from any kind of annoyance produced by smog, humidity, smell and other forms of air-pollution. Public interest, in that field, was not only assured by private reaction, but also and principally, by the municipal magistrates. Such legislation shows a certain awareness for a clean environment within Roman mentality; moreover, these laws were an attempt to protect not only nature, but also public health.

The second section of the volume ('Environment and pollution in literary and epigraphic evidence') focuses on the evidence in historical, literary, philosophical, and in epigraphic sources. The main aim is the analysis of the intellectual thought and the ancient environmental awareness about the specific agents and forms of environmental pollution and depletion of resources.

Via an investigation of the ancient Greek terminology about problems related to ecology and pollution in the Greek and Roman world, particularly within the Hippocratic treatises, CINZIA BEARZOT highlights the importance of fixing the precise ancient lexicon. Especially the ideas of 'balance' and 'disproportion', as well as 'stability' and 'change' are investigated and related to the definition of 'ideal environment'. The paper of GIAN FRANCO CHIAI deals with further aspects of ancient mentality: the collected epigraphic evidence shows how religion and particularly the fear of divine punishment were used in order to protect the purity of water, and the cleanliness of roads and public and private spaces. As evidenced by ARNALDO MARCONE, Frontinus' and Plinius the Younger's testimonies, as well as Statius and other literary sources are particularly helpful to reconstruct the specific housing strategy of the Roman elite, who aimed at avoiding the environmental inconvenience of city life. We can observe that very soon, together with practical needs, ideological motivations also shaped the projects of architects, who created underground areas in villas. This was a noteworthy change of sensitivity. It is even possible that the creation of a more protected space inside one's mansion, which was in some way immune from external contagion, also played a role. EDOARDO BIANCHI analyses the management of Tiber's floods with 'scientific' method by the Julio-Claudians emperors and focuses on the ancient debate about the possibility to change the natural order of nature in order to protect the life of the inhabitants of Rome. Such a debate shows a view of nature that was not only considered to be a divine *kosmos*, but also a world that human civilization can modify. ORIETTA CORDOVANA's paper demonstrates that the entire treatise of Pliny's *Naturalis Historia* is permeated by environmental awareness: stoic philosophical principles mainly inspired the books, in which he shows the natural order of the world and its delicate balance that is constantly under threat. By focusing on Medieval culture, LUCA MONTECCHIO compares the difference between the Roman perception of the environment and that of the medieval monastic culture. After the Late Antiquity and during the Medieval age, environmental awareness seems disappear and the view of the nature was deeply affected by the religious concepts of the Roman Church.

The third part ('Environment, pollution, and diseases') deals with the material evidence of pollution and diseases connected to polluting agents. Via the analysis of the Hippocratic text, ELIZABETH CRAIK shows that the connection between malaria and marshes or stagnant water, a typical breeding ground for mosquitoes, was well recognized in antiquity. Local cult epithets of certain deities indicating a particular association with marshy ground suggest disease prophylaxis. Likewise, medical papyri from Hellenistic and Roman Egypt show that people were affected by diseases related to polluted air and water, as evidenced in ISABELLA ANDORLINI's study. The essay offers comparative materials and combines the methods of archaeology with an admixture of the history of medicine. Papyri and the material records are also used to counterbalance the literary evidence. Moreover, the paper evaluates the contribution of ancient medical writers to the environmental diseases.

In the fourth section ('Pollution and the environment in ancient life: material evidence') the approach to the archaeological and literary evidence characterizes ALAIN BRESSON's contribution, which also deals with the presence of diseases caused by a polluted environment especially among the inhabitants of ancient cities. His study focuses on the transformation of the environment by human activity that could have a negative impact on the health and welfare of human populations. Two main aspects are considered: domestic and non-domestic pollution. Domestic pollution was mainly linked to the absence of chimneys in ancient houses that would allow the evacuation of smoke. Non domestic pollution was determined, for instance, by emissions of ammonia of the *fullonicae*, contamination by metals in the mining districts and metal-processing workshops. According to J. DONALD HUGHES' reconstruction, deforestation was a major environmental trend in the ancient Mediterranean world in the period from the Bronze Age through the Roman Empire. The result of continual use for construction and dependence on wood and charcoal for fuel, the degradation of the landscape through deforestation became lasting, where the rate of use exceeded the rate of forest regeneration and growth. Ancient societies took measures to ameliorate these problems through forest plantations, the reservation of areas as sacred groves, and oversight of royal and public lands. Government officials were appointed to regulate forest contracts, and guards with stations and regular patrols limited illegal cutting. Water supervisors directed construction by means of purification such as settling tanks for urban water supplies. Even with these measures, powerful consortia had political influence, and the sheer demand for fuel and wood products exceeded the supply that Mediterranean forests could provide. JOCELYNE NELIS-CLÉMENT's paper considers the relevance of the organisation of mass spectacle in the Roman world. The surviving sources provide us with a mass of information that suggest a great deal about the practicalities involved in putting on games in the amphitheatres and circuses of the Roman world: hunting on a massive scale, deforestation for building purposes, intensive transportation, mass slaughter and the elimination of waste materials. The evidence reveals the impact of the organization of mass spectacle on the natural world and considers both contemporary Roman reactions to these processes and also suggests how modern scientific methods of measuring ecological impact could be brought to bear on the ancient evidence.

Bibliographical note

A remarkable number of studies concerning the relationships between environment and human life has been published in recent years. A useful introduction to this subject is the important monograph of Donald Hughes.[28] This is a significant contribution to understand the relationship between Classical Civilizations in the Mediterranean basin and its contemporary natural environment during the period from 800 BC to AD 600. This book incorporates much of Hughes's ongoing investigations published in journals and essay collections from disciplines as diverse as religion, forestry, archaeology, and environmental studies proper. The monography of Karl-Wilhelm Weeber represents another useful introduction to the topic of the environmental problems in the Greek and Roman world.[29] He analyses the case of the Athenian mines which polluted and destructed the nature in Attica, showing how this human activity changed the face of the Attic landscape. He emphasizes that we can find many critical references to the transformation of this landscape in the ancient sources (for example in Plato). Paolo Fedeli's book makes another important contribution.[30] He collected many passages of Latin authors that deal with the destruction of the nature by human activities in order to reconstruct the concept of environment in Roman mentality. This is an important reference work about this topic for the Roman world, though archaeological and epigraphic evidence could have completed the picture conveyed by the literary sources.

Two books by Holger Sonnabend deal with natural catastrophes in the Greek and Roman world.[31] Based on classical sources, he analyses how such catastrophes were seen and interpreted in the ancient mentality. Another important work is that of Giangiacomo Panessa, who has collected the sources on the climate and associated climatic catastrophes in the Greek and Roman world.[32] This book attempts to reconstruct the ancient mentality and behaviour with regards to the climatic changes.

Lukas Thommen's work represents another important contribution to this subject.[33] It is an attempt to examine the relationship between man and environment in the Greek and Roman world. Thommen shows that this relationship takes place on two levels. The first one is that of nature, viewed as a physical place dominated by divine and non-human powers. Man, however, can control such non-natural powers through ritual, according to the traditional view. The second one is that of the scientific knowledge, according to which nature can be observed, studied, and rationally understood. Thommen also recognises that the ancient sources on this topic are not only scattered, but also disagree with each other. Within the landscape descriptions, classical authors focussed mostly their attention on the natural features. The information about pollution and destruction of the environment is thereby often neglected. Like the literary evidence, the archaeological evidence about this topic has neither been systematically collected nor studied in comparison to the literary sources and the inscriptions. The finds from Roman sewers, for example, have not yet been comprehensively examined with the aim to reconstruct not only the everyday life of a Roman city, but also the impact of the rubbish on the environment and on human health (the book by Mark Bradley and Kenneth Stow about the hygienic problems in Rome from Antiquity to Modernity represents a good exception). This is surprising, considering for example the eulogy of the *locus amoenus*, characterized by *salubritas*, which is a widely known *topos* within the classical literature. Indeed, the archaeological evidence as well as a lot of references by ancient authors, notably from the Roman imperial period, shows that rubbish and malodorous streets represented an urgent problem in many cities.

---

[28] Hughes 2014[2].
[29] Weeber 1990.
[30] Fedeli 1990.
[31] Sonnabend 1999; 2013.
[32] Panessa 1991.
[33] Thommen 2009 (cf. the critical review of Sonnabend on this book in *Sehepunkte*, 14.20.2010).

A study about this subject is, for example, the short book of GÜNTHER THÜRY, which deals with the problem of the polluted environment in the Greek and Roman world, considering the archaeological evidence from sewers and fountains.[34] Through a detailed analysis of the archaeological material, he shows how pollution of waters and public places was an everyday problem in the Greek and Roman areas. His research is also a good attempt to study the archaeological evidence related to the literary and epigraphic sources. He compares, for example, the interesting Pompeian inscriptions against *cacatores* with epigraphic documents concerning the prohibition to discharge rubbish within the city.

Bibliography

BEARD, M. and J. HENDERSON 2000. *Classics: a Very Short Introduction*. Oxford.

BLÁZQUEZ, J. M. 1997. 'El grifo en mosaicos africanos y su significado', *Antiquités Africaines*, 33, 155–63.

BRADLEY, M. and K. STOW (eds.). 2012. *Rome, Pollution and Propriety. Dirt, Disease and Hygiene in the Eternal City from Antiquity to Modernity*. Cambridge.

CAPOGROSSI COLOGNESI, L. 1996. *Ai margini della proprietà fondiaria*. Roma.

CAPOGROSSI COLOGNESI, L. 2014. *La lex rivi Hiberiensis e gli schemi delle servitù d'acqua in diritto romano*, in L. MAGANZANI and C. BUZZACCHI (eds.), *Lex rivi Hiberinensis. Diritto e tecnica in una comunità d'irrigazione della Spagna romana*. Napoli, 75–91.

CARANDINI, A., A. RICCI and M. DE VOS 1982. *Filosofiana: la villa di Piazza Armerina. Immagine di un aristocratico romano al tempo di Constantino*. Palermo.

CORDOVANA, O. D. 2007. *Segni e immagini del potere. I Severi e la provincia Africa Proconsularis*. Catania.

CRACCO RUGGINI, L. 1980. 'La Sicilia tra Roma e Bisanzio', in: R. ROMEO (ed.), *Storia della Sicilia* 3. Napoli, 3–96.

DELPLACE, C. 1980. *Le griffon de l'archaïsme à l'époque impériale. Étude iconographique et essai d'interprétation symbolique*. Bruxelles-Rome.

DI PORTO, A. 2015. *Salubritas e forme di tutela in età romana. Il ruolo del* civis. Torino.

DUNBABIN, K. M. D. 1978. *The Mosaics of Roman North Africa. Studies in Iconography and Patronage*, Oxford.

FEDELI, P. 1990. *La natura violata. Ecologia e mondo romano*. Palermo.

FERNÁNDEZ GALIANO, D. 1995. 'La Filosofia de Filosofiana. Piazza Armerina y los juegos de Anfiteatro', in J. M. ALVAREZ MARTÍNEZ, J. J. ENRÍQUEZ NAVASCUÉS (eds.), *El Anfiteatro en la Hispania Romana*, Coloquio Internacional, Mérida, 26–28 de Noviembre de 1992. Badajoz, 45–67.

FOUCHER, L. 1969. 'À propos d'un griffon', in J. BIBAW and M. RENARD (eds.), *Hommages à Marcel Renard* 3. Bruxelles, 232–8.

GEUS, K. and M. THIERING (ed.) 2014. *Features of Common Sense Geography. Implicit knowledge structures in ancient geographical texts* (Antike Kultur und Geschichte 16). Wien-Berlin.

JAMES, P. 2013–2014. 'People, Planet, and the Anthropocene. Spectators of our own demise?', *Arena* 41/42, 1–6.

KARANASTASSI, P. 1992. 'Nemesi', *LIMC* 6.1, 1992, 733–62.

LABRUNA, L. 2008. 'Rome et le droit de l'environnement', in E. HERMON (ed.), *Vers une gestion intégrée de l'eau dans l'Empire romain*. Roma, 277–80.

LINANT DE BELLEFONDS, P. 1992. 'Nemesis (in Peripheria Orientali)', *LIMC* 6.1, 770–3.

MANGANARO, G. 1959. 'Aspetti pagani dei Mosaici di Piazza Armerina', *Archeologia Classica*, 11, 241–50.

MANGANARO, G. 1960. 'Grifo' s. v., *EAA* 3.

MARROU, H. I. 1978. 'Sur deux mosaïques de la villa romaine de Piazza Armerina', in Christiana tempora. *Mélanges d'histoire, d'archéologie, d'épigraphie et de patristique* (Publications de l'École Française de Rome 35). Rome, 253–95.

---

[34] THÜRY 2001.

MAZZARINO, S. 1953. 'Sull' *otium* di Massimiano Erculio dopo l'abdicazione', *RAL* s. 8, 417–21.

NEWMYER, S. 2006. *Animals, Rights and Reason in Plutarch and Modern Ethics*. London.

PANESSA, G. 1991. *Fonti greche e latine per la storia dell'ambiente e del clima nel mondo greco*. Pisa.

PENSABENE, P. 2009. 'I mosaici della Villa Romana del Casale: distribuzione, programmi iconografici, maestranze', in M. C. LENTINI (ed.), *Mosaici mediterranei*. Caltanissetta, 87–116.

RAUSA, F. 1992. 'Nemesi a Roma e nelle province occidentali', *LIMC* 6.1, 762–70.

SETTIS, S. 1975. 'Per l'interpretazione di Piazza Armerina', *MEFRA* 87, 873–994.

SETTIS FRUGONI, C. 1975. 'Il grifone e la tigre nella "Grande Caccia" di Piazza Armerina', *Cahiers archéologiques: fin de l'Antiquité et Moyen Âge* 24, 21–32.

SFAMENI, C. 2013. 'La villa del Casale e l'archeologia delle ville tardoantiche oggi: problemi e prospettive', in F. P. RIZZO (ed.), *La villa del Casale ed oltre: territorio, popolamento, economia nella Sicilia Centrale tra Tarda Antichità ed Alto Medioevo*, Piazza Armerina, 29–30 settembre 2010, SEIA 15–16, 159–79.

SIMON, E. 1962. 'Zur Bedeutung des Greifen in der Kunst der Kaiserzeit', *Latomus* 21, 749–80.

SONNABEND, H. 1999. *Naturkatastrophen in der Antike. Wahrnehmung – Deutung – Management*. Weimar.

SONNABEND, H. 2013. *Katastrophen in der Antike*. Darmstadt.

STEFFEN, W., J. GRINEVALD, P. CRUTZEN and J. MCNEILL, 2011. 'The Anthropocene: conceptual and historical perspectives', *Philosophical Transactions of the Royal Society A: Mathematical, Physical and Engineering Sciences* 369, 842–67.

THOMMEN, L. 2012. *An Environmental History of Ancient Greece and Rome*. Cambridge.

THÜRY, G. E. 2001. *Müll und Marmorsäulen. Siedlungshygiene in der römischen Antike*. Mainz.

TOYNBEE, J. M. C. 1963. *A silver casket and strainer from the Walbrook Mithraeum in the City of London*. Leyden.

TUTRONE, F. 2012. *Filosofi e animali in Roma antica. Modelli di animalità e umanità in Lucrezio e Seneca*. Pavia.

TUTRONE, F. 2012–2013. '*Commune Ius Animantium* (*Clem.* 1.18.2): Seneca's Naturalism and the Problem of Animal Rights', in *Phasis* 15–16, 511–50.

WACKE, A. 2002. 'Protection of the Environment in Roman Law?', in *Roman Legal Tradition* 1, 1–14.

WEEBER, K.-W. 1990. *Smog über Attika. Umweltverhalten im Altertum*. Zürich.

WITTS, P. 1994. 'Interpreting the Brading "Abraxas" Mosaic', *Britannia* 25, 111–7.

ZALASIEWICZ, J., M. WILLIAMS, W. STEFFEN, and P. CRUTZEN, 2010. 'The New World of the Anthropocene', *Environmental Science & Technology* 44.7, 2228–31.

Environment, pollution, and legal sources

Cristina Simonetti

Uso e gestione delle acque in Mesopotamia nel Secondo Millennio a. C.

1. La Mesopotamia: terra tra due fiumi

Parlare di inquinamento nel mondo antico, e nel Vicino Oriente antico in particolare, non è semplicissimo, perché bisogna valutare la consapevolezza che gli uomini antichi avevano dell'ambiente e del ruolo più o meno nocivo che essi potevano giocare in relazione ad esso. Sappiamo che l'uomo, dal momento in cui è comparso sulla terra, ha dovuto interagire con l'ambiente che lo circondava: uno dei principali fattori che distinguono lo sviluppo umano da quello animale consiste proprio nel fatto che mentre gli animali tendono ad adattarsi all'ambiente in cui vivono, gli uomini tendono a modificarlo per adattarlo alle proprie esigenze. Ovviamente nel corso del tempo la capacità di incidenza dell'uomo sull'ambiente è andata aumentando, ma è innegabile che sin dalla preistoria l'uomo abbia interagito con esso e sfruttato ogni cosa che potesse rendergli più semplice la vita. Si pensi, ad esempio, all'uso del fuoco per scaldarsi, proteggersi, cuocere i cibi, ma anche liquefare i metalli e costruire strumenti sempre più perfezionati e adatti alle proprie esigenze: esso ha avuto un impatto non certo irrilevante sull'ambiente. Ma anche le più ecologiche delle attività umane, introdotte dall'età neolitica, come l'agricoltura e la pastorizia, che hanno fatto da volano al successivo sviluppo storico umano (divisione del lavoro, complessità sociale e scrittura), hanno avuto un impatto non trascurabile sull'ambiente.[1] Non è facile dire se gli uomini, nel corso della storia, abbiano avuto la consapevolezza che il loro intervento rischiasse di distruggere l'ambiente in cui vivevano. Però è probabile che si rendessero conto di quanto il loro intervento modificasse l'ambiente: incendiare un bosco o bonificare delle paludi per ampliare le aree coltivabili, oppure disboscare intere aree per procurare combustibile, difficilmente potevano essere considerate pratiche naturali.

Volendo parlare di inquinamento e ambiente in Mesopotamia, un aspetto fondamentale riguarda lo straordinario intervento fatto sulle acque dei due grandi fiumi che la delimitano, il Tigri e l'Eufrate. Sia il Tigri sia l'Eufrate hanno regimi stagionali molto accentuati, e in determinati periodi dell'anno essi aumentano di molto la propria portata, provocando

---

[1] Cfr. Asouti e Fairbairn 2010, 165: 'Domestication cannot be considered as a static end state, but should be seen rather as an ongoing process of human-induced transformation of plant genotype and phenotype which is co-determined by cultivation harvesting practices, plant behavior and environmental circumstances. Cultivation/harvesting practices necessarily take place in specific places (plants being dependent on particular environments) that may be managed in order to improve food productivity for human communities'.

straripamenti ed esondazioni, tali da modificare anche permanentemente il loro corso.[2] Per ovviare a questo problema, piuttosto dannoso per l'agricoltura, gli antichi abitanti della Mesopotamia hanno provveduto a irreggimentarne il corso, costruendo dighe e canali.[3] In questo modo è stato possibile trasportare l'acqua in aree più lontane dai letti dei fiumi, ed ampliare la superficie coltivabile, ma si è anche potuto mitigare la forza dell'acqua nei periodi di maggiore afflusso. La rete dei canali, visibile sulle mappe idrogeologiche ricostruite da vari gruppi di studiosi, è complessa, quasi illeggibile per un non esperto, anche perché nel corso del tempo alcuni canali sono stati abbandonati, a favore di altri più nuovi. C'è da dire, inoltre, che almeno per i primi millenni i lavori di canalizzazione non erano coordinati tra loro, perché il territorio non era unificato. Molti erano i luoghi abitati, tutti indipendenti gli uni dagli altri, anche dopo l'emersione delle prime realtà urbane, cioè a partire da Uruk IVa (3500 a. C.).[4] Per iniziare ad osservare una realtà politica più ampia, che inglobasse buona parte della Mesopotamia meridionale bisogna attendere l'impero di Akkade (2350–2225 a. C.) e soprattutto, quello della III Dinastia di Ur (2150–2004 a. C.). In realtà, però, lavori davvero incisivi e ben coordinati sono attestati a partire dalla Prima Dinastia di Babilonia, dall'unificazione operata da Ḫammurapi (1792–1750 a. C.) in poi.[5] All'incirca allo stesso periodo (inizio del secondo millennio a. C.) risale l'unificazione della zona centro-settentrionale della Mesopotamia ad opera degli Assiri, che però furono meno impegnati in opere di canalizzazione, data la specificità del loro territorio più montuoso e con un clima più piovoso rispetto a quello meridionale. Bisogna ricordare, inoltre, che la parte più meridionale della piana, che si estende all'incirca dalla confluenza dei due fiumi fino allo sbocco sul Golfo Persico, è molto diversa dal punto di vista paesaggistico, perché è molto ricca d'acqua in ogni periodo dell'anno, e la coltura delle palme da dattero è stata ed è tuttora florida. Non a caso, in quella zona l'intervento umano è stato pressoché nullo.

2. Lo sfruttamento agricolo delle acque

Gli abitanti della piana mesopotamica, sin dalla più lontana antichità, hanno utilizzato l'acqua dei due fiumi per vari scopi: per irrigare i campi e per trasportare uomini e mezzi, innanzitutto, ma anche per la pesca e l'allevamento del bestiame. Le scarse piogge e il clima torrido rendevano necessari lavori di canalizzazione, che non solo ampliavano l'area coltivabile, estendendola oltre la fascia lungo il corso naturale dei fiumi, ma fornivano acqua

---

[2] Riguardo al Tigri, sappiamo da AbB 1.141 che la piena avveniva verso Febbraio. A tal proposito, cfr. LEEMANS 1968: 171–226. Sul suo corso si veda anche ADAMS 1981: 3, 6–7. In relazione all'Eufrate si veda BESANÇON-GEYER 2003: 26, dove si dice che le variazioni stagionali della portata dell'Eufrate, sono di due tipi: le acque basse durano dalla fine di giugno alla fine di settembre (modulo settembre portata di 216m³/s), mentre le acque alte da marzo a giugno (modulo aprile 3422m³/s). Queste ultime sono alimentate dalle piogge primaverili e dal coincidente disgelo dei massicci della Turchia orientale.

[3] Cfr. BURINGH 1960: 52; BRINKMAN 1984: 175; KLENGEL 1993: 150.

[4] KRAELING e ADAMS 1960: 280 s., ritengono che 'before the Early Dynastic period the watercourse were purely local concern'.

[5] Si pensi al 33° anno di regno di Ḫammurabi, durante il quale fece scavare un canale per approvvigionare d'acqua Nippur, Eridu, Ur, Larsa, Uruk e Isin, oppure al suo 43°, quando fece costruire una diga a Sippar (Abu Abbah), oppure a quella costruita a Sippar Amnanum (Tell ed-Der) da Samsuiluna.

sufficiente per evitare e ritardare l'innalzamento delle falde freatiche, che avrebbero fatto affiorare i sali minerali e reso improduttivo il terreno. Sappiamo, infatti, grazie alle Georgiche sumeriche, un testo molto antico di carattere letterario ma con contenuto molto tecnico, che per coltivare i cereali era necessario allagare il terreno prima ancora di seminare; dopo la semina facevano seguito altri quattro allagamenti del terreno tra aprile e giugno.[6] Questo sistema, che implicava l'uso di fossati d'irrigazione gestiti con piccole chiuse aperte su ogni appezzamento di terreno coltivato, è attestato per millenni: almeno fino al decimo secolo d. C.[7] Secondo alcuni studiosi, inoltre, sarebbero attestate anche altre pratiche atte ad attenuare il rischio di danneggiare irrimediabilmente l'equilibrio ambientale: l'abbandono dei campi per alcuni anni, ad esempio destinandoli al pascolo, consentiva alle piante spontanee di ripopolare la zona. Si trattava per lo più di leguminose, che pure aiutavano a rallentare il processo di salinizzazione.[8]

Il fatto che i lavori di canalizzazione non fossero ben coordinati, ma, anzi, del tutto indipendenti gli uni dagli altri, non ha inciso molto sull'efficienza del sistema stesso: probabilmente la portata dei fiumi era tale da sopportare derivazioni di notevole entità in tutto il corso dei fiumi.

Nel periodo paleo-babilonese sappiamo che ci furono tentativi di gestione unitaria delle canalizzazioni, proprio a causa dell'unificazione del paese da parte di una sola dinastia. Tuttavia la gestione unitaria non poteva che attenere alle grandi vie d'acqua (*naru*), perché in realtà ogni singolo appezzamento di terreno aveva necessità di essere raggiunto dall'acqua, e una così fitta rete di piccoli canali non poteva che essere a gestione locale.

Sappiamo che esisteva un sistema basato sugli *ugaru*, distretti attraversati da un canale secondario, detto *atappum*: l'*ugarum* era a sua volta suddiviso in *eqlu*, campi, bagnati da fosse d'irrigazione, provvisti di chiuse, dette *iku*.[9] Sembrerebbe che la responsabilità della manutenzione di tali corsi d'acqua ricadesse sui gestori e gli utenti: i fossati erano gestiti dai singoli contadini, gli *atappu* dall'insieme dei proprietari dei campi che formavano l'*ugarum*. Il sistema degli *ugaru*, caratteristico della Babilonia centro settentrionale, venne introdotto da Ḫammurapi anche nelle zone del sud da lui stesso conquistate.[10] In questo periodo, inoltre, sono attestati degli accorgimenti che servivano a proteggere i campi dalle piene dei due fiumi, da un lato, e dall'insabbiamento dei fossati d'irrigazione, dall'altro. Nel primo caso si utilizzavano delle dighe, realizzate con legno, paglia, bitume e, in alcuni casi, anche pietra: esse proteggevano da eccessi d'acqua, ma in caso di necessità vi si aprivano delle brecce, per far defluire l'acqua. Inoltre, si innalzavano gli argini dei fiumi e dei canali, cercando di contenere la massa d'acqua. Tuttavia dalle molte lettere degli epistolari paleo-babilonesi, sappiamo che non sempre tali accorgimenti erano sufficienti a scongiurare catastrofi.[11] C'era

---

[6] Salonen 1968.

[7] Cfr. Adams 1981.

[8] Cfr. a questo proposito Gibson 1974: 7–20 e Cocquerillat 1968.

[9] Arnaud 1982: 45–59.

[10] Cfr. Stol 1982: 351–7.

[11] Dalle lettere sappiamo che lo straripamento dei fiumi distrusse nel tempo vari quartieri, come a Sippar-Amnanum, cfr. Jannsen, Gasche e Tanret 1994: 110–1. Pare che questa inondazione fosse stata preceduta da un'altra, sotto il regno di Abi-esuḫ (secondo gli studiosi, da 10 a 40 anni prima), che in seguito a questa alluvione richiese la mobilitazione di tutti gli operai disponibili nella regione di Sippar per rinforzare gli argini del canale Irnina e del Purattum (l'Eufrate) perché le acque della

poi la manutenzione dei fondi dei canali e dei vari corsi d'acqua secondari, che andavano dragati periodicamente, per evitare l'innalzamento del letto.

Per l'altro caso, cioè la mancanza d'acqua, bisognava prestare una continua manutenzione dei canali e delle dighe e soprattutto dragare regolarmente i fiumi e i canali.[12] Ma uno dei principali rimedi per riportare acqua in un canale insabbiato era costituito dall'uso delle pompe idrauliche dette *šiknum*.[13] Secondo JØRGEN LOESSØE: 'The function of the machine šiknu seems be raising of the water level, a sort of amplifier in the canalization system whereby a current was produced. If so, however, the šiknu must have required a power source and cannot be identified with the naʾūra; but the letters are held in terms too general to allow any conclusion as to the exact nature of the structure'.[14]

In realtà, sono attestati diversi dispositivi idraulici, che poi, in età neo-babilonese, consentiranno le rigogliose coltivazioni a terrazza sugli edifici della capitale Babilonia, come attestato ben oltre l'epoca di Erodoto.[15]

3. La gestione delle acque nella prima metà del secondo millennio a. C.

È innegabile che il modello mesopotamico abbia resistito per molti secoli, nonostante lo sfruttamento capillare del terreno, ma sappiamo che ci sono stati periodi di maggiore e minore resa agricola, e che alle fasi più ricche e abbondanti si sono alternate fasi di devastazione e declino. Forse proprio queste fasi alterne hanno reso più lenta la desertificazione, in atto certamente già a partire dall'epoca medievale (con le eccezioni della zona montuosa a nord e di quella palustre a sud). In questo caso l'intervento umano, così evidente agli occhi di tutti coloro che vi arrivavano, per quanto massiccio esso fosse, era riuscito a trovare una sorta di equilibrio ecosostenibile *ante-litteram*, che i sovrani, a partire almeno dal secondo millennio, avevano cercato di preservare.

La documentazione paleo-babilonese, infatti, ci fornisce una serie di informazioni molto interessanti riguardo la gestione e la regolamentazione dei corsi d'acqua. Si tratta, in particolare, di alcuni archivi provenienti da Larsa e Isin, studiati molto attentamente a partire dagli anni Sessanta, e che ci permettono di distinguere almeno tre livelli di gestione dei corsi d'acqua.[16] Oltre alle lettere celeberrime di Ḫammurapi[17], indirizzate ai luogotenenti di Larsa, infatti, in quegli anni erano stati identificati gli archivi relativi a due alti funzionari: Šamaš-magir, ufficiale del re Rim-Sin di Larsa (1822–1763 circa a. C.), che si trovava ad Ešn-

---

[ ] piena avevano superato il livello degli argini. Cfr. AbB 2.70, a Nippur (GIBSON ET AL. 1983: 177) e Uruk (COLE e GASCHE 1998: 8–9).

[12] AbB 4.19 (manutenzione di canali e dighe); AbB 2.4, 5 e 55 (dragaggio).

[13] Il termine ricorre in AbB 4.34, lettera inviata da Ḫammurapi a Šamaš-ḫaṣir, un suo luogotenente a Larsa, in cui il re ordinava al governatore di provvedere all'irrigazione di alcuni campi per mezzo di questa pompa, in grado di rifornire il fossato d'acqua.

[14] LOESSØE 1953: 24.

[15] Cfr. ad esempio il *nartabum*; secondo CARDASCIA 1966: 153–64, si tratterebbe di una noria; o il nag.ku₅ cfr. SALONEN, 1968: 223–8. Notizie sulla Mesopotamia e sui lavori idraulici sono presenti, oltre che in Hdt. 1.185–6 e 191; Polyb. 5.48.1–10 (Tigri), 9.43.1–6 (Eufrate). Tac., *ann.* 6.37; Strab., 9.17; Plut., *Lucullus* 24.2–3; Ptol. 5.12; Plin., *nat. hist.* 5.83–5 (Eufrate); 6.127–9 (Tigri).

[16] ROWTON 1967: 267–75; WALTERS 1970.

[17] Note sin dagli anni Venti grazie all'edizione di THUREAU-DANGIN 1924.

unna, e Lu-igi-sa, ufficiale del re Sumuel di Larsa (1894–1866 a. C.).[18] In sostanza, abbiamo informazioni su una zona piuttosto ristretta della Babilonia meridionale (tra Isin e Larsa) nel periodo in cui Larsa era indipendente (Sumuel e Rim-Sin) e in quello della sua conquista da parte del sovrano babilonese.

Senza entrare troppo nel dettaglio, si può vedere come ci fossero sostanzialmente tre livelli di gestione dei corsi d'acqua: uno attinente ai sovrani, diciamo di livello internazionale; uno di livello amministrativo di alto livello, gestito da ufficiali preposti, e infine uno più prettamente locale, attinente a gruppi di piccoli proprietari terrieri, raggruppati in distretti.

Per quanto riguarda il primo livello di gestione, si tratta di dispute su diritti d'acqua tra sovrani confinanti. Su questo punto, in realtà, la documentazione mesopotamica è molto abbondante (si vd. per esempio le iscrizioni reali sumeriche, la Stele degli avvoltoi etc.), ma tra i documenti dell'archivio di Šamaš-magir ci sono due testi in cui Rim-Sin rivendica dei diritti d'acqua contro un altro sovrano non meglio identificato (il re di Ešnunna?).[19]

Il secondo, invece, di cui ci informano più dettagliatamente gli archivi prima ricordati, oltre a quello di Ḫammurapi, riguarda i canali più importanti, che approvvigionano i centri urbani e le aree agricole. Sono funzionari palatini ad occuparsene, organizzando il lavoro di manutenzione, di sorveglianza e di protezione. Il lavoro effettivo lo svolgevano gli uomini che dovevano periodicamente assicurare le *corvées* (l'*ilkum* in particolare), una sorta di tassa dovuta al palazzo, che consisteva in giornate di lavoro obbligatorio, da effettuare periodicamente a vantaggio del palazzo. Dagli archivi sappiamo che erano impiegati centinaia di uomini alla volta, che dragavano i canali, riparavano le brecce nelle dighe e nei ponti, rafforzavano gli argini. L'amministrazione centrale si occupava delle loro razioni giornaliere. In molti casi erano i sovrani stessi ad intervenire, impartendo ordini specifici, ed erano previste sanzioni in caso di negligenza.

Infine, l'ultimo livello, quello più capillare, era gestito dalla popolazione locale: i distretti e i singoli proprietari terrieri, che erano tenuti alla manutenzione dei fossati e dei canali secondari, e anche all'uso corretto di essi. L'uso delle chiuse, infatti, presupponeva una precisa regolamentazione della distribuzione dell'acqua per allagare i campi al momento opportuno: trascurare la chiusura della piccola diga, oppure allagare il campo dei vicini poteva costare il raccolto. Nei codici babilonesi abbiamo qualche articolo che parla di questi aspetti.

Già nel codice di Ur-Namma, che risale alla fine del terzo millennio, è previsto il caso del paragrafo 30: 'Se un uomo causa l'allagamento del terreno di un altro, sia condannato a pagare 3 *gur* di orzo per ogni *iku* di terreno allagato'. Si prevede il caso di una responsabilità per colpa, che viene punita con il risarcimento del danno causato, calcolato con una resa presunta del campo danneggiato.

Un altro testo che presenta contenuti legislativi, il prisma FLP 1287, redatto in sumerico, ma datato da MARTHA ROTH intorno al 1700 a. C., cioè in piena età paleo-babilonese, nel paragrafo racchiuso in IV 35–41 recita: 'Se un uomo causa l'allagamento del terreno di un altro, già preparato per la coltivazione, rifonderà basandosi sul raccolto dei vicini'.[20] Anche in questo caso si disciplina la responsabilità per colpa, ma il risarcimento è calcolato sulla base della produzione degli altri campi vicini.

---

[18] Lettere di Ḫammurapi: AbB 2.4, 5, 55, 70; AbB 4.19, 39, 80, 85; AbB 13.5.
[19] Si tratta di A7535 e A7537. Cfr. a questo proposito ROWTON 1967: 267–75; SAPORETTI 2002: 235–8.
[20] Pubblicato da ROTH 1979. I passi dei codici citati sono nella traduzione di SAPORETTI 1998.

Il Codice di Ḫammurapi, infine, dedica quattro paragrafi all'argomento:

> *CH* 53: Se un uomo ha trascurato di rinforzare l'argine del suo terreno, per cui si sono verificate delle falle, con conseguenti inondazioni, egli dovrà pagare una quantità di orzo pari a quella di cui ha causato la perdita.
>
> *CH* 54: Se non è in grado di risarcire i danni, sarà venduto con tutti i suoi beni, e il ricavato sarà diviso tra i proprietari dei campi il cui raccolto è andato distrutto.
>
> *CH* 55: Se un uomo per irrigare il suo campo ha aperto le sue riserve d'acqua e poi ha trascurato il lavoro e le acque hanno invaso il terreno del vicino, egli dovrà risarcire con una somma corrispondente al raccolto di questo vicino.
>
> *CH* 56: Se un uomo allaga il campo del vicino, dovrà risarcirlo con 10 *gur* d'orzo per ogni bùr di terreno. Il raccolto dell'orzo spetterà al proprietario.

I primi tre paragrafi parlano di responsabilità per colpa (negligenza nella manutenzione e nella sorveglianza dell'acqua) e prevedono il risarcimento del mancato raccolto, calcolato probabilmente su quello dei vicini. L'ultimo, invece, prevede il caso in cui il danneggiatore non agisca con negligenza, e la pena è più severa, prevedendo un risarcimento calcolato con una quota fissa. In questi casi si vede come la gestione delle acque fosse affidata ai singoli agricoltori, che dovevano sorvegliare l'acqua, quando irrigavano i campi, facendo attenzione a non danneggiare i campi vicini. I codici, in questo caso, prendono in considerazione soltanto il danneggiamento del singolo contadino, che deve essere risarcito per l'incuria e la negligenza di un vicino.

Da altre fonti sappiamo, inoltre, che erano conosciute anche le servitù di passaggio e d'acqua: la compravendita di terreni agricoli, infatti, molto diffusa specialmente nella Babilonia centro-settentrionale, doveva necessariamente prevedere, in caso di parcellizzazioni di terreni originariamente più grandi, la possibilità di accedere ai fossati d'irrigazione, anche attraversando fondi altrui.[21]

4. L'uso delle acque come mezzo di trasporto

Fiumi e canali, però, in Mesopotamia, avevano anche la funzione del trasporto di uomini e merci. Esistevano chiatte e barche, per lo più di forma rotonda, fatte di giunchi e poi incatramate col bitume, affiorante naturalmente in alcune località come Hit. Esse venivano usate sia scendendo verso sud, che risalendo verso nord: quando si andava controcorrente era necessario trainarle con delle funi dagli argini dei fiumi e dei canali.

Il luogo più frenetico delle città mesopotamiche era il *karum*, il porto fluviale, a ridosso del quale erano i magazzini e gli uffici dei mercanti, che trasportavano tutto, dalle derrate alimentari, vera ricchezza della Mesopotamia centro-meridionale, alle stoffe, fino ai metalli, alle pietre e al legname: tutte materie prime, queste ultime, carenti nella zona, e provenienti dall'Anatolia, dall'Iran, dall'Egitto, dalla Siria, ma anche dal Golfo Persico. I commerci di questi beni sono attestati da epoca remota, ben prima dell'età protostorica, e l'utilizzazione delle barche deve essere stata assai antica anch'essa. Le imbarcazioni, però, erano anche utilizzate per il trasporto delle truppe: pure gli eserciti si muovevano sulle barche, che certamente dovevano essere più veloci delle carovane, e forse anche gli approvvigionamenti e le

---

[21] Cfr. a questo proposito Simonetti 2010: 379–81. Cfr. anche *CAD* M/2 s. v. mūṣû: 247–9.

tassazioni che venivano recuperate durante le campagne militari giungevano via fiume. Se pensiamo che le imbarcazioni potevano facilmente affondare e che erano incatramate, possiamo dire che anche così poteva avvenire l'inquinamento delle acque: ma forse la sensibilità degli antichi per queste fonti di inquinamento doveva essere assai scarsa. Nei codici mesopotamici, ad esempio, la maggior parte dei paragrafi dedicati all'argomento prendono in esame soltanto la responsabilità del nocchiero nei confronti del proprietario della nave e del carico. Anche in questo caso, cioè, i codici proteggono gli interessi lesi dal comportamento negligente di un altro, e non quelli generali della collettività.

> *CLI* 8: Se un uomo provoca l'affondamento di un'imbarcazione, egli sarà condannato a risarcire il danno.
> *CLI* 9: Se un uomo prende a nolo un'imbarcazione ma non segue le disposizioni sulla rotta e causa la perdita dell'imbarcazione facendola incagliare sul fondo sabbioso, dovrà risarcire il valore dell'imbarcazione.

I due paragrafi del Codice di Lipit-Ištar riguardanti la perdita delle navi, vedono il risarcimento del danno come pena in caso di perdita dell'imbarcazione.

In un testo considerato scolastico e redatto in sumerico, YOS 1.28, risalente, pare, al 1800 a. C., cioè in piena età paleo-babilonese, nel paragrafo 3, si legge: 'Se un uomo, mutando la rotta di un'imbarcazione, ne causa la perdita, sia condannato a pagarne il noleggio finché non sarà restaurata'. In questo caso noi possiamo pensare che spesso le barche potevano essere recuperate e rimesse in condizione di navigare.

Anche nel prisma FLP 1287 ci sono dei paragrafi dedicati all'argomento:[22]

> iv 42-v 11: Se un uomo prende a nolo un'imbarcazione ma non segue le disposizioni sulla rotta e ne causa la perdita facendola incagliare sul fondo sabbioso, dovrà risarcire il valore dell'imbarcazione
> v 21–26: Se (un uomo prende a nolo un'imbarcazione e) la barca affonda dovrà rimborsarne il valore e pagare ugualmente il noleggio.
> v 27–31: Se un'imbarcazione che va controcorrente causa l'affondamento di un'altra barca che segue la corrente, il suo proprietario/nocchiero dovrà risarcire il valore dell'imbarcazione perduta.
> v 32–36: Se una barca che segue la corrente causa l'affondamento di un'altra che va controcorrente, il suo proprietario/nocchiero non sarà tenuto a risarcire.

Il Codice di Ḫammurapi ci offre una più ricca casistica.

> *CH* 235: Se il lavoro del calafato non è stato accurato ed entro l'anno l'imbarcazione comincia a pendere e a denunciare difetti, il calafato dovrà demolirla e rifarla solida.

In questo caso si parla di un lavoro mal eseguito, che l'artigiano dovrà provvedere a rifare a sue spese.

> *CH* 236: Se un nocchiero causa l'affondamento o la perdita di un'imbarcazione avuta a nolo, dovrà risarcire il proprietario con un'altra imbarcazione.
> *CH* 237: Se un uomo ha preso a nolo un'imbarcazione ed un nocchiero, che provoca l'affondamento e la perdita dell'imbarcazione e del carico (orzo, lana, olio, datteri o qualsiasi altro bene), il nocchiero dovrà risarcire imbarcazione e carico.

---

[22] Cfr. Roth 1979.

CH 238: Se un nocchiero provoca il naufragio di un'imbarcazione e poi riesce a recuperarla, dovrà risarcire la metà del valore dell'imbarcazione.

CH 240: Se un'imbarcazione che naviga controcorrente sperona un'imbarcazione che naviga secondo corrente provocandone il naufragio, il padrone dell'imbarcazione affondata dovrà dichiarare davanti al dio il valore del carico, e dovrà essere risarcito della perdita del carico e dell'imbarcazione dal nocchiero che lo ha speronato.

In tutti questi casi, la responsabilità viene attribuita al nocchiero, cioè a chi è responsabile della navigazione: tale responsabilità sarà per la sola barca, se vuota, per la barca e il carico se carica, e ridotta alla metà qualora il nocchiero sia in grado di recuperare la barca affondata. L'ultimo paragrafo, che pure attribuisce la responsabilità della perdita di una barca che segue la corrente al responsabile della navigazione della barca che, navigando controcorrente, l'abbia speronata, invece, ci induce a chiederci cosa sarebbe accaduto in caso contrario. Probabilmente era più facile manovrare la barca controcorrente (doveva essere sospinta e spesso trainata), rispetto a quella che seguiva la corrente e che più difficilmente poteva scegliere perfettamente il punto del fiume o del canale da percorrere. È probabile che si seguisse la norma contenuta alle righe V 32–36 del Prisma FLP 1287.

Osservazioni conclusive

In conclusione, quindi, possiamo dire che lo sfruttamento dei due fiumi da parte dei popoli che abitarono la Mesopotamia nel corso del secondo millennio a. C. fu piuttosto massiccio. Se da un lato le grandi opere di canalizzazione hanno sfruttato il più possibile il terreno racchiuso dai due fiumi, tuttavia dall'altro alcune pratiche agricole e la continua manutenzione dei fiumi e dei canali hanno preservato per molti secoli l'efficienza del sistema. Certamente la consapevolezza di modificare l'ambiente in cui vivevano era forte, specialmente quando i sovrani ordinavano di scavare brecce nelle dighe in caso di forti piene, o di ripararle, una volta passato il pericolo, oppure di dragare il letto dei canali, per aumentarne la portata in tempi di magra. Che questa consapevolezza fosse tale da pensare di poter inquinare, e cioè distruggere, l'ambiente in cui vivevano ed operavano, è più difficile da appurare. Così come difficilmente gli antichi si saranno resi conto dei danni che l'affondamento delle imbarcazioni, con i loro vari carichi, potesse arrecare ai fiumi: sembrerebbe che la loro preoccupazione fosse rivolta ai soli danni materiali creati dalla cattiva gestione della nave da parte del nocchiero.

Insomma, l'intervento umano sulle acque dei due grandi fiumi c'è stato ed è stato massiccio. L'impatto sulla vasta piana mesopotamica anche. Gli uomini erano certamente consapevoli del loro intervento, e di quanto esso potesse influire sull'ecosistema naturale. Più difficile, invece, è valutare quanto fossero consapevoli degli effetti potenzialmente nocivi di tale intervento. Da quel che possiamo vedere, anche dai pochi esempi riportati circa la regolamentazione dei comportamenti dei singoli in questi ambiti, l'interesse era posto più sulla buona gestione del sistema antropico (risarcire i danni direttamente ai vicini danneggiati o ai proprietari delle imbarcazioni e delle merci naufragate), piuttosto che sulla repressione del danno ambientale.

Forse qualcosa in più possono dirci altri tipi di fonti, che hanno un carattere più ideologico che storico, e cioè le benedizioni e maledizioni incluse all'interno delle iscrizioni reali. In esse, i sovrani da un lato fanno affidamento sulla benevolenza o malevolenza divina, ma dall'altro sanno anche che il benessere del popolo dipende dalla condotta giusta del sovrano che lo governa. Gli dèi donano abbondanza a chi agisce bene, a chi è pio e fa quello che bisogna fare, mentre colpiscono con carestie, alluvioni e devastazioni chi si comporta male. Ovviamente questa ideologia evolverà nel periodo seguente, riportando l'intervento degli dèi e i capricci della natura a una volontà esterna e diversa da quella umana, del tutto indipendente e indifferente all'operato umano (teodicea babilonese, il giusto sofferente, ecc.). In ogni caso, in ambito religioso si può vedere una sorta di specchio della natura così come la vedevano gli antichi abitanti della Mesopotamia: da un lato c'è il dio Ea (il sumerico Enki), il dio delle acque dolci sotterranee, il dio benevolo, che protegge e preserva sempre le peculiarità degli uomini (nel Diluvio, nel mito di Adapa, ecc.), dall'altro c'è Enlil (e i suoi simili Adad, Ninurta, ecc.), dio della tempesta, della pioggia, dell'atmosfera, che porta la pioggia, la tempesta, gli uragani e che fa straripare i fiumi e i canali. Un dio nemico degli uomini (è lui che invia il Diluvio, che vuole annientare gli uomini). Ci sono acque buone e acque nocive: bisogna curare bene le prime (i fiumi e i canali) affinché le altre non prendano il sopravvento.

Bibliografia

Abbreviazioni

AbB 1   = KRAUS, F. R. 1964. *Briefe aus dem British Museum* (CT 43 und 44). *Altbabylonische Briefe* 1. Leiden

AbB 2   = FRANKENA, R. 1966. *Briefe aus dem British Museum* (LIH und CT 2–33). *Altbabylonische Briefe* 2. Leiden.

AbB 4   = KRAUS, F. R. 1968. *Briefe aus dem Archive des Šamaš-ḫāzir in Paris und Oxford* (TCL 7 und OECT 3). *Altbabylonische Briefe* 4. Leiden.

AbB 13 = VAN SOLDT, W. H. 1994. *Letters in the British Museum. Altbabylonischen Briefe* 13. Leiden.

CAD   = *Chicago Assyrian Dictionary*. Chicago.

CH   = *Codice di Ḫammurapi*

CLI   = *Codice di Lipit-Ištar*

YOS 1 = Yale Oriental Series, CLAY, A. T. 1915. *Miscellaneous Inscriptions in the Yale Babylonian Collection* (Yale Oriental Series, Babylonian Texts). New Haven.

ADAMS, R. MC C. 1981. *Heartland of Cities*, Chicago-London.

ARNAUD, D. 1982. 'La legislation de l'eau en Mésopotamie du IIIe au Ier millénaire', in *L'Homme et l'eau en Méditerranée et au Proche Orient. Volume II. Aménegement Hydrauliques, état et legislation.* Lyon, 45–59.

ASOUTI, E. e A. S. FAIRBAIRN 2010. 'Farmers, Gatherers or Horticulturalists? Reconstructing Landscapes of Practice in the Early Neolithic', in B. FINLAYSON e G. WARREN (eds.), *Landscapes in Transition.* (Levant Supplementary Series. Vol. 8). Oxford, 161–72.

BESANÇON, J. e B. GEYER, 2003. 'La géomorphologie de la basse vallée de l'Euphrate syrienne', in B. GEYER, e J. Y. MONCHAMBERT (eds.), *La Basse vallée de l'Euphrate Syrienne du Néolithique à l'avènement de l'islam: géographie, archéologie et histoire.* Beyrouth, 7–59.

BRINKMAN, J. A. 1984. 'Settlements Surveys and Documentary Evidence: Regional Variation and Secular Trend in Mesopotamian Demography', *Journal of Near Eastern Studies* 43, 169–80.

BURINGH, P. 1960. *Soil and Soil Condition in Iraq.* Baghdad.

CARDASCIA, G. 1966. 'Faut-il limine giš apin = nartabu: une machine d'irrigation?', *Revue d'Assyriologie et d'Archéologie Orientale* 60, 153–64.

COCQUERILLAT, D. 1968. *Palmieraies et cultures de l'Eanna d'Uruk.* Berlin.

COLE, S. W. e H. GASCHE 1998. 'Second and First Millennium BC. Rivers in Northern Babylonia', in H. GASCHE e M. TANRET (eds.), *Changing watercourses in Babylonia: Towards a Reconstruction of the Ancient Enviroment in Lower Mesopotamia*. Leuven, 1–64.

GIBSON, McG. 1974. 'Violation of fallow and engineered disaster in Mesopotamian civilization', in TH. DOWNING e MC G. GIBSON (eds.), *Irrigation's Impact on Society*. Tucson, 7–20.

GIBSON, McG., R. L. ZETTLER e J. A. ARMSTRONG 1983. 'The Southern Corner of Nippur. Excavation during the 14th and 15th seasons', *Sumer* 39, 170–90.

JANNSEN, C., H. GASCHE e M. TANRET 1994. 'Du chantier à la tablette. Ur-Utu et l'histoire de sa maison à Sippar-Amnānum', in H. GASCHE, M. TANRET, C. JANNSEN, e A. DEGRAEVE (eds.), *Cinquante-deux reflexions sur le Proche-Orient Ancient offertes en homage à Léon De Mayer* (MHEO 2). Leuven, 91–123.

KLENGEL, H. 1993. *Il re perfetto. Hammurabi e Babilonia*. Bari.

KRAELING, C. H. e R. MC C. ADAMS 1960. *City Invincible*. Chicago.

LEEMANS, W. F. 1968. 'Old Babylonian Letters and Economic History', *Journal of Economic and Social History of Orient* 11, 171–226.

LOESSØE, J. 1953. 'Reflexion on modern and ancient oriental water works', *Journal of Cuneiform Studies* 7, 5–26.

ROTH, M. T. 1979. *Scholastic Tradition and Mesopotamian Law: a Study of FLP 1287, a Prism in the Collection of the Free Library of Philadelphia*. Ann Arbor.

ROWTON, M. B. 1967. 'Watercourses and water rights in the official correspondence from Larsa and Isin', *Journal of Cuneiform Studies* 21, 267–75.

SALONEN, A. 1968. *Agricultura Mesopotamica nach sumerisch-akkadischen Quellen*. Helsinki.

SAPORETTI, C. 1998. *Antiche leggi*. Milano.

SAPORETTI, C. 2002. *La rivale di Babilonia*. Roma.

SIMONETTI, C. 2010. 'Note in margine ad alienazioni immobiliari d'età paleo-babilonesi', in M. G. BIGA e M. LIVERANI (eds.), *Ana turri gimilli*. *Studi dedicati al padre Werner R. Mayer, S. J. da amici e allievi*. Roma, 379–81.

STOL, M. 1982. 'A Cadastral Innovation by Hammurabi', in G. VAN DRIEL (ed.), *Zikir šumim. Assyriological Studies Presented to F. R. Kraus*, Leiden, 351–7.

THUREAU-DANGIN, F. 1924. 'Correspondance de Ḫammurapi avec Šamaš-ḫaṣir', *Revue d'Assyriologie et d'Archéologie Orientale* 21, 1–58.

WALTERS, S. D. 1970. *Water for Larsa: Old Babylonan Archive Dealing with Irrigation*. New Haven-London.

Luigi Capogrossi Colognesi

Roman Rural Landscape and Legal Rules

1. Natural resources in Roman times and the legal point of view

The idea that in ancient Rome natural resources were even *too* abundant, given the insufficient human capacity to exploit them, is suggestive, but also misleading. It was a challenge
for the Romans to learn how to exploit this abundance in a sustainable and defendable way.
In early Roman times, the natural elements on which human life depends, the water and
land needed to be changed from their original, untouched and untamed nature and transformed in workable land for husbandry and agriculture, or brought in places where water
could become a useful element for human settlements.

It is sort of a paradox that the problems of the Romans (as well as other ancient societies) were to a certain extent the opposite of ours, in modern times. Now, human life and
our social organization can be, in many ways, dangerous to nature; whereas at that time,
human life, particularly in the rural settlements, was endangered by the savage elements of
nature. Cultivated land, pasturage, and the *silvae caeduae* were products of human work and
transformed from their natural conditions. To the Romans, these lands were object of such
a kind of conquest no less important than military conquest. Just as with military conquest,
the new areas brought under human control needed to be defended, not only from re-en-
croachment by nature, but also from the results of human conflict.

The consequence was that, from the very beginning, legal rules and social institutions
were aimed to coordinate and discipline the forms in which natural resources were distributed and exploited by individuals, as well as by the community as a whole. In this regard, we
find in the most ancient legal history, codified in the XII Tables, a system of rules concerning private relationships between individuals, primarily aimed at protecting their property.
Additionally, we can observe that the same legal system was also concerned with the activity
of public magistrates and institutions in land management, where the private interests were
strictly connected and integrated in the public interest of the whole community. It is still
more interesting that, in many cases, this public interest was achieved through the prosecution of the aims and interests of private individuals.

Rather frequently, in fact, we observe that jurists and the Roman praetor governed individual conflict between private landowners, not only by giving victory to one and defeat to
the other, but also by pushing the litigants toward a new form of cooperation. As a result of
this latter form of conflict resolution, certain general advantages, concerning protection and
further exploitation of natural resources, were attained for the whole community through
inter-individual conflict. Private interests and public utility were the two aims – sometimes
conflicting – pursued by Roman legal system. Any attack against property rights and, in

particular, the boundaries of any *fundus* was punished very severely. On the other hand, the autonomous exercise of the owner's powers on his own property was limited, if his behavior would damage a neighbor's property. Already, in the XII Tables, an important *actio aquae pluviae arcendae* limited the freedom of the owner to undertake work on his land designed to limit damage from waterfalls, if these innovations would damage his neighbor's property.[1]

In fact it is in the legal commentaries of classical Roman jurists regarding the *actio aquae pluviae arcendae* that we find the most meaningful evidence of this parallel pursue of individual and common benefits. They frequently make references to *fossae* and *aggeres*: they did consist in drainage works by which the Roman landowners did try to protect their properties from damages produced by water rainfall. I think that, in general, these works were realized through the cooperation and the agreement of the neighboring landowners, under the influence of the relationship established by the *actio aquae pluviae arcendae*. In that way, through cooperation (we should refer here to the *pagi* and their local governance), not only the individual's property, but also the whole rural district was protected against natural dangers. Furthermore, we must consider that, in the system of XII Tables, there were also various legal tools aimed at maximizing the exploitation of natural resources of limited availability (e. g. water for drinking or for irrigation and the system of local roads connecting multiple agrarian properties with the public roads). The rights to use or the ownership of these resources were shared in common between various landowners, and the law regulated their common exploitation.

2. Roman management of territory

From the middle of the fourth century BC, all these elements were combined in a vast project for managing Roman territory. After the Latin League's dissolution in 338 BC, Rome's power to found new Latin colonies not only ensured it would control new territory, but also allowed it to pursue a policy of demographic and economic growth.[2] Rome now had access to a constantly renewed supply of land to distribute to its citizens (as well to the Latins, and to the inhabitants of the *civitates foederatae*), and thanks to the colonies, it could move its excess population into the newly occupied areas, to the benefit especially of the poorer classes of Romans, and of the cities that had the closest ties to Rome. In that way, many Romans, Latins, and citizens of other allied towns came to settle in areas that had either seen little development, or from which the previous inhabitants had been driven out, leading to their gradual urbanization and, most important, hastening the full agrarian exploitation of the entire Italian peninsula. This happened especially in those recently conquered areas that

---

[1] See Pomp., *l. 7 ex Plaut.*: D. 40.7.21*pr. Labeo libro posteriorum ita refert (…) sic et verba legis duodecim tabularum veteres interpretati sunt 'si aqua pluvia nocet', id est 'si nocere poterit'*; Ulp., *l. 53 ad ed.*: D. 39.3.1*pr. Si cui aqua pluvia damnum dabit, actione aquae pluviae arcendae avertetur aqua. aquam pluviam dicimus, quae de caelo cadit atque imbre excrescit, sive per se haec aqua caelestis noceat, ut Tubero ait, sive cum alia mixta sit.* Ibid., § 1: *Haec autem actio locum habet in damno nondum facto, opere tamen iam facto, hoc est de eo opere, ex quo damnum timetur: totiensque locum habet, quotiens manu facto opere agro aqua nocitura est, id est cum quis manu fecerit, quo aliter flueret, quam natura soleret, si forte immittendo eam aut maiorem fecerit aut citatiorem aut vehementiorem aut si comprimendo redundare effecit. quod si natura aqua noceret, ea actione non continentur.* See also Paul., *l. 16 ad Sab.*: D. 43.8.5.

[2] Capogrossi Colognesi 2014b: 93–7.

were best suited to large-scale agricultural investment; the landscape of Picenum and Gallia Cisalpina was greatly altered by colonization in this way, with lasting effects on the history of Italy as a whole. The *limitatio* was directly connected to the expansion of Roman power on the Italian peninsula.

It is well known the extraordinary Roman project of reshaping a great part of Italian territory through *centuriatio*. Organizing the physical configuration of rural landscape was a crucial stage in the foundation of a colony. From the fourth century BC, the Romans employed a system that subdivided the colony's land into uniform, same-size lots. Under the supervision of the magistrates in charge, a team of land-surveyors (*agrimensores*) would select a central point through which they traced two perpendicular lines, the *cardo* and *decumanus maximus*, drawing the central axis of the future city. At equal distances from these, other parallel lines were then traced (further *cardines* and *decumani*), intersecting at right angles to form identical quadrangles known as *centuriae*. In Italy, the *cardines* and *decumani* – referred to more generally as *limites* ('boundaries') by surveyors – were broad enough to serve as rural roads and formed an extensive network ensuring that all landed estates were linked to the most important public routes. Each *centuria* was then divided by the Roman magistrates, with the help of land-surveyors, into many lots, each of them being given as property (*adsignatus*) to a member of the colony. Along with the *limitatio* (this division) of the land, a large system of regular roads, drainage canals was also built. In that way, a geometrical design shaped a great part of landscape, in Roman Italy, first, and then in the whole Roman Empire.[3]

The practice of *centuriatio* was rooted in the most ancient Italian and Roman traditions, particularly Etruscan ideas of space as a dimension of a religious cosmos and the town-planning experience of the Greek cities. From a technical standpoint, the practice evolved into a highly advanced science comprising elements of geometry, astronomy, geology, as well as the law. Ancient sources provide excellent documentation on *centuriatio*, but the most important evidence we have is the enduring presence, throughout the area once occupied by the empire, of the traces left by this massive alteration of the landscape. In many cities today, especially in the plains of northern Italy, the urban grid is superimposed on the ancient colonial grid that once organized the urban, as it did the rural, portions of the colonies.

This new system developed parallel to the efficient spread of rural settlements (*coloniae civium Romanorum* and *latinorum*, but also *municipia*) throughout the whole Peninsula. The principal aim achieved was a very regular and clear land distribution plan allowing to the Roman landowners, an easy and sure knowledge of the boundaries of each individual land-allotment. But not less important was the efficient and continuous management, by public authorities, of the best possible conditions for agrarian exploitation of land. For that reason, this enormous cultural edifice was not limited to a geometrical design, but did change the whole physiognomy of the natural element constituted by land. Through the *limitatio*, the Romans did submit the whole natural landscape – the wild forests and marshes, as well the cultivated lands – to a system of rules and regulations.

In many municipal laws and in various writings of Roman land-surveyors, there is clear evidence that Roman magistrates and land surveyors, in managing the system of the *limitatio*, pursued also a more comprehensive policy. A policy aimed to preserve land integrity, wild-life (eg. the *silvae* and the peculiar form of *silva caedua*) in and near the agrarian settle-

---

[3] Capogrossi Colognesi 2012: 96–8.

ments, water resources, and the integrity of territory from natural (and sometimes also human) dangers.[4] In the *leges coloniariae* we find many references to the building and maintenance of *fossae*, channels and agrarian roads as a general duty of municipal magistrates. This is confirmed by the land-surveyor reports and is clear evidence of the pursuit of the best conditions for rural life and economy, in parallel to the protection of the individual interests and legal claims.

3. Measures for protection of natural environment

It was through this second element – individual action – that many of the general aims of land management and protection of natural environment were attained. I have already quoted the *actio aquae pluviae arcenda*, which became one of the foundations of all private relationships between landowners. But, many other legal tools, primarily the *interdicta*, were shaped by the praetor to the same end. Many of them, as well as the *actiones populares*, were created in order to allow anyone in the political community to pursue a public interest by suing someone who did attempt to the public trust. In that way, not only the direct action of Roman magistrates, but also this private opposition could contribute to preserve many of the public resources such as roads, harbours, rivers, lakes, beaches (*litora maris*) etc.

This cooperation of private activity with Roman magistrates is a characteristic feature of the Roman republican government. In this way, individual interests were mobilized to obtain protection also for the whole community. Without recognizing the real relevance of this peculiar kind of cooperation, it would be impossible for us to understand the way in which many of the public aims of the Roman government were pursued. In particular, this must be kept in mind when considering how Romans protected natural resources from pollution and other human predation.

Even during the last period of the kings, the political community was already directly involved in this policy. On this point, it is sufficient to recall one of the principal enterprises of the first centuries of Roman history: the *Cloaca maxima*. Undoubtedly, this great collective work was mainly designed to drain the marshes of the Forum, but at the same time it served to drain away any rain-water or flood water that accumulated in the city.[5] In

---

[4] See e.g. *Lex col. Genetivae, cap.* 77 (*FIRA* I, p. 184; CRAWFORD 1996, 1: 404): *Si quis vias fossas cloacas IIvir aedilisve publice facere immittere commutare aedificare munire intra eos fines, qui coloniae Iuliae erunt, volet, quot eius sine iniuria privatorum fiet, it is facere liceto; cap.* 104 (*FIRA* I, p. 191, CRAWFORD 1996, 1: 409): *Qui limites decumanique intra fines coloniae Genetivae deducti facti que erunt, quaecumque fossae limitales in eo agro erunt, qui iussu C. Caesaris dictatoris imperatoris et lege Antonia senatusque consultis plebique scitis ager datus atsignatus erit, ne quis limites decumanosque opsaeptos neve quit immolitum neve quit ibi opsaeptum habeto, neve eos arato, neve eis fossas opturato neve opsaepito, quo minus suo itinere aqua ire fluere possit. si quis adversus* ea quit fecerit, is in res singulas, quotienscumque fecerit, HS ∞ colonis coloniae Genetivae Iuliae dare damnas esto, eiusque pecuniae cui volet petitio persecutioque esto.* See also *Lex Mamilia, Roscia Peducaea et Allia, cap.* 4 (*FIRA* I, p. 139; CRAWFORD 1996, 2: 763): *…quecumque fassae limitales in eo agro erunt, qui ager hac lege datus adsignatus erit, ne quis eos limites decumanosque obsaeptos neve quid in eis molitum neve quid ibi opsaeptum habeto, neve eos arato, neve eis fossas opturato neve opsaepito, quominus suo intinere aqua ire fluere possit.*

[5] Dion. Hal., *ant. rom.* 3.67.5; 4.44.1: He also began the digging of the sewers, through which all the water that collects from the streets is conveyed into the Tiber – a wonderful work exceeding all description. Indeed, in my opinion the three most magnificent works of Rome, in which the greatness of her empire is best seen, are the aqueducts, the paved roads and the construction of the

addition to its primary aims, this enormous work of drainage had a positive impact on the sanitary conditions of early Rome. We should also recall that, in the fourth cent. BC, the first important aqueduct was built, bringing the healthy waters from hills that were many miles from the city.[6]

With the city development, the opportunities for mutual disruption and pollution multiplied and this reflected especially in the relations among individuals. The peculiar rules concerning private sewers, for example, are very ancient, and surely date back to the Republic. However, there are doubts that the messy rebuilding of Rome after the Gallic fire produced the peculiarity that, since then, many sewers crossed private properties.[7]

Whatever the historical origin and initial dating of these private channels might be, it is likely that already in late Republican Rome they had taken a major importance for the purpose not only of public but also of private hygiene. And this explains the configuration of a specific easement allowing the owner of a building to dig his sewer under another person's property in order to connect with the public system. This explains also the distinctive character of the interdict conceded to the same owner of *reficere cloacam* in that property. Unlike the other interdicts concerning the exercise of a private aqueduct or a private street, this peculiar interdict did not depend on the existence of an effective private *ius cloacae immitendi per fundum tuum*, or of its prior use. The owner disturbed by the restoration works could not avoid this interference in his property: he had as only guarantee the *cautio damni infecti*.

Nevertheless, what attracts our attention is, with the latitude of this protection, the reasons advanced by the jurists. We can read them in Ulpian in his commentary to the text of the interdict: first of all in *D.* 43.23.1.2 (71 *ad ed.*), where he explains that the intervention of the praetor with this interdict in favour of the restoration and cleaning of private sewers is due to the fact that *utrumque et ad salubritatem civitatium et ad tutelam pertinet*. This policy is reiterated in paragraph 7, where we read that the *cloacarum refectio et ad utilitatem publicam spectare purgatio videtur.*[8] But the idea of this *publica utilitas* constituted by the *salubritas civitatium* is not accepted only by Imperial jurists: being present at least at the end of the Republic, as it is witnessed by Labeo's interpretation – in the Augustan period – with which

---

sewers. I say this with respect not only to the usefulness of the work (concerning which I shall speak in the proper place), but also to the magnitude of the cost, of which one may judge by a single circumstance, if one takes as his authority Gaius Acilius, who says that once, when the sewers had been neglected and were no longer passable for the water, the censors let out the cleaning and repairing of them at a thousand talents. (Engl. transl. by E. Carey, Cambridge MA 1939, 239–41).

[6] Front., *aq.*, 5; Livy 9.29; Diod. Sic. 20.36.

[7] Livy 5.55.4: *Antiquata deinde lege, promisce urbs aedificari coepta. Tegula publice praebita est; saxi materiaeque caedendae unde quisque vellet ius factum, praedibus acceptis eo anno aedificia perfecturos. Festinatio curam exemit vicos dirigendi, dum omisso sui alienique discrimine in vacuo aedificant. Ea est causa ut veteres cloacae, primo per publicum ductae, nunc privata passim subeant tecta, formaque urbis sit occupatae magis quam divisae similis.*

[8] See Ulp., *l.* 71 *ad ed.*: *D.* 43.23.1 *pr. Praetor ait: 'quo minus illi cloacam quae ex aedibus eius in tuas pertinet, qua de agitur, purgare reficere liceat, vim fieri veto. damni infecti, quod operis vitio factum sit, caveri iubebo'.* Ulp., *l.* 71 *ad ed.*: *D.* 43.23.2. *Curavit autem praetor per haec interdicta, ut cloacae et purgentur et reficiantur, quorum utrumque et ad salubritatem civitatium et ad tutelam pertinet: nam et caelum pestilens et ruinas minantur immunditiae cloacarum, si non reficiantur.* Ulp., *l.* 71 *ad ed.*: *D.* 43.23.7 *Quia autem cloacarum refectio et purgatio ad publicam utilitatem spectare videtur, idcirco placuit non esse in interdicto addendum 'quod non vi non clam non precario ab illo usus', ut, etiamsi quis talem usum habuerit, tamen non prohibeatur volens cloacam reficere vel purgare.* On this general theme see the fundamental works of: Di Porto 1989–90: 271–309; Di Porto 2014: 51–81; Solidoro Maruotti 2009. See also Capogrossi Colognesi 1976: 307–9.

he extended the protection of the interdict also to those who wanted *privatam cloacam in publicam immittere,* including therefore a new *cloaca* built to serve one's home.[9]

4. Water management between public and private welfare

An important consequence of the availability of water brought into the city by an increasing number of aqueducts was the growth of *balnea* and public *termae,* with the consequent improvement in public health. But, these baths would also damage in some ways the private and public welfare: moisture, heat, and noise were the inconveniences of which the inhabitants, living near these baths, ever more frequently complained. These problems remained partially unsolved all through the Imperial age: it is a case for which, I think, we could speak of a kind of a difficult and always changing balance between opposite needs and values: of the community on one side, of private inhabitants on the other.

There is no doubt that, in general, for what concerns water, through the important complex of water easements, there was a general purpose of maximizing the exploitation of this scarce resource, through a very articulated allocation of rights of use. Yet, both the historical antecedents dating back to XII Tables legislation, than the subsequent transformation of proprietary forms relating to the ancient *iura aquarum* in the scheme of individual rights remain largely beyond our topic.

The legal instruments aimed at protecting the proper functioning of private aqueducts, were, not only the Roman civil process, but – and may be principally – also the system of interdicts. Although these appear slightly less effective than those destined to protect private sewers. Here the priority is to ensure the regular supply of running water to all the entitled people, but also to keep in good conditions private springs, fountains and waterworks and to avoid any kind of damages or pollution. In fact, the problems of pollution mix and blend in the more general guarantee of the proper functioning of the group of structures intended to elevate the quality of urban as well as country life.[10]

The *servitus aquae ductus,* to which these rules refer to, is essentially designed to regulate agrarian relations. In the city, by contrast, the judiciary has a priority role since it holds the responsibility of the public water supply. A role destined to increase significantly – as for breadth of powers and intervention capacity – with the *curatores aquarum* of imperial age, on which we have an enlightening testimony offered by Frontinus' treaty.

For what concerns city water the intertwining of public roles and private law situations increases greatly. The same water conduits in private homes, widely attested by both literary and archaeological evidence, present themselves in an ambiguous form: they are modelled on the civil regime of a private *servitus* but not always carried out through this precise legal figure. And even here, except for the *refectio* of public aqueducts, it is difficult to single out the precise intervention area where is possible to detect the specific protection against pollu-

---

[9] Venul., *l.* 1 *interd.*: D. 43.23.2. *quamquam de reficienda cloaca, non etiam de nova facienda hoc interdicto comprehendatur, tamen aeque interdicendum Labeo ait, ne facienti cloacam vis fiat, quia eadem utilitas sit: praetorem enim ssic interdixisse, ne vis fieret, quo minus cloacam in publico facere liceret: idque Ofilio et Trebatio placuisse.*

[10] Capogrossi Colognesi 1976: 501–16.

tion of this precious commodity: although a public interest in that field is indirectly attested, if nothing else, from Frontinus.[11]

In general, the management of water resources was a central issue for Roman government, not only the Roman aqueducts testify it, but also legal system. Anyhow, referring to the means to protect this precious element from pollution, we must admit that, in a rural settlement, they are less obvious than in the urban sphere. In any case, there is a passage of the Digest that weighed a great deal on the attempts of scholarship to understand the precise Roman notion of private waters and their pollution. I am referring to *D.* 43.24.11 pr., on which an impressive amount of literature has concentrated, especially on the case considered by Labeo consisting of someone *qui aliquid effuderit in puteo vicini, ut hoc facto aquam corrumperet*, rightly sanctioned with the interdict *quod vi aut clam*.

Surely this principle was applied also to protect springs and water conduits. But I don't think that, for the protection of all individual interests connected to the availability of private and public waters, should have been sufficient this interdict alone. I am sure that, at least from the time of another great jurist of late Republic, Servius Sulpicius Rufus, also the *actio negatoria* could be employed against any attempt to pollute or damage springs, fountains, aqueducts and so on.[12]

---

[11] *Lex Quinctia de aquae ductis* (Front. 129): *T. Quintus Crispinus consul *** populum iure rogavit populusque iure scivit in foro pro rostris aedis Divi Iulii pr. K. Iulias. Tribus Sergia principium fuit. Pro tribu Sex. L. f. Virro primus scivit. Quicumque post hanc legem rogatam rivos, specus, fornices, fistulas, tubulos, castella, lacus aquarum publicarum, quae ad urbem ducuntur, sciens dolo malo foraverit, ruperit, foranda rumpendave curaverit peiorave fecerit, quo minus eae aquae earumve quae pars in urbem Romam ire, cadere, fluere, pervenire, duci possit, quove minus in urbe Roma et in eis locis, aedificiis, quae loca, aedificia urbi continentia sunt, erunt, in eis hortis, praediis, locis, quorum hortorum, praediorum, locorum dominis possessoribusve aqua data vel adtributa est vel erit, saliat, distribuatur, dividatur, in castella, lacus immittatur, is populo Romano HS centum milia dare damnas esto. Et qui D. M. qui eorum ita fecerit, id omne sarcire, reficere, restituere, aedificare, ponere et celere demolire damnas esto sine dolo malo atque omnia ita ut quicumque curator aquarum est, erit, si curator aquarum nemo erit, tum is praetor qui inter cives et peregrinos ius dicet, multa, pignoribus cogito, coercito, eique curatori aut si curator non erit, tum ei praetori eo nomine cogendi, coercendi, multae dicendae sive pignoris capiendi ius potestasque esto. Si quid eorum servus fecerit, dominus eius HS centum milia populo Romano D. D. E. Si qui locus circa rivos, specus, fornices, fistulas, tubulos, castella, lacus aquarum publicarum, quae ad urbem Romam ducuntur et ducentur, terminatus est et erit, ne quis in eo loco post hanc legem rogatam quid opponito, molito, obsaepito, figito, statuito, ponito, conlocato, arato, serito, neve in eum quid immittito, praeterquam eorum faciendorum, reponendorum causa, quae hac lege licebit, oportebit. Qui adversus ea quid fecerit, siremps lex, ius causaque omnium rerum omnibusque esto, atque uti esset esseve oporteret, si is adversus hanc legem rivum, specum rupisset forassetve. Quo minus in eo loco pascere, herbam, fenum secare, sentes [tollere liceat, eius hac lege nihilum rogatur]. Curatores aquarum, qui nunc sunt quique erunt, faciunto ut in eo loco, qui locus circa fontes et fornices et muros et rivos et specus terminatus est, arbores, vites, vepres, sentes, ripae, maceria, salicta, harundineta tollantur, excidantur, effodiantur, excodicentur, uti quod recte factum esse volent; eoque nomine eis pignoris capio, multae dictio coercitioque esto; idque eis sine fraude sua facere liceat, ius potestasque esto. Quo minus vites, arbores, quae villis, aedificiis maceriisve inclusae sunt, maceriae, quas curatores aquarum causa cognita ne demolirentur dominis permiserunt, quibus inscripta insculptaque essent ipsorum qui permisissent curatorum nomina, maneant, hac lege nihilum rogatur. Quo minus ex eis fontibus, rivis, specibus, fornicibus aquam sumere, haurire eis, quibuscumque curatores aquarum permiserunt, permiserint, praeterquam rota calice, machina liceat, dum ne qui puteus neque foramen novum fiat, eius hac lege nihilum rogatur.* See PALMA 1987: 439–57; MAGANZANI, 2004: 185–220. The study of VALLOCCHIA 2012, vol. 1 esp. chapt. 1–2, as well as vol. 2 chapt. 1–3 is just a thematic overview, without any analysis or focus on problems, about which, by contrast, previous legal doctrine deeply had investigated, although irregularly.

[12] CAPOGROSSI COLOGNESI 1976: 501–16.

Nor, one should be brought to think that everything was limited to Rome or Italy: within the provinces, we have today a precious evidence showing how the entire local community was involved in the exploitation and maintenance of aqueducts for local uses. I refer in particular to the so-called *lex Rivi Hiberiensi* that adds information to what we know from the *lex aquae* of Lamasba concerning a derivation of water from the Ebro river in Spain, meant to serve a relatively large territory and several smaller communities, belonging to different administrative structures within the province. The text suggests the presence of two different logics, a private and a public one, as demonstrated by the presence of a provincial envoy to sanction the agreement between the parties recorded in our document. DIETER NÖRR speaks, in this respect, of a 'Wassergenossenschaft', although he rules out with reason the hypothesis that this discipline of water conduits had roots in a pre-Roman reality[13].

The point here is that the approved regulation concerned first of all a capillary subdivision between all the persons burdened with the maintenance and cleaning costs of the derivation implants. These costs were divided in proportion to the extent of the rights of participation in the water: the text speaks of genuine *ius aquae*. This obligation is all the more important because the text implies how water was not only used for agriculture, but could also be used for domestic purposes as drinking water.

Of course, we're dealing with provincial realities: but how can we not think that the Romans on the peninsula had already adopted such schemes? The regime of the *dominium* certainly was not an obstacle to that effect. And even more so if we think of how the ancient colony and municipal regulations could have operated in that sense. A pale echo remains in the surviving *leges coloniariae*. In fact they widely attest the power of intervention of the municipal magistrates for the construction and maintenance of sewers and public water conduits, without possibility for individuals to interfere in their action. At least, in that occasion, the direct tuition of *publica utilitas* by municipal magistrates seems to be preferred to the protection of private property.

5. Cities and air pollution

But pollution, in the cities, did concern, not only waters, but also another *res communis*: air. The Chapter 76 of the *lex Coloniae Genetivae* states that: 'no one is to have tile works with a capacity of more than 300 tiles or tile like objects in the town of the *colonia* Iulia. Whoever shall have had (one), that building and place is to be public property of the *colonia* Iulia,

---

[13] Lex Rivi Hiberiensi (BELTRÁN LLORIS 2006: 147–97): *Id omne magistri pagi in commune redigunto | Cuius eorum qui operas aliutue quid-praestare debebit magistri pagi curatoresue praesentiam | habere non potueri<n>t, domo familiaeue eius de-|nuntie<n>t et cuius domo familiaeue eius denu[n]t[i]|atum erit ut s(upra) s(cripta) est non dederit feceritue, [ean]|dem poenam quae s(upra) s(cripta) est praestare debeat. Adri|uom Hiberiensem Capitonianum purgandum| reficiendumue ab summo usque ad molem i-|mam quae est ad Recti centurionis omnes pa-|gani pro parte (vacant 4) sua quisque praestare debe|ant. | Riuos quibus utentur communiter purgent re|feciant ita ut qua fine quisque aquam habet | usque eo operas praestet; perfectis riuis,[a]b ea| mole qua quisque aquam deriuat ad proxuma[m] | molem purgare… anno bis cum ei magistri pa|gi diem dixerint denuntiauerint; id adsidue | fieri debeat quod ipsius dolo malo non fiat. | Item si quis canalem aut pontem positum habet, | tamquam moles obseruabitur et eum locum is | tueri et purgare debebit et quantum ab ea re | riuus impeditus erit quominus aqua iusta per-|fluat. Magistri pagi magisterium gerent ex k(alendis) Iun(is) | in k(alendas) Iunias sequentes et ex quo magistri suffec|ti erunt diebus quin.* See NÖRR 2008: 108–88; BELTRÁN LLORIS 2014: 54–73; CAPOGROSSI COLOGNESI 2014a: 75–91.

and whoever shall be in charge of jurisdiction in the *colonia Genetiva Iulia* is to pay into the public hands that sum (derived) from that building'.[14] I believe this prohibition is clearly inspired to the need of avoiding in the city centre the presence of strong centrals for burning, incompatible with the dense human settlement. Starting from the late Republican age, Rome in the first place and the great city centres required the production of great quantities of bricks and lime, for building purposes. It was a material difficult to transport, considering the relation between unitary value and transportations costs. Thus, the importance of those suburban villas at the outskirts of the city, being able to supply these products, is directly related to the relevance of the ban of mass production of this material *within* the city (also for lime, the jurists speak of *calcem coquere*).

This prohibition is all the more significant because in its generality, it impacted directly on one of the few industrial sectors that the Roman-Italic economy had developed since the end of the Republican age: the building industry, as well as the production of fine ceramic artefacts throughout the Mediterranean basin. In fact any 'mass' production fell within the limit of three hundred tiles laid down for urban activities.

Although we still do not know the time frame in which the production was limited to that amount (one day, one month, one year?), the production capacities of many small terracotta factories had to remain below this size. Even today, after all, in various towns of central Italy one can find still functioning small traditional ceramics factories: until the middle of the last century they met a substantial local demand but now, closing one after the other for the lack of craftsmen, they respond mainly to the requests of amateurs and tourists.

This rule raises a serious and almost insoluble problem as to the possibility of a generalized intervention of laws and city administrations to regulate polluting activities carried out in the city. We are immediately reminded of two different types of industrial processes able to affect very seriously the context in which they operated: the tannery of leather and of wool. Here we must deal with a cryptic silence: first of all for the incomplete character of our knowledge of municipal laws, but also because in many cases the limitations must have been entrusted to the local magistrates, as in Rome to the competence and to the edicts of the *aediles*.

Despite the doubtful value of these silences, I tend to believe that, in Roman cities, the policy of containment of polluting activities had a limited impact: and after all one must consider as still in the Italian towns, throughout the middle ages and beyond, so much of the activities we are talking about was strength and pride of individual corporations operating within them. As an alternative to a general public policy, the praetor, as I have already mentioned, was entrusted with the power to arbitrate and mediate between opposing needs: on one side between the right to exploit economically the buildings and, on the other the right not to be disrupted from outside in the peaceful enjoyment of property.

The pollution caused by the *fullonicae* actually appears in a passage of Ulpian, but only in an indirect and episodic way[15]. Not only that, but it basically considers a rural situation:

---

[14] Lex col. Genetivae, *cap.* 76: *Figlinas teglarias maiores tegularum CCC tegulariumque in oppido coloniae Iuliae ne quis habeto. Qui habuerit it edificium isque locus publicus coloniae Iuliae esto, eiusque aedificii quicumque in colonia Genetiva Iulia iure dicundo praerit, sine dolo malo eam pecuniam in publicum redigito.*

[15] Ulp. *l.* 53 *ad ed.*: D. 39.3.3 pr. *Apud Trebatium relatum est eum, in cuius fundo aqua oritur, fullonicas circa fontem instituisse et ex his aquam in fundum vicini immittere coepisse: ait ergo non teneri eum aquae pluviae arcendae actione. si tamen aquam conrivat vel si spurcam quis immittat, posse eum impediri plerisque placuit.*

relatively less significant than the urban realities to which I was referring. The opinion of Trebatius, reported in the text, refers to the owner of a piece of land where *aqua oritur*, who, having implanted a *fullonica*, drained water to the neighbouring property. For Trebatius, the neighbour could defend himself, although not with the *actio aquae pluviae arcendae*, but with another procedural instrument, referring moreover to *aqua spurca*: wherein my opinion he was referring to the polluted water from the working of the *fullonica*.

This logic based on self-defence and not on a widespread intervention of the city, is the same we find in the series of cases, perhaps the most complex we have in this field, included in the Digest (*D.* 8.5.8.5–7) taken from the seventeenth book *ad edictum* by Ulpian, but based on material of jurists of the late Republic and early first century AD. The text addresses essentially harmful material interferences in land property or in a building, and investigates the limits in which one can oppose to them: this is primarily the case of fumes produced by a dairy farm *in superiora edificia*, since the owner of this building could act against the author of the emissions, claiming that *ius illi non esse ita facere*. The opinion of a jurist of the first principate, Aristo, recalls a real case, in which he explores in more detail the mutual game of rights: of he who denies that others can *immittere* in his properties, but also of he who denies that one can prevent him from *facere in suo*: in this case producing cheese. The solution suggested in the next paragraph 6 by Pomponius, and founded on the idea of the possibility to *immittere in alienum* a *fumum non gravem*, emphasizes the absence of an objective criterion and defines a conflict between opposing proprietary spheres, whose optimal exercise results incompatible. And this, it must be stressed, must have been the general criterion of tolerance for *immissiones*. It is crucial what we know about the inconvenience caused to the residents living close to the public baths who could do little to defend themselves.

Because this is the limit of delegating rights of self-defence to individuals: eventually the boundary line between the opposing claims must be entrusted to the empirical evaluation of the judge. In conclusion the fact that such rights were protected and tempered one another on the basis of practical interests claimed in court, does not rule out that various city areas could set specific requirements. And this especially during the imperial age when both the government and the protection systems of Rome and of the countless cities of the empire became better organized and systematic. Anyhow it is difficult, may be impossible, to affirm that Romans did have an 'ecological consciousness' and there is no evidence that they did conceive a general policy on that matter. But it is sufficient to refer to Cato's and Varro's agrarian treatise to affirm that they have a serious interest in the 'bel paesaggio' of agrarian Italy. Also in that case, protection of natural landscape, of wild life as of the 'coltivo', was essentially the final result of individual activity and interests of a large number of land-owners and peasants.

Bibliography

Beltrán Lloris, F. 2006. 'An Irrigation Decree from Roman Spain: *the Lex Rivi Hiberiensis*', *JRS* 96, 147–97.

Beltrán Lloris, B. 2014. 'La *lex rivi Hiberiensis* nel suo contesto: i pagi e l'organizzazione dell'irrigazione in Caesar Augusta', in L. Maganzani and C. Buzzacchi (eds.), Lex rivi Hiberiensis. *Diritto e tecnica in una comunità d'irrigazione della Spagna romana*. Napoli, 54–73.

Capogrossi Colognesi, L. 1976. *La struttura della proprietà e la formazione dei 'iura praediorum' nell'età repubblicana*, 2. Milano.

Capogrossi Colognesi, L. 2012. *Padroni e contadini nell'Italia romana*. Roma.

CAPOGROSSI COLOGNESI, L. 2014a. 'La *lex rivi Hiberiensis* e gli schemi delle servitù d'acqua in diritto romano', in L. MAGANZANI and C. BUZZACCHI (eds.), Lex rivi Hiberiensis. *Diritto e tecnica in una comunità d'irrigazione della Spagna romana.* Napoli, 75–91.

CAPOGROSSI COLOGNESI, L. 2014b. *Law and Power in the Making of the Roman Commonwealth.* Cambridge.

CRAWFORD, M. 1996. *Roman Statutes* (BICS Suppl. 64). London.

DI PORTO, A. 1989–1990. 'La tutela della *salubritas* tra Editto e giurisprudenza. Il ruolo di Labeone. II. Cloache e salubrità dell'aria', *BIDR* 92–93, 271–309.

DI PORTO, A. 2014. Salubritas *e forme di tutela in età romana. Il ruolo del* civis. Torino.

MAGANZANI, L. 2004. 'L'approvvigionamento idrico negli edifici urbani nei testi della giurisprudenza classica: contributi giuridici alle ricerche sugli acquedotti di Roma antica', in M. ANTICO GALLINA (ed.), *Acque per l'*utilitas *per la* salubritas *per l'*amoenitas. Milano, 185–220.

NÖRR, D. 2008. 'Processuales (und mehr) in der lex rivi Hiberiensis', *ZSS* 125, 108–88.

PALMA, A. 1987. 'Derivazioni di acqua *"ex castello"*', *Index* 15, 439–57.

SOLIDORO MARUOTTI, L. 2009. *La tutela dell'ambiente nella sua evoluzione storica. L'esperienza del mondo antico.* Napoli.

VALLOCCHIA, F. 2012. *Studi sugli acquedotti pubblici romani*, 1. Napoli.

Environment and pollution in literary and epigraphic evidence

Cinzia Bearzot

Ancient Ecology: problems of terminology

'Ecology' is a Greek word resulting from the combination of *oikos* and *logos*; however, as well known, ecology is a modern word, not an ancient one. The ancient world did not know the concept of ecology and, therefore, lacked corresponding terminology. In 1866, ERNST HAECKEL defined ecology 'the body of knowledge concerning the economy of nature – the total relations of the animal to both its inorganic and organic environment, including in the broader sense all "conditions of existence"'. His definition as the 'interdisciplinary scientific study of the interactions between organisms and their environment' is still valid.[1]

The ancients, however, in some way posed the problem of the mutual relation between man and the environment. Basic references on this topic can be found in the short monograph by LUKAS THOMMEN, *Umweltgeschichte der Antike* (2009). It is no accident that this book, like others about the same subject, does not include any reflection on terminology.[2] I should like to contribute with this paper to the identification of at least some ancient elements, be they terminological or conceptual, which may have some 'ecological' interest.

I shall begin with the pseudo-Hippocratic treatise *On Airs, Waters and Places*, which could be considered the 'manifesto' of environmental determinism, which holds that 'the characteristics of a natural element affect other elements of nature, the body of the inhabitants in its normal and pathological aspects, and the psychic form of man'.[3] The author of the treatise, by exploring the influence that the environment exerts on man, suggests a 'médecine environnementale' (quoting JACQUES JOUANNA). He is keen to define the ideal environment and considers several factors: the course of the seasons, climate and wind regime, and conditions of air and water, as well as soil characteristics (Chapter 1).[4] In his analysis of these factors, he introduces that allow us to draw concepts and terms which are crucial for understanding ecological thought in the ancient context.[5]

---

[1] Further details in THOMMEN 2009: 11–6; quotation in THOMMEN 2012: 6. See MURRAY 1992.

[2] English translation THOMMEN 2012.

[3] BOTTIN 1996: 12 ('le proprietà di un elemento della natura condizionano altri elementi della natura, il corpo degli abitanti nei suoi aspetti normali e patologici, la forma psichica dell'uomo').

[4] JOUANNA 1996: 7–82, esp. 33.

[5] RÜST 1952 is about the language of the treatise; however, it focuses on phonetic, grammar, and syntactic features while ignoring lexical items.

1. Environmental balance in Pseudo-Hippocrates' *On Airs, Waters and Places*

In Chapter 5 of *On Airs, Waters and Places*, while treating the topic of exposure to winds, the author discusses the condition of cities which face the east. They are healthier (ὑγιεινότεραι) than those facing the north, which are exposed to cold winds, and than those exposed to hot winds, even though they lie very close nearby (these were examined in the two previous chapters). These cities are healthier because they enjoy 'more moderate' heat and cold (μετριώτερον ἔχει), which guarantees clear and delightful waters (λαμπρά, εὐώδεα, μαλακά) and, in the absence of fog, good air quality. As a result, people have better complexions, clear voices and quick intelligence, and everything in these cities is better as well (τὰ ἄλλα τὰ ἐμφυόμενα ἀμείνω ἐστίν). The author concludes that a city so situated 'is just like spring, because the heat and the cold are tempered (κατὰ τὴν μετριότητα τοῦ θερμοῦ καὶ τοῦ ψυχροῦ)'.[6]

Here the notion of μετριότης (also present in Chapters 10, 12 and 18) surfaces again, reinforced by the striking image of spring, the temperate season to which the city facing the east bears resemblance.[7] A comparison of the two previous chapters, which accumulate 'extreme' terms related to heat and cold, as well as damp and dryness, shows that wholesomeness and moderation are parallel concepts. Indeed, the quality of environmental 'healthfulness' (ὑγιεινός) is synonymous with temperance (μέτριος), as emerges clearly in the discussion of 'noisome' airs and waters (νοσερός and νοσώδης in Chapters 6 and 7).[8] The opposition wholesome/unwholesome is framed from the point of view of man and his relationship with the environment, and the same can be said for the opposition healthy/unhealthy (ἐπιτήδειος/ἀνεπιτήδειος), introduced at the beginning of Chapter 7 with respect to the winds. From the latter viewpoint, this theme re-emerges in Chapter 10, which is about the seasons. It is from their course that one can assess whether the year will be 'noisome' or 'healthful' (νοσερός or ὑγιηρός). A 'very healthful' year (ὑγιεινότατος) will result 'if there be rains in autumn, if the winter be moderate (μέτριος), neither too mild nor unseasonably cold, and if the rains be seasonable in spring and in summer'. Moderation and absence of excess in the climate contribute to a healthy environment for mankind. The concept of 'tempering' appears to be a key point in the definition of a man-friendly environment.

In Chapter 12, which is devoted to the climate of Asia, the concept of μετριότης resurfaces again. The author begins by saying that 'everything in Asia grows to far greater beauty and size; the one region is less wild than the other, the character of the inhabitants is milder and more gentle'. He continues stating that the excellence of Asia originates from its temperate climate (ἡ κρῆσις τῶν ὡρέων). The term κρῆσις expresses the idea of a balanced and measured mix, and it is related to Asia's geographical location, namely midway (ἐν μέσῳ) between the risings of the sun and removed from the cold.[9] The author then explains the reasons for the excellence of the best part of Asia, the one that is, in its own turn, ἐν μέσῳ, and is therefore characterised by richness of fruits and trees, genial climate, purest waters, abundance in both spontaneous and cultivated plants, cattle of very good quality, as well as beautifully shaped and well-nourished inhabitants. Finally, he concludes that it is natural that 'this region, both in character and in the mildness of its seasons, might fairly be said to

---

[6] Ps.-Hipp., *aer.* 5.21. (Engl. transl. by JONES 1923).
[7] JOUANNA 1996: 197, n. 5.
[8] Lt. *pestiferus/salūber.* BOZZI 1982: 49–50, 66–7.
[9] JOUANNA 1996: 294–5.

bear a close resemblance to spring (κατὰ τὴν φύσιν καὶ τὴν μετριότητα τῶν ὡρέων)'. Here the juxtaposition between spring and μετριότης is proposed again in the context of discussion about the theme of tempering and lack of excesses (heat or frost, drought or excessive rainfall).

Besides the concept of μετριότης, Chapter 12 also introduces the notion of 'balance/equality', ἰσομοιρίη. After recalling the position of Asia as ἐν μέσῳ – 'because it lies towards the east midway between the risings of the sun, and farther away than is Europe from the cold' –, the author adds: 'growth and freedom from wildness are most fostered when nothing is forcibly predominant (μηδὲν ᾖ ἐπικρατέον βιαίως), but equality in every respect prevails (ἀλλὰ παντὸς ἰσομοιρίη δυναστεύῃ)'.[10]

This idea of balance is developed in Chapter 13, which is devoted to the characteristics of the peoples who live on the right hand of the summer risings of the sun, as far as the Palus Maeotis. Territory and population are characterised in this area by significant differences, which are connected with the course of the seasons (μεταβολαί). The adjective διάφορος expresses the idea of difference, but also of conflict (see the noun διαφορά, 'difference', but also 'clash'). Relevant and frequent changes in the climate make the land very wild and uneven (ἡ χώρη ἀγριωτάτη καὶ ἀνωμαλωτάτη ἐστίν), and this is reflected in the inhabitants. By contrast, where the seasons are not strongly differentiated, there is a lack of variation in the landscape and in its people (ἡ χώρη ὁμαλωτάτη ἐστίν).

An emphasis on 'evenness' and 'stability' *versus* 'difference' and 'change' is already present in the contrast between cities facing the east and those facing the west, as well as between spring – the season of tempering and measure – and autumn, the season of contrasts and changes (in Chapters 5 and 6). Such an emphasis is connected with the concepts of measure and balance, which exists when there is no violent predominance of one element over another, and which contrasts with the notions of disproportion, breach of harmony, tension, and conflict. The term ἰσομοιρίη occurs only once in the text; in contrast the term μεταβολή is frequent in a generally negative sense, as it underlines the dangers posed to health by changes of the seasons (Chapters 1, 2, 3, 4, 10, 11) and changes in lifestyle (δίαιτα: Chapter 2). A positive effect of μεταβολή is emphasised only during the discussion of the characteristics of the peoples of Europe, where (in contrast with Asia), the changeability of climate keeps the intelligence/judgement (γνώμη) of man awake and does not allow it to remain inactive. It should be noted, however, that the author assigns an important role to institutions too, in determining the different characteristics of European and Asian peoples.[11]

2. Alteration of balance: water and air pollution

In this framework, which highlights the idea of a natural balance that needs to be preserved, the intervention of man on the environment may raise concerns that can be broadly defined as 'ecological', especially when it endangers the environmental balance. Nevertheless,

---

[10] ἰσομοιρίη conveys the idea of 'equality between parts'. See JOUANNA 1996: 295–6, who notes some affinity with the doctrine of health as harmony of the elements and with the political and moral fields. The comparison with the opposition ἰσονομία/μοναρχία is also interesting, in the medical and not constitutional sense found in Alcmaeon of Croton (F 24 b 4 DK).

[11] BORCA 2003: 15–37.

human impact on nature was considered inevitable by the ancient tradition and, therefore, substantially legitimate.

Opening the *On Airs*, Pseudo-Hippocrates pinpoints the elements of a healthy environment: climate, exposure, wind regime, and water quality.[12] His approach looks at the influence of the environment on man and, therefore, at the situation of the natural resources. The ancient tradition does not ignore that such a balance can be subverted by man, as he more or less consciously intervenes on the environment. In Italy pioneering studies on these topics have been proposed by ODDONE LONGO on Greece, *Ecologia antica. Il rapporto uomo/ambiente in Grecia* (1988), and by Paolo FEDELI on Rome, *La natura violata: ecologia e mondo romano* (1990). The violation of nature – to borrow from the title of FEDELI's work – can occur in different ways. With respect to the Greek world, LONGO cited: (a) exploitation of animal, vegetable, and mineral resources; (b) release into the environment of organic and inorganic waste from consumption and production; (c) alterations to the characteristics of natural species (animals and plants) and of the environment for the benefit of man. I shall limit my discussion to the topic of pollution of water and air. The pseudo-Hippocratic author of *On Airs* considers the wholesomeness of these natural resources to be crucial to man's health. Any damage done to them alters, inevitably, the ideal equilibrium on which (what we would call) a good ecosystem depends.

It is worth remembering at this stage that, if Pseudo-Hippocrates is attentive to the natural availability of good air and water, Aristotle emphasises the importance of their quality also considering the contribution of human action (*polit.* 7.1330b, 4–14). By stating that the health (ὑγίεια) of the inhabitants 'depends upon the place being well situated both on healthy ground and with a healthy aspect, and secondly upon using wholesome water-supplies (ὕδατα ὑγιεινά)', he fully aligns himself with the *On Airs*. Moreover, the philosopher adds that 'in wise states' if there is a difference is in the quality of the waters available and the supply is not plentiful, a distinction is made between drinking water (which is used for τροφή) and waters which are put to any other use. He concludes:

> Those things which we use for the body in the largest quantity, and most frequently, contribute most to health; and the influence of the water-supply and of the air is of this nature (ἡ δὲ τῶν ὑδάτων καὶ τοῦ πνεύματος δύναμις τοιαύτην ἔχει τὴν φύσιν).[13]

So water and air must have and, especially, preserve those optimal qualities that make them healthy for man.[14] Yet, man can alter these qualities affecting the natural balance (the ἰσομοιρίη of Pseudo-Hippocrates); and he may even have to undo damage done and restore the balance.

---

[12] On this last point, see JOUANNA 1994; JORI 1994.

[13] Arist., *pol.* 7.10. (1330b 14–5. Engl. transl. by RACKHAM 1944).

[14] See HUGHES 2014[2]: 176 (on water pollution), 177–8 (on air pollution).

2.1 Water

In 2002, Livio Rossetti drew scholarly attention to a very interesting piece of epigraphic evidence regarding waters.[15] Inscription *IG* I³ 1.257 is a regulation – whether it is a local one or, more probably, a decree issued by the Athenian βουλή and δῆμος is controversial – dating from around 430 BC. In the preserved part we can read:

με/δὲ δέρματα σέπεν ἐν τõ/ι hιλισõι καθύπερθεν / τõ τεμένος τõ hερακλέ/[ο]ς· μεδὲ βυρσοδεφσἓν μ/[εδὲ καθά]ρμα[τ]α <ἐ>ς τὸν π/[οταμὸν βάλλεν ...]

(It is not allowed) to put the skins to rot in the Ilissos upstream of the temple of Hercules or tanning (skins) or dispose of waste (leather work) in the river.

The decree prohibits the use of the waters of the river Ilissos, upstream of the temple of Heracles, for soaking skins and eliminating waste. Evidently, the habits of skin and leather craftsmen, who operated in the area, were heavily impinging on the environment. They polluted the river waters, worsened air quality, deprived the sanctuary of pure waters, and posed hygienic and health risks. The promulgation of the decree shows considerable environmental sensitivity; even more so, it targeted an important economic activity and, therefore, strong interests. It is well known, for instance, that Cleon the demagogue was an entrepreneur in this industry, and was at the peak of his career in those years. The decree produced the desired effects. This is suggested by the fact that the Ilissos appears completely untainted by environmental damage in the description of Plato's *Phaedrus* (227a–230e, especially 230b–c), which depicts a very pleasant environment with clear and fresh water (ἥ τε αὖ πηγὴ χαριεστάτη ὑπὸ τῆς πλατάνου ῥεῖ μάλα ψυχροῦ ὕδατος), as well as delightful and scented air (εὐωδέστατον; τὸ εὔπνουν τοῦ τόπου ὡς ἀγαπητὸν καὶ σφόδρα ἡδύ). An explicit connection has been proposed between this excellent environmental situation and the application of the decree *IG* I³ 1.257. They may refer both to the date of the setting of *Phaedrus* (late fifth century) and to its date of composition (in the 370s).[16]

This document attests to a typical case of alteration (and then restoration) of the environmental balance. It is remarkable evidence of what we would describe in terms of ecological awareness. The Eridanus, another river in Attica, underwent a similar alteration, the reasons for which are, however, unknown. According to Strabo (9.1.19), Callimachus stated in his Συναγωγῆ τῶν ποταμῶν (F 458 Pfeiffer) that a line about Athenian virgins drawing water from the 'pure' stream of the Eridanus (ἀφύσσεσθαι καθαρὸν γάνος Ἠριδανοῖο) made him laugh, since even the cattle kept away from it.[17] Strabo adds that the sources of the river provided pure and potable water (καθαροῦ καὶ ποτίμου ὕδατος). Additionally, he states that in the past there was also a fountain of human construction which yielded abundant and excellent water (πρότερον δὲ καὶ κρήνη κατεσκεύαστό τις πλησίον πολλοῦ καὶ καλοῦ ὕδατος). Later, some sort of change (μεταβολή) occurred, which corrupted the nature of those pure and abundant waters. Although Strabo does not speculate about the nature of this 'change', he declares that such events are unsurprising: 'and even if the water is not so now, why should it be a thing to wonder at, if in early times the water was abundant and pure,

---

[15] Rossetti 2002.

[16] Lind 1987.

[17] The author of the line cited by Callimachus is unidentified.

and therefore also potable, but in later times underwent a change?' (εἰ δὲ μὴ νῦν, τί ἂν εἴη θαυμαστόν, εἰ πάλαι πολὺ καὶ καθαρὸν ἦν ὥστε καὶ πότιμον εἶναι, μετέβαλε δὲ ὕστερον;).[18]

Both sources, though different in nature, draw attention to the issue of the wholesomeness of the waters. Even when the natural conditions are good, water wholesomeness is subject to changes due to human intervention or to other unspecified causes. From a lexical point of view, an opposition can be highlighted between the verb σήπειν (to 'rot' skins in the Ilissos) and the adjective καθαρός, referred to water pureness.

Elsewhere, in a context that is conceptually similar, we find an analogous opposition between διαφθείρειν, 'to corrupt', also 'to pollute', and καθαίρειν, 'to purify', 'to clean'. A long passage in Plato's *Laws* (6.761a–c) prescribes the treatment of rainwater and spring waters, to be entrusted to the ἀγρονόμοι, the country counterparts of the ἀστυνόμοι, who were in charge of the urban environment. It is noteworthy that a passage in Plutarch's *Life of Themistocles*, 31.1, assigns the responsibilities of ἐπιστάτης τῶν ὑδάτων (overseer of water reservoirs) to the archon and *strategos*, but Aristotle attests the existence of the specific office of ἐπιμελητὴς τῶν κρηνῶν (*Ath. Pol.* 43.1).[19] At 8.844, Plato refers to 'excellent old laws' that regulate water supply to the farms.[20] He explains that in drawing water, one should not cut off the spring or do harm to the channel. In the event of droughts, it is appropriate to share water with neighbours and if, as is likely, a dispute should arise over the influx of rainwater, one must consult the ἀστυνόμοι and ἀγρονόμοι. Finally, the most interesting passage is 8.845, which precisely addresses the problem of water pollution. Water is 'easy to spoil' (εὐδιάφθαρτον). Indeed, Plato adds,

> For while soil and sun and wind, which jointly with water nourish growing plants, are not easy to spoil (φθείρειν) by means of sorcery or diverting or theft, all these things may happen to water; hence it requires the assistance of law. Let this, then, be the law concerning it: – if anyone want only spoil another man's water (Ἄν τις διαφθείρῃ ἑκὼν ὕδωρ ἀλλότριον), whether in spring or in pond, by means of sorcery, digging, or theft, the injured party shall sue him before the ἀστυνόμοι, recording the amount of the damage sustained; and whosoever is convicted of damaging by poisons shall, in addition to the fine, clean out (καθηράτω) the springs or the basin of the water, in whatever way the laws of the interpreters declare it right for the purification (κάθαρσις) to be made on each occasion and for each plaintiff.[21]

Here we find a contrast between the terminology of corruption (φθείρειν, διαφθείρειν, εὐδιάφθαρτος) and the terminology of pureness (καθαίρειν, κάθαρσις), which is similar to the language of the Athenian decree about river pollution, where the verb σήπειν (to 'rot', used of the skins in the Ilissos) contrasts with the adjective καθαρός, which referred to water pureness. The terminology of 'pureness' (adjective καθαρός, verb [ἐκ]καθαίρειν) is also present in the inscriptions studied by Reinhard Koerner in the 1970s, which related to the care of fountains and structures of public water supply, e. g. prohibitions against bathing, washing clothes and vessels, throwing garbage, watering cattle, and throwing objects of any

---

[18] English translation by Jones 1927.

[19] On this theme, important piece of evidence is the inscription of the *astynomoi* of Pergamon, dated to the second century BC (*SEG* 13.521). Lines 160–90 record the duties of the *astynomoi* with respect to water treatment. See on this aspect Klaffenbach 1953; Saba 2012.

[20] See Plut., *Sol.* 23.6–7.

[21] Engl. transl. by Bury 1926, 181.

kind into the fountains, as well as measures for the cleaning of basins and pipes.[22] I should like to quote especially an inscription from Kea, from the beginning of the third century BC (*IG* 12.5.569 + *IG* Suppl. 114, ll. 3–5). It mentions the duties of an Epimeletes in charge of the cleaning of pipes and fountains for the preservation of pure water flowing to the sanctuary of Demeter (ὅπως ἂν εἶ κ[α][θ]α[ρὸς ὁ ὀχ]ετὸς ὁ κρυπτός, ἐπιμελεῖ[σθαι καὶ τῆς κά] τω κρήνης, ὅπως [ἄ]ν μήτ[ε λό]ωνται μήτε πλύ[ν]ωσιν ἐ[ν ταῖς κρήναις, ἀ]λλὰ κα[θα]ρὸν τὸ ὕδωρ εἴσεισιν ἐς τὸ ἱε[ρὸν] τῆς Δήμητρο[ς: 'in order to maintain the covered water-pipe clean, he has to take care of the lower spring, so that people neither wash nor clean anything in the fountain, but the water flows pure in the temple of Demetra'). It is remarkable that here, as well as in the case of pollution of the Ilissos above mentioned, the main concern seems to be the availability of pure waters for the sanctuary. We can plausibly infer that those in charge of the sanctuary must have reported possible sources of water pollution to the authorities and asked for their removal, and that environmental restoration, therefore, was encouraged by the influential cult centre.

2.2 Air

With respect to air, the issue is less clear. Pseudo-Hippocrates' *On Airs* attributes great importance to air quality, and it is well known that many illnesses were traced back to contaminated air (μίασμα).[23] However, in the above mentioned passage of Plato's *Laws* (8.845 d) it is stated that it is not easy to spoil (φθείρειν) soil, sun, and wind (πνεύματα). Thus, it was considered even more difficult to affect air quality, even by those who were aware of the risks of polluting water. However, just as there was concern for the care of water, measures were similarly taken to preserve the wholesomeness of air. Aristotle, in the *Constitution of the Athenians* (50.2), reports that some Athenian officials called ἀστυνόμοι had the task of ensuring that the collectors of sewage should deposit their waste at least ten stadia from the city-walls.[24] They also prevented people from blocking up the streets by building, stretching barriers across them, and making drain-pipes in mid-air with a discharge into the street. Furthermore, Aristotle says, they also removed the corpses of those who died in the streets.[25] It is evident that the first and the last of these measures were intended to ensure good air quality and preserve the urban environment from contamination. An atmosphere that was 'clean-smelling' (εὐώδης) and 'pleasant to breathe' (εὔπνους), to use the terminology of *Phaedrus*, reflected the environmental balance that Pseudo-Hippocrates believed to be a requisite for ensuring good health.

---

[22] Koerner 1974. On the theme of the public treatment of waters see also Angelakis and Koutsoyiannis 2003; Krasilnikoff 2002; Angelakis, Koutsoyiannis, Tchobanoglous and Zarkadoulas 2008.

[23] Gourevitch 1995.

[24] Rhodes 1981: 573–5.

[25] The issue of air quality is better attested for Rome. See Hughes, 2014[2]: 177–8. See Thommen 2009: 65–8, on air pollution caused by mine extraction activity.

Conclusions

My contribution derives from a lexical interest. I conclude by assessing what we have been able to recover through this approach.

First of all, it was possible to identify a number of terms related to the concept of environmental 'balance'. In a positive sense, this includes terms that refer to the notions of 'measure, tempering, moderation', such as μέσον, μετριότης; of 'balance' between parts, such as ἰσομοιρίη; of 'reconciliation', such as κρῆσις. These are opposed to negative terms referring to the idea of 'change', such as μεταβολή; of 'difference' and 'contrast', such as διαφορά (διάφορος); of 'violence', such as βία (βιαίως).

Moreover, one should also consider terminology related to health, which contrasts what is 'proper' to human settlement (ἐπιτήδειος), as well as wholesome (ὑγιεινός, ὑγιηρός), with what is 'improper' (ἀνεπιτήδειος) and, therefore, unhealthy (νοσερός, νοσώδης). Similarly, we find specific terminology with regard to the characteristics of natural resources. Their good quality is expressed by adjectives like λαμπρός, καθαρός, ψυχρός, εὐώδης, μαλακός for water, and εὐώδης, εὔπνους for air.

Finally, there emerged a clear pattern of terminology relating to human intervention, both of a polluting and a repairing nature: σήπειν, φθείρειν, διαφθείρειν (with the adjective εὐδιάφθαρτος) and καθαίρειν, κάθαρσις respectively.

FEDELI opened his *La natura violata* with these words: 'The history of the relations between man and the environment in the ancient world has not been written yet. In the absence of systematic research, the only way to proceed is by reading the texts directly'.[26] The same holds true, and perhaps even more so, for the lexicon of ecology. It is clear that such a lexicon must be gathered from those texts, both literary and documentary, that deal with ecological issues. Then, over time it can be enriched through research in this field, which has indeed been expanding recently. My present attempt at a preliminary definition of ancient ecological terminology may appear unsystematic. Yet, it is also proof of the method that I believe should be followed in this field of research. I hope that others may continue the work, and that we may reconstruct a fuller lexicon of ancient ecology over time.

Bibliography

ANGELAKIS, A. N. and D. KOUTSOYIANNIS. 2003. 'Urban Water Engeneering and Management in Ancient Greece', in B. A. STEWART and T. HOWELL (eds.), *The Encyclopedia of Water Science*. Dekker-New York, 999–1007.

ANGELAKIS, A. N., D. KOUTSOYIANNIS, G. TCHOBANOGLOUS, and N. ZARKADOULAS. 2008. 'Urban Water Management in Ancient Greece. Legacies and Lessons', *Journal of Water Resources Planning and Management* 134.1, 45–54.

BORCA, F. 2003. *Luoghi, corpi, costumi. Determinismo ambientale ed etnografia antica*. Roma.

BOTTIN, L. (ed.), 1996. *Ippocrate, Arie, acque, luoghi*. Venezia.

BOZZI, A. 1982. *Note di lessicografia ippocratica. Il trattato sulle arie, le acque, i luoghi*, Roma.

BURY, R. G. 1926. *Plato. Laws, 2: Books 7–12*. Cambridge MA.

BURY, R. G. 1967–1968. *Plato in 12 Volumes, Vols. 10 & 11*. Cambridge MA-London.

---

[26] 'La storia dei rapporti fra uomo e ambiente nel mondo antico è ancora tutta da scrivere. In assenza di ricerche sistematiche, l'unico modo di procedere è la lettura diretta dei testi'. FEDELI 1990: 17.

FEDELI, P. 1990. *La natura violata. Ecologia e mondo romano.* Palermo.

GOUREVITCH, D. 1995. 'La medicina ippocratica e l'opera "Delle arie, acque, luoghi". Breve storia della nascita e del potere di un "inganno" scientifico', *MedSec* n. s. 7(3), 425–33.

HUGHES, D. J. 2014². *Environmental Problems of the Greeks and Romans. Ecology in the Ancient Mediterranean.* Baltimore.

JONES, H. L. 1927. *Strabo, Geography,* Volume IV: Books 8–9. Cambridge MA.

JONES, W. H. S. 1923. Hippocrates. *Ancient Medicine. Airs, Waters, Places. Epidemics 1 and 3. The Oath. Precepts. Nutriment.* Cambridge MA.

JORI, A. 1994. 'L'acqua di Ippocrate. Physis e ordine politico in Arie acque luoghi', in O. LONGO and P. SCARPI (eds.), *Letture d'acqua,* Atti del II Colloquio *Homo Edens* 21–22 Settembre 1991. Padova, 171–210.

JOUANNA, J. 1994. 'L'eau, la santé et la maladie dans le traité hippocratique des "Airs, eaux, lieux"', in R. GINOUVÈS, A.-M. GUIMIER-SORBETS, J. JOUANNA and L. VILLARD (eds.), *L'eau, la santé et la maladie dans le monde grec* (*Bulletin de correspondance hellénique,* Supplément 28). Paris, 25–40.

JOUANNA, J. 1996. *Hippocrate, Airs, eaux, lieux.* (Texte établi et traduit par). Paris.

KLAFFENBACH, G. 1953. 'Die Astynomeninschrift von Pergamon', *Abhandlungen der Deutschen Akademie der Wissenschaften zu Berlin. Klasse für Sprachen, Literatur und Kunst* 6, 3–25.

KOERNER, R. 1974. 'Zu Recht und Verwaltung der griechischen Wasserversorgung nach den Inschriften', *Archiv für Papyrusforschung und verwandte Gebiete* 22–23, 155–202.

KRASILNIKOFF, J. A. 2002. 'Waters and Farming in Classical Greece: Evidence, Method and Perspectives', in K. ASCANI, V. GABRIELSEN, K. KVIST and A. H. RASMUSEN (eds.), *Ancient History Matters. Studies presented to Jens Erik Skydsgaard on His Seventieth Birthday.* Rome, 47–62.

LIND, H. 1987. 'Sokrates am Ilissos. IG I³ 1 257 und die Eingangsszene des platonisches "Phaidros"', *Zeitschrift für Papyrologie und Epigraphik* 69, 15–9.

LONGO, O. 1988. 'Ecologia antica. Il rapporto uomo/ambiente in Grecia', *Aufidus* 6, 3–30.

MURRAY, O. 1992. 'The Ecology and Agrarian History of Ancient Greece', *Opus* 11, 11–21.

RACKHAM, H. 1944. *Aristotle* in 23 Volumes, Vol. 21. Cambridge MA-London.

RHODES, P. J. 1981. *A Commentary on the Aristotelian Athenaion Politeia.* Oxford.

ROSSETTI, L. 2002. 'Il più antico decreto ecologico a noi noto e il suo contesto', in T. M. ROBINSON and L. WESTRA (eds.), *Thinking about the Environment. Our Debt to the Classical and Medieval Past.* Lanham MD, 44–57.

RÜST, A. 1952. *Monographie der Sprache des Hippokratischen Traktates 'Peri aeron, hydaton, topon'.* Freiburg.

SABA, S. 2012. *The Astynomoi Law of Pergamon. A New Commentary.* Mainz.

THOMMEN, L. 2009. *Umweltgeschichte der Antike.* München (English translation: THOMMEN, L. 2012. *An Environmental History of Ancient Greece and Rome.* Cambridge).

Gian Franco Chiai

Rivers and Waters Protection in the Ancient World: how religion can protect the environment

Introduction

Nos et flumina inficimus et rerum naturae elementa, ipsumque quo vivitur in perniciem vertimus.

We taint the waters of the stream, and we infect the various elements of Nature; indeed, the very air even, which is the main support of life, we turn into a medium for the destruction of life.

(Pliny *NH* 18.1. Engl. transl. by H. Rackham 1936)

This quotation from Pliny sounds like a modern-day criticism against the pollution of rivers and natural resources (*rerum naturae elementa*) by human action.[1] But why was Pliny cautioning the Romans? Did the pollution and depletion of the environment really represent a widespread problem in ancient times? It is possible that the exploitation of natural resources was considered as an act of violence (*hybris*) against nature. But then again, Pliny's warning could have been an isolated voice?[2] Was the ancient world really a dirty world, in contrast to the literary representation of the bucolic and idyllic landscape, which has inspired the literary and artistic European culture?

In lamenting the destruction and pollution of the environment, Pliny's claim is by no means an isolated voice. Indeed, especially in the Latin literature, we come across an impressive number of critical statements against the exploitation and destruction of the nature. Horace (*Carm.* 3.1.33–37) protests, for instance, against the construction of gigantic villas on the Italian coast that caused the destruction of the natural habitat of fish and fauna and made the sea smaller. Sallust (*Cat.* 13.1–2) disapproves the excessive luxury of Lucullus, who constructed a sumptuous residence by levelling mountains and covering the sea. Varro (*rust.* 3.17.9) remembers that Lucullus again ordered a passage to be dug through a mountain near Naples, in order to conduct the water from the sea to his property. Seneca (*Ep.* 89.21) raised his voice against such interventions that modified the natural order (*kosmos*) and destroyed the environment, emphasizing (*Ep.* 122.8) that such people were living against nature (*con-*

---

* I am greatful to my friends Daniela Kleine Burhoff, Matteo Olivieri, Orietta Cordovana and Ralph Häussler for correcting my English and for helping with useful criticism and suggestions.

1 On this text see Mazzarino 1969–1970: 643–5; Fedeli 1990: 69–70, and the contribution by Cordovana in this volume.

2 Fedeli 1990: 51–8, collected ancient sources for cases of exploitation of natural resources in the Greek and Roman world. A good overview is in Hughes 2014; Pucci 2015. Archaeological evidence is in Thüry 1995 and 2001.

tra naturam). All these literary testimonies, though only briefly mentioned here, not only demonstrate a concern for the problems and the destruction of the environment in the Roman world, but also show the existence of an ongoing debate on the subject, associated with the issue of the decadence of the traditional *mores* within the Roman society. Unfortunately, the literary evidence allows only a rough reconstruction of such a discussion. Furthermore, the Roman attitude towards the environment should also be considered in the perspective of the social changes that took place since Augustus' reign. Strabo (5.3.8), for example, considered the Romans to have been the first to construct roads, aqueducts, and sewers in order to improve the quality of urban life, while the Greeks preferred the beauty of their cities.[3] Frontinus (*De aquaeductu Urbis Romae* 88) praises the utility of Roman aqueducts, because they had improved everyday quality of life in Rome, the air of which was reeking with the exhalations of the wastewaters. Despite the undoubted benefits of progress, we can recognise the stirring of an 'ecological' awareness over the damage caused by indiscriminate exploitation of the natural resources and environmental depletion. Going back in time, we can briefly allude to Plato's disappointment over the terrible deforestations of Attica[4], large resulting from the indiscriminate exploitation of the silver mines at Laureion that caused a radical transformation of the landscape.[5] In this case, economic profits were preferred over preservation of nature. Another significant source of evidence are the *leges sacrae*, which reveal the concern for preserving a clean environment by prohibiting the pollution of waters or the cutting of trees throughout the sacred spaces, as emphasized for instance by MATTHEW P. J. DILLON.[6]

I am persuaded that the examples I present here indicate that the sensibility for ecologic problems was more diffused and debated in the ancient world more than one commonly assumes.

On the basis of epigraphic, literary, and archaeological sources, this paper aims at reconstructing what the sensibility for environmental issues was, particularly with regard to the pollution of rivers and waters in the Greek and Roman world. I shall begin by showing the importance of rivers for the formation of local cultural identities in the ancient world, especially in the Greek East. Subsequently, I shall analyse archaeological and epigraphic sources regarding the pollution of rivers. Finally, I shall look to the measures that were adopted in order to protect the rivers, particularly via an analysis of the religious influences in the preservation of the environment against human pollution.

---

[3] In Strabo's times, Vipsanius Agrippa was *curator aquarum* in Rome; he conducted the restoration of the *Cloaca maxima*.

[4] On deforestation in the ancient world see HUGHES 2014: 68–87 and his contribution to this volume with further literature.

[5] Pl., *Criti.* 111b–c; on industrial technology and environmental damage see HUGHES 2014: 129–49. On the exploitation of the silver mines and the destruction of the environment in Attica see WEBER 1990: 63–84 with a collection of literary sources.

[6] DILLON 1997.

1. Rivers and city identity

City names, such as *Laodicea ad Lycum* or *Antiochia ad Orontem,* emphasized the geographical position of a *polis* on the riverbank; they show the importance of the river in the choice of the city name, and represent a central element of the local identity.[7]

Rivers also played an important role for a city's eulogy, especially in the context of the panegyrical literature during the time of the Second Sophistic. In his handbook, for example, the rhetorician Menander advises to praise a city by mentioning its *eudria* (the richness of water), and to stress its importance for urban life. The praise of a city was intrinsically bound to that of its *chora*.[8] In his eulogium of Antiochia on Orontes, Libanios (*or.* 11.244–7) emphasizes the polis' *eudria*, which represents a central element of its beauty (*kallos*) and greatness (*megethos*). Great and clean rivers are very important for the life and economy of a city, of course; the people can drink the water and the landscape becomes fertile. During the Second Sophistic, therefore, rivers were an important element of urban identity, and they were praised in literary eulogies and epigraphic dedications to the local river god, but they were also represented on the reverse of local coins.[9] This attitude towards the landscape can be recognised already in the early coinages of the Greek cities in the colonial territories; numerous representations of river- and fountain gods are figured on the coins.[10] These deities were representation of paramount features of the sacred landscape, divine personifications of natural elements, created by Greek settlers in those regions. They served the purpose of protecting the territory of the *polis* and bestowed prosperity to the inhabitants.

We may take into consideration the case of Laodicea on Lycos. The *chora* of Laodicea was characterised by the presence of three rivers, Lycos, Asopos, and Kapros, as well as by the presence of Mount Kadmos, from which the rivers Lycos and Asopos originate. A great monumental aqueduct, which connected the river Kapros to the urban centre and the monumental thermal complex, must have been object of city praise. Literary sources testify to the importance of the local rivers also for the treatment of the wool, which was famed

---

[7] Very important elements of the link between rivers and city-identities, particularly in the Greek East during Roman times, are in Nollé 1994; Peter 2005 (about the river god Hebros and the city identity of Traianopolis); Nollé 2006: 50–60, 66–8, with numismatic evidence from Roman Asia Minor.

[8] Men. Rhet. 1.347–9: τὸ γὰρ εὔυδρον εἶναι τὴν πόλιν ἢ ποταμοῖς περιειλῆφθαι ἔφην τῶν περὶ χώραν εἶναι τὸ γὰρ εὔυδρον εἶναι τὴν πόλιν ἢ ποταμοῖς περιειλῆφθαι ἔφην τῶν περὶ χώραν εἶναι. αὐτῶν δὲ τούτων ἕκαστον καὶ πρὸς ἡδονὴν καὶ ὠφέλειαν κατὰ τὴν ἄνω γεγονυῖαν διαίρεσιν. [...] (349) ὑδά-των δὲ φύσεις τριχῆ δεῖ διαιρεῖν, ἢ ὡς πηγῶν, ἢ ὡς ποταμῶν, ἢ ὡς λιμνῶν. κριτέον δ' αὐτὰ ὥσπερ καὶ τὰ ἄλλα, πρὸς ἡδονὴν καὶ ὠφέλειαν, καὶ ἔτι πρὸς ταύτῃ τῇ διαιρέσει πρὸς πλῆθος καὶ αὐτοφυῖαν· ἐνιαχοῦ γὰρ καὶ θερμαὶ πηγαὶ εὑρίσκονται. That a city is well-watered or surrounded by rivers is what I have called a 'country topic'. Each of these should be considered with reference both to pleasure and to utility, according to the division made above. [...] Resources of water should be divided into three: springs, rivers, lakes. Like everything else, these are to be judged on grounds of pleasure and utility; a further division may be made in terms of abundance and natural occurrence. In some places hot springs are also found. (Engl. translation by D. A. Russel and N. G. Wilson, Oxford 1981).

[9] A fundamental collection of the representations of river-gods on coins is by Imhoof-Blumer 1923; Weiss 1984 is about the cults of river-gods before the Hellenistic age.

[10] Representation of nymphs and river-gods on Sicilian archaic Greek coins are in Günther 2009 with further literature. On the river names see Curbera 1998.

across the Roman Empire and represented a staple of Laodicea's economy. Furthermore, the Lycos was also an important communication route since it connected the city with other Phrygian urban centres and facilitated commerce and transport of local products.

In the local iconographical tradition, the rivers Lycos and Kapros are often depicted respectively as a wolf and as a boar on the coins of Laodicea, since Hellenistic times.[11] These rivers are perceived as part of the cityscape and their representations are placed beside the city goddess on the coins. Furthermore, the two rivers played a central role also in the elaboration of local mythical traditions, as well as in the creation of a sacred landscape. According to the local traditions (*patria*), the territory of Laodicea was the sacred landscape where Zeus was born and spent the first years of his childhood, as portrayed on coins. A medal, minted under the emperor Caracalla, represents a central cloaked female figure who is holding a child in her left arm; we may safely recognize Adrasteia, Zeus' wet nurse, surrounded by four armed Korybantes who are dancing to shelter the child, while the river gods Lycos and Kapros lay on the left and right sides.[12] This coin image, as Louis Robert conjectured, may allude to a local tradition, according to which Zeus was born not in Crete, but in the territory of Laodicea.[13] Another local emission bears, on the obverse, the bust of Zeus Aseis, a local Phrygian god identified with the Greek Zeus; on the reverse, a representation of the goat Amaltheia together with a child (Zeus) is visible.[14] The coin legend on the obverse is ΖΕΥΣ ΑΣΕΙΣ, on the reverse ΛΑΟΔΙΚΕΩΝ. The coin legends also emphasize the strong relation of the god to the city: Zeus is the god of the Laodiceans, because he was born and nurtured in their territory during his childhood. This coin is a local emission meant for the local market and for circulation in Phrygia only; the images on it can also be interpreted as a response to the coins minted by neighbouring cities, by Aizanoi, for instance, which displayed claims of being the birth place of Zeus in competition with the Laodiceans.

The rise of these traditions, that constructed an ancient Greek and noble origin for the Phrygian *poleis*, is connected to the creation of sacred landscapes, where the myths would have taken place, i. e. the natural backdrop for the *patria*. The natural features of the landscape (rivers, hills, mountains etc.) are considered sacred and cannot be violated by human activity.

The depictions on coins of the *polis* of Aizanoi show a case in point of local tradition according to which Zeus was born there and spent his childhood with the local river god Echeloos.[15] In this case, we also have the creation (and manipulation) of a sacred landscape, where rivers and mountains are the natural setting for the traditions. The reverse of another local bronze coin of Aizanoi depicts the Meter Steunene as Rhea, who holds Zeus in her right arm and is surrounded by the dancing armed Korybantes;[16] this iconographic layout was also adopted in Laodicean emissions. Another Aizanian coin shows the goat Amaltheia on the reverse, to represent Aizanoi as the birth place of Zeus.[17] This iconography could be

---

[11] On these local emissions see Huttner 1997, who collects numismatic evidence for the Laodicea's river gods; Chiai 2012: 57–9 with further literature.

[12] On this emission see Lindner 1994: 176–8, Taf. 21, 1; Nollé 2003: 636–7; Chiai 2012: 61–2.

[13] Robert 1969: 359; Lindner 1994: 176–8, Taf. 21, 2; Nollé 2003: 636–7.

[14] *BMC Phrygia* 124; *SNG Kopenhagen Phrygia* 580; *SNG Aulock* 3849; *BMC Phrygia* 258; *SNG Aulock* 3866.

[15] On the mythical traditions of Aizanoi see Jes 2007 with further literature. On the numismatic evidence see Robert 1981, and Chiai 2012: 59–64 with further literature.

[16] See von Aulock 1979, Taf. 77, n. 62; Robert 1981: 353–7; Lindner 1994: 172.

[17] See von Aulock 1979, Taf. 77, n. 66; Robert 1981: 357–8.

explained as a response to the Laodicean emissions that claimed the status of birth place of Zeus for the *polis*. The reverse of another specimen represents the river god Penkalos, holding a child in his right arm; this should be identified as Zeus bathing in the river, according to the interpretation of Robert.[18] The territory of Aizanoi was established as a sacred landscape, the backdrop of the myth of Zeus' birth and the place where the father of the gods spent his infancy.

Altogether these coin images represent local traditions and testify to the importance of natural features and their sacralisations. River gods are of course worshiped in the city and, due to their sacralisation, the city and the citizens must protect the rivers and the mountains of their territory, the sacred landscape where the gods were born. In this way religion could become a means for the protection of the environment.

2. River pollution: archaeological evidence, literary and epigraphic sources

River pollution was a widespread problem in the ancient world, as evidenced by abundant archaeological data.[19] For example, in Kastellvicus von Butzbach, in Hessen (Germany), archaeologists have discovered a well containing the remains of animals, plants, wood, glasses, and leather.[20] The well was situated within the Roman settlement and was used by its inhabitants. Chemical analyses have dated these materials to the 2nd century AD; furthermore, botanical researches, conducted by Karl-Heinz Knörzer, have proven that a large part of the organic remains had decomposed on the surface of the well; sedimentation also indicates that this place was used for a long period, although the water must have been noxious.[21] Similarly, the wells discovered in *Londinium's* Queen Street, provide another very interesting case: filled with waste, they had probably been used as dumps.[22]

Another case of pollution of public wells and sources of water is attested in Chalon-sur-Saône. On the bank of the river near the Roman settlement, 75.000 fragments of pottery have been found; these findings encourage identifying this place as the dump of the *vicus*.[23] In the river harbour of Cologne, a very impressive sedimentation of rubbish (metal, glass, wood, leather, animal remains, etc.) has been discovered, indicating that the place was used as dump for the city waste.[24] A similar case has been discovered on the bank of the Chiem-

---

[18] Von Aulock 1979: 91, Taf. 76, n. 51; Robert 1981: 350–2; Nollé 2003: 637; Chiai 2012: 60–3.

[19] I do not know of any monograph on the history of river pollution in Greek and Roman antiquity. In her PhD thesis, Nicole Albrecht (Albrecht 2014) has investigated the remains from Roman wells in the Germanic provinces. The archaeological evidence shows that the water of most of these wells was polluted by plant and animal remains. Thüry 2001: 45–52, and Thommen 2009: 112–4 offer a useful overview on the topic.

[20] Archaeological evidence from this Roman settlement in Knörzer 1973.

[21] The botanist Karl Heinz-Knörzer remarks (Knörzer 1973: 106): 'Es dürfte sich … nicht um eine einmalige Ablagerung handeln, denn es sind Reste von sehr verschiedenartigen Kulturpflanzen mit unterschiedlichen Reifezeiten (Obst, Gemüse, Gewürze) vorhanden. Der Brunnen hat offenbar eine längere Zeit zur Aufnahme solcher Abfälle gedient'.

[22] On the archaeological discoveries in Queen Street see Wilmott 1982 with an analysis of the remains found in the grounds of the wells.

[23] Törbrügge 1970/1971: 54–60.

[24] Doppelfeld 1953: 102–6.

see, in the district of the *vicus Bedaium*, equally situated near a settlement. The archaeologists report that: 'All das, was in der Siedlung … unbrauchbar geworden ist, wurde ins Wasser geworfen, darunter Tausende von Tierknochen'.[25]

These examples, albeit here merely summarised, provide some convincing evidence that water pollution must have constituted a problem in small settlements and villages of the Roman Empire. Furthermore, the sedimentation of the material would point to the fact that the local authorities hardly adopted any measures to clean out or expurgate these water sources, and that the rivers near these settlements were likely heavily polluted by human waste. The situation in the larger towns could not have been any better. Archaeological remains and numerous literary sources attest to the problems of pollution in Rome: narrow and reeking streets, rubbish along the roads, and the murky water of the Tiber.[26]

Among the numerous ancient authors who took concern for river pollution, we can consider the testimony of Pliny the Younger regarding the contamination of the city river in Amastris.[27]

C. PLINIUS TRAIANO IMPERATORI

Amastrianorum civitas, domine, et elegans et ornata habet inter praecipua opera pulcherrimam eandemque longissimam plateam; cuius a latere per spatium omne porrigitur nomine quidem flumen, re vera cloaca foedissima, ac sicut turpis immundissimo aspectu, ita pestilens odore taeterrimo. Quibus ex causis non minus salubritatis quam decoris interest eam contegi; quod fiet si permiseris curantibus nobis, ne desit quoque pecunia operi tam magno quam necessario. (*Ep.* 10.98.1–2)

The elegant and beautiful city of Amastris, Sir, has, among other principal constructions, a very fine street and of considerable length, on one entire side of which runs what is called indeed a river, but in fact is no other than a vile common sewer, extremely offensive to the eye, and at the same time very pestilential on account of its noxious smell. It will be advantageous, therefore, in point of health, as well as decency, to have it covered; which shall be done with your permission: as I will take care, on my part, that money be not wanting for executing so noble and necessary a work.

The very short answer of the emperor was:

TRAIANUS PLINIO

Rationis est, mi Secunde carissime, contegi aquam istam, quae per civitatem Amastrianorum fluit, si intecta salubritati obest. Pecunia ne huic operi desit, curaturum te secundum diligentiam tuam certum habeo. (*Ep.* 10.99)

---

[25] Czysz, Keller 1981: 18.

[26] Panciera 2000, presents a careful discussion of the epigraphic and literary sources related to Rome and its pollution problems. See also Lo Cascio 2001; Morley 2005 and the contributions in Bradley and Stow 2012 with further literature. Silvio Panciera (p. 97) asserts that 'appare certo che l'aspetto di Roma antica non dovesse essere propriamente quello di un ridente villaggio svizzero'. Suet. *Vesp.* 5.4, enumerates the omens that presaged the Vespasian rule as future emperor: 'once, when Vespasian was at breakfast, a stray dog brought in a human hand from the crossroads, and threw it under the table'. For long time, this passage has been considered the paradigm for the abject hygienic conditions in Rome, where a dog could find cadavers lying in the street: Scobie 1986: 419. Cic., *Ver.* 2.5.27, tells of Verres, who used a rose crown to protect himself from the stench, when he went out with his *lectiga*, because of the reeking exhalations of the roads.

[27] A short collection in Fedeli 1990: 58–72.

It stands to reason, my dearest Secundus, that the water in question which flows through the city of Amastris should be covered up, if in its uncovered state it is in injurious to health (*si intecta salubritati abest*). As to money not failing for the work, that I am confident you will see to with your customary diligence. (Engl. transl. by B. RADICE 1975)

The choice of vocabulary is noteworthy. According to the common rhetorical terminology for the praise of cities, Amastris is described as *elegans* and *ornata* by a very beautiful and long boulevard, and on one of its sides there is a river. Yet, this is not a river, but a sewer. This is probably a small river and we have to infer that it was so overwhelmed by waste and rubbish to be described as a *cloaca*. The terms *turpis, immundissimus, pestilens,* and *taeterrimus* emphasize the danger posed to the public health (*publica salubritas*). This river, *cloaca*, is not only harmful to the citizens' health, but also detrimental to the city's beauty (*decor*). Pliny does not propose to cleanse and purify the river, but simply to cover up the sewer. The problem concerns above all the city's beauty: the public *salubritas* is a secondary issue.

The emperor agrees too, and does not order to clean the river, but to cover up the *cloaca* as soon as possible, providing the necessary funds.

This evidence shows how pollution problems in the ancient world could be perceived and understood also as an act of violence against the city's beauty. In this context, I would use the concept of *decorum,* elaborated by TONIO HÖLSCHER.[28] According to the German scholar, *decorum* is the appropriate image (statue, picture etc.) for a given space. If *decorum* refers to what is appropriate for a given place, waste, rubbish, and all forms of pollution are 'matter out of place' and, therefore, they are not appropriate for any private, public, and sacred space.[29] Rubbish reeks, is detrimental to health and pollutes the environment. It is not *decor* and must be eliminated.

Another interesting source, dealing with the possible causes for water pollution in the ancient world, is Vegetius (2.12) who admonished:

Si autumnali aestivoque tempore diutius in hisdem locis militum multitudo consistat, ex contagione aquarum et odoris ipsius foeditate vitiatis haustibus et aere corrupto perniciosissimus nascitur morbus, qui prohiberi non potest aliter nisi frequenti mutatione castrorum.

If numerous army continue long in one place in summer or in the autumn, the waters become corrupt, and the air infected. Malignant and fatal distempers proceed from this and can be avoided only by frequent changes of encampments. (Engl. transl. by J. CLARKE 2012)

Among the causes of river pollution was also the sojourn of an army for a long time in the same place.[30] This is unsurprising. In the *castra*, thousands of men lived and attended to their daily needs; furthermore, horses and other animals had to be fed, ministered, and washed. The cleanness of the river and the respect for the environment were certainly not the army's foremost concern. Vegetius uses the expression *perniciosissimus morbus,* whereas he does not specify the name of the illness, the outbreak of which is caused by the contamination of air and water. Here the author is drawing a 'scientific' correlation between

---

[28] On the concept of *decorum* see BRAVI 2012: 18–20, who quotes the unpublished TONIO HÖLSCHER's Sather Classical Lectures 2007, 14 March: 'Images and the Order of Things: Towards a Theory of Decorum'.

[29] DOUGLAS 1966: 36.

[30] On the environmental impact of warfare see HUGHES 2014: 150–62 with further literature.

the polluted environment and the outbreak of an infection that was hazardous to all. The statement owes perhaps to the empirical observation of the consequences of a long sojourn of an army in the same place. We may take into consideration the permanent settlement of Roman armies at the frontiers (for example along the German *limes*), which must have often originated similar problems. Other causes of river pollution must have been the discharges of the workshops: we may think – particularly – of the *fullonicae* that tanned animal skins to produce leather, as well as of the use of chemical substances in textiles manufacturing.[31] Another cause of pollution was also the blood of the sacrificial victims when sacrifices were performed near or at the shores of rivers.[32] A passage by Vitruvius (8.6.10–11) mentions the toxicity of the lead water plumbing, another health hazard to people.[33]

Rubble and rubbish, often discharged into the rivers, could cause their clogging, as Suetonius testifies by reporting the Augustus' measures to remove the rubbish that was blocking the flow of the Tiber.[34] The adverb *olim,* in the text, may refer to the fact that the rubbish had lain in the river for a long time, causing floods and other problems to the city. Augustus was the first to take care of the safety of the population living near the banks of the Tiber.

A similar initiative is attributed to Nero, who had the Tiber cleansed from the putrid corn that had been dumped into the waters.[35] But it seems that not all Roman emperors showed the same concerns for respecting the environment.[36] Claudius is said to have tossed into the sea a container of poison, which belonged to his predecessor Caligula, hence provoking the death of large quantity of fish that eventually were washed ashore.

Concerning the dangers of eating animals from a polluted river, Galen warns against fish from noxious waters. He was aware of the fact that the quality of the fish caught in the Tiber was very bad and that it was not fit for human consumption.[37] Similar concerns over the fish from the Tiber is expressed by Juvenal (5.103–5), who defines the Roman river as *torrens cloaca*, because the *cloaca maxima* flows into it. Though the Tiber was a polluted river, the inhabitants fished and drunk its waters for centuries. Moreover, the important city cult of the *Tiberinus pater* brought the citizens of Rome in contact with the polluted waters of the river. Archaeological evidence even suggests that this god was considered a healing god, like most of the river gods. A sacred deposit of approximately 200 clay kegs was discov-

---

[31] See the cases discussed by Isabella Andorlini in her contribution to this volume.

[32] Fedeli 1990: 67–9, with a collection of literary sources; the battles by rivers and lakes were also another possible cause for water pollution. Silius Italicus (*Pun.* 1.42–54) describes the waters of the Trebbia and Trasimenus polluted by the blood of the Roman soldiers. Cicero (*Tus. Disp.* 5.97) mentions the Asiatic waters that were polluted by the blood of the Persians fallen during the struggle against Alexander. Vitruvius (1.2.7) suggested, for example, to construct temples (particularly the Asklepieia) near rivers and sources.

[33] See Hughes 2014: 176 with further literature.

[34] Suet., *Aug.* 30.1: *ad coercendas inundationes alveum Tiberis laxavit ac repurgavit, completum olim ruderibus et aedificiorum prolatibus coartatum.*

[35] Tac., *Ann.* 15.18.

[36] Suet., *Cal.* 49: *Inventa et arca ingens variorum venerorum plena, quibus mox a Claudio demersis infecta maria traduntur non sine piscium exitio, quos aestus in proxima litora eiecit.* On this tradition see also Or., *Hist.* 7.5.10.

[37] Gal., *de alim. fac.* 3.25.709; 29.718; 30.721.

ered on the Tiber bank not far from the *cloaca maxima*.[38] Joël le Gall interpreted this evidence in terms of a drinking cult. In his opinion, that place, though near the city dump, was indeed a holy place at which healing drinking rituals were performed. Another similar case is attested in Roman Egypt, where, according to Athenaeus (2.42a), many people died after having drunk water from the Nile in a drinking ritual, indicating that this water must have also been heavily polluted in certain places. These examples show the value attributed to the holiness of rivers and the influence of religious beliefs that could lead people to drink toxic waters. Rivers were held sacred throughout the ancient world and the belief in river gods as gods of healing was widespread.[39] It is not a coincidence, for example, that all the sanctuaries of Asklepios were near rivers and springs. Yet, pollution of rivers and other water sources is often attested in the Greek and Roman world; hence we may suppose that these kinds of hygienic problems were an issue in many other cities and settlements lying on a river bank.

A notorious modern-day case in point is the river Ganges, the water of which, though heavily polluted, is drunk by thousands of people every day.

The above mentioned literary testimonies show, on one hand, that there was a concern and consciousness over these issues among ancient thinkers. On the other hand, the city authorities rarely adopted any particularly severe and effective measures to combat the pollution and the toxicity of rivers. This can help us to understand Pliny's protest over the destruction of the environment and the poisoning of water and air. Nevertheless, the civic authorities attempted to defend the *decor* of the public spaces by banning, for example, the transport and deposit of *stercus* within the bounds of the cities, as epigraphic evidence testifies.[40]

Concerning the pollution of public fountains and wells, numerous epigraphic documents not only testify that this was a serious problem in the ancient world, but they also inform on the measures adopted against such hazards to public health.[41]

A case in point is provided by a 5th century BC Delian inscription that prohibited washing clothes and swimming in the fountain.[42]

Μὴ πλύνεν ἐπὶ τὲν κρή[νε]ν / μηδέν, μηδὲ κολυμ[βᾶν ἐν τ]/ει κρήνει, μηδὲ [βάλ]λ[εν] κ[α]/τὰ τὴν κρήν[εν κόπρον μηδ]/[έ τι ἄλλ]ο. ἐπιζήμια δραχμ/αὶ ╞╞ [ἱ]ερ[α]ί.   (*LSCG* Nr. 50)

It is forbidden to clean something in the fountain, to swim in the fountain and to throw something into the fountain. Two drachms fine.

---

[38] See Hopkins 2012 on the importance of this monumental construction in Rome's religious traditions with further literature.

[39] A good overview for the Roman world is Campbell 2012: 331–68.

[40] On this subject Panciera 2000, collects and analyses the epigraphic and literary sources. Other examples (from *Herculaneum*) *AE* 1962, 234: *M(arcus) [Alf]icius Pa[ul]us / aedil[is] / [si qu]is velit in hunc locum / stercus abicere monetur n[on] / facere si quis adver[sus hoc] / i(n)dictum fecerit liberi dent / [dena]rium n(ummum) servi verberibus / in sedibus atmonentur(!).* (from Rome) *CIL* 6.31614: *L(ucius) Sentius C(ai) f(ilius) pr(aetor?) / de sen(atus) sent(entia) loca / terminanda coer(avit) / b(o-num) f(actum) nei quis intra / terminos propius / urbem ustrinam / fecisse velit nive / stercus cadaver / iniecisse velit. CIL* 6.40885: *L(ucius) Sentius C(ai) f(ilius) pr(aetor) / de sen(atus) sent(entia) loca / terminanda coer(avit) / b(onum) f(actum) nei quis intra / terminos propius / urbem ustri-nam / fecisse velit nive / stercus cadaver / iniecisse velit.*

[41] A good overview is in Klingenberg 1979: 291–301.

[42] Delos, *LSCG* Nr. 50: Μὴ πλύνεν ἐπὶ τὲν κρή[νε]ν / μηδέν, μηδὲ κολυμ[βᾶν ἐν τ]/εῖ κρήνει, μηδὲ [βάλ]λ[εν] κ[α]/τὰ τὴν κρήν[εν κόπρον μηδ]/[έ τι ἄλλ]ο. ἐπιζήμια δραχμ/αὶ ╞╞ [ἱ]ερ[α]ί.

A similar ban, to the purpose of protecting the cleanliness of a public fountain, is found in another Hellenistic inscription from Keos (*IG* 12.5.569): 'It is forbidden to wash (clothes) in the fountains'. These two epigraphic documents reveal the concerns of the civic authorities for the pollution of public fountains, vital features to the citizens' daily life, as well as for the spread of disrespectful behaviours against such public monuments. The so-called Astynomoi law (*OGIS* 483.159–189 = *SEG* 13.521.172–202) from Pergamon is the most important epigraphic text containing a detailed regulation about the protection of public wells and fountains.[43] It is forbidden to let animals drink from public fountains (κτῆνος … ἐπὶ τῶν δημοσίων κρηνῶν) and to wash clothes (μήτε ἱμάτια πλύθειν) there. Trespassers were fined 50 drachms. Furthermore, it established that the *astynomoi* were responsible for the fountains within and near the city, protecting their purity, important for the city life and for public health.

These epigraphic documents, dating to Hellenistic times, attest that the civic authorities of some Greek cities took care of protecting the purity of public waters, which were drunk by all citizens, by adopting measures and punishing acts of pollution as crimes against public health. In this framework, we can shortly mention the statement of Plato on the importance of protecting public waters against human pollution.[44] He suggests the institution of a body of five *agoranomoi*, responsible for the civic κρῆναι, and of three *astynomoi* competent for the general water supply of the city. We find similar considerations also in the *Politics* of Aristotle (1330b 9–18), who emphasizes the importance of preserving the cleanliness of the polis' public fountains in order to preserve the citizens' healthy.[45] Furthermore, clean

---

[43] On this important epigraphic document see KLAFFENBACH 1954; see now the comment of SABA 2012: 69–86.

[44] Pl., *Leg.* 6.763d 1–4. ἵνα κατὰ νόμους γίγνωνται πᾶσαι, καὶ δὴ καὶ τῶν ὑδάτων, ὁπόσ᾽ ἂν αὐτοῖς πέμπωσι καὶ παραδιδῶσιν οἱ φρουροῦντες τεθεραπευμένα, ὅπως εἰς τὰς κρήνας ἱκανὰ καὶ καθαρὰ πορευόμενα, κοσμῇ τε ἅμα καὶ ὠφελῇ τὴν πόλιν. δεῖ δὴ καὶ τούτους δυνατούς τε εἶναι καὶ σχολάζοντας τῶν κοινῶν ἐπιμελεῖσθαι· διὸ προβαλλέσθω μὲν πᾶς ἀνὴρ ἐκ τῶν μεγίστων τιμημάτων ἀστυνόμον ὃν ἂν βούληται, διαχειροτονηθέντων. To see that all these conform to the requirements of the law; and they shall also have charge of all the water-supplies conveyed and passed on to them by the guards in good condition, to ensure that they shall be both pure and plentiful as they pour into the cisterns, and may thus both beautify and benefit the city. (Engl. transl. by R. G. BURY, Cambridge MA 1926).

Pl., *Leg.* 6.764b 1–c2: πᾶσιν ἔκ τινος ἀνάγκης συνιέναι. τοὺς δὲ δὴ ἀγορανόμους τὸν περὶ τὴν ἀγορὰν κόσμον διαταχθέντα ὑπὸ νόμων φυλάττειν, καὶ ἱερῶν καὶ κρηνῶν ἐπιμελεῖσθαι τῶν κατ᾽ ἀγοράν, ὅπως μηδὲν ἀδικῇ μηδείς, τὸν ἀδικοῦντα δὲ κολάζειν, πληγαῖς μὲν καὶ δεσμοῖς δοῦλον καὶ ξένον, ἐὰν δ᾽ ἐπιχώριος ὤν τις περὶ τὰ τοιαῦτα ἀκοσμῇ, μέχρι μὲν ἑκατὸν δραχμῶν νομίσματος αὐτοὺς εἶναι κυρίους διαδικάζοντας, μέχρι δὲ διπλασίου τούτου κοινῇ μετὰ ἀστυνόμων ζημιοῦν δικάζοντας τῷ ἀδικοῦντι. The market-stewards must see to it that the market is conducted as appointed by law: they must supervise the temples and fountains in the market, to see that no one does any damage; in case anyone does damage, if he be a slave or a stranger, they shall punish him with stripes and bonds, while if a native is guilty of such misconduct, they shall have power to inflict a fine up to a hundred drachmae of their own motion, and to fine a wrongdoer up to twice that amount, when acting in conjunction with the city-stewards. (BURY 1926).

[45] Arist., *pol.* 1330b 9–18: ἐπεὶ δὲ δεῖ περὶ ὑγιείας φροντίζειν τῶν ἐνοικούντων, τοῦτο δ᾽ ἐστὶν ἐν τῷ κεῖσθαι τὸν τόπον ἔν τε τοιούτῳ καὶ πρὸς [10] τοιοῦτον καλῶς, δεύτερον δὲ ὕδασιν ὑγιεινοῖς χρῆσθαι, καὶ τούτου τὴν ἐπιμέλειαν ἔχειν μὴ παρέργως. οἷς γὰρ πλείστοις χρώμεθα πρὸς τὸ σῶμα καὶ πλειστάκις, ταῦτα πλεῖστον συμβάλλεται πρὸς τὴν ὑγίειαν· ἡ δὲ τῶν ὑδάτων καὶ τοῦ πνεύματος δύναμις τοιαύτην ἔχει τὴν φύσιν. διόπερ ἐν [15] ταῖς εὖ φρονούσαις δεῖ διωρίσθαι πόλεσιν, ἐὰν μὴ

fountains and water supply could represent an important element of the city's beauty; they were to be preserved and praised in the context of inter-poleis antagonism: a city rich in pure water was a beautiful and enviable city.

3. Gods against human pollution

Based on epigraphic evidence, I wish to analyse how religion and the gods could provide incentive for the protection of the environment, and how the provisions regarding ecology in the sacred laws point to a sensibility for uncontaminated nature and environment in the Greek and Roman world. The collected epigraphic material testifies how theodicy – the widespread in antiquity belief that the gods can punish mortals for their sins with death or disease so restoring the justice – played a central role in these inscriptions. The confession inscriptions from the rural sanctuaries of Roman Phrygia narrate how the gods are aware of all human actions from heaven, and how the divine powers can punish human sins. This evidence provides a pertinent case in point for theodicy. Moreover, ANGELOS CHANIOTIS in particular has emphasized that these epigraphic documents reveal how religion and the belief in the divine powers could be used as a tool for social control in the Greek and Roman world.[46] The Egyptologist JAN ASSMAN has proposed the concept of 'konnektive Gerechtigkeit' to name and define the relationship between physical conditions and social behaviour.[47] According to this theory, the cause of an illness can be traced to a misdeed, a violation of social norms, laws or divine order. This idea can be applied to the study of our inscriptions, which contain threatening admonishments against actions that could result in the pollution of a place that was under the protection of the gods. In addition, according to a common practice often attested in the prayers for divine justice, the property of a place or of an object can be transferred to a god, who becomes responsible for its purity and cleanliness against human pollution. Inscriptions and images of gods, therefore, can signal that places were under divine protection.[48]

On the importance of religion in the preservation of the purity of a place, the philologist OTTO JAHN observed in 1855: 'Wenn man im Altertum einen Ort nicht verunreinigt haben wollte, diente statt Schutzmann und Polizeiplacat ein Götterbild'.[49] This image could be, for instance, the image of the *genius loci*, the god who protected the place. The satirist Persius recorded that the depiction of two snakes signalled the divine protection of a place and the

---

πάνθ᾽ ὅμοια μηδ᾽ ἀφθονία τοιούτων ἢ ναμάτων, χωρὶς τά τε εἰς τροφὴν ὕδατα καὶ τὰ πρὸς τὴν ἄλλην χρείαν. And since we have to consider the health of the inhabitants, and this depends upon the place being well situated both on healthy ground and with a healthy aspect, and secondly upon using wholesome water-supplies, the following matter also must be attended to as primary importance. Those things which we use for the body in the largest quantity and most frequently, contribute most to health; and the influence of the water supply and of the air of this nature. Hence in wise cities if all sources of water are not equally pure and there is not an abundance of suitable springs, the water-supplies for drinking must be kept separate from those of other requirements. (Engl. transl. by E. BARKER, Oxford 1946).

[46] See especially CHANIOTIS 1997; 2007; 2012.
[47] ASSMANN 1990: 58–91.
[48] About the prayers for justice see VERSNEL 1991, 2009.
[49] JAHN 1855: 87.

prohibition of urinating there.[50] Inscriptions of this sort, which provide interesting testimonies for people's sacral practices in religion, have been found in Rome and particularly in
Pompeii.

4. Pollution of Public spaces

The first inscription discussed was found in Rome at the Baths of Titus, where it was painted
on a wall. It bears the image of two snakes with the following threatening admonishment:

> *Duodecim deos et Deanam et Iovem/optimum maximu[m] habeat iratos/quisquis hic mixerit aut*
> *cacarit.*   (*CIL* 6.13740)

> Whoever shall urinate or shit in this place will anger the twelve gods, Diana and Jupiter, the best
> and greatest.

This epigraphic document aims at protecting the cleanness of this public place, and invokes the power of the twelve gods, Diana, and the most important Roman god *Iupiter optimus maximus*. Divine wrath will strike anyone who urinates or shits there, consequently
contaminates the cleanness of the public space and pesters the other visitors. On the other
hand, this inscription testifies that some guests did heed to their bodily needs and did not
respect basic hygienic norms of a public bath. This is an exemplar case of a public religious
inscription against human pollution which employed the respect towards the gods to the
safeguard of the cleanness of a public space.

The next testimony is an inscribed relief from Salona (Split, Dalmatia) with a representation of Hecate. It also cautions against *cacatores*, protecting the place against human
pollution.

> *Quis(que) in eo vico stercus non posu/erit aut non cacaverit aut non me/iaverit, habeat illas propi-*
> *tias;/si neglexerit, viderit.*   (*CIL* 3.1966)

> Whoever will not dump waste or will not shit or will not urinate in this residential district, he will
> have their (i. e. Hecate's) goodwill, if he disobeys, he will see.

The prohibition against human pollution involves the powerful goddess of the underworld,
Hecate. In this case, the relationship between text and image is noteworthy: the onlooker
sees at first sight the image of the terrible goddess, connected with the underworld and
often appealed to in *defixiones*. But Hecate was also the protector of the crossroads and it
is possible that this inscription was placed at a crossroads to defend the street of the urban
settlement against anyone who might sully the street or the district. The cleanness of roads
and *fora*, and generally of public spaces, represented a problem in the ancient world since
archaic times. This is testified by an inscription from Paros (*IG* 12.5.107), dated to the early
5th century BC.[51] The epigraphic text contains provisions for the cleanness of public roads
and attests the means by which civic authorities attempted to keep clean the public spaces
of the polis. With regard to the Roman world, we can make use of the numerous epigraphic
documents that prohibited dumping *stercus* or other rubbish within the city's boundaries.[52]

---

[50] Pers., *Sat.* 1.113: *Pinge duo anguis: pueri, sacer est locus, extra/meite.*
[51] KOERNER 1993: 215–7 n. 57.
[52] An exhaustive overview in PANCIERA 2000.

The inscription under discussion here, probably set up at a crossroads, would seem to testify an attempt by the civic authorities to instrumentalise religion in order to protect the cleanness of the roads and urban districts.

Similar examples include two graffiti from Ephesos, inscribed on the wall of the city gate.

> Εἴ/τις ἂν ὧδε / οὐρήσι, ἡ Ἑκά/τη αὐτῷ κε/χώλωται. (*I.Ephes.* 2.567)

> If someone urinates in this place, Hecate will be angry against him.

Again the mentioning of Hecate alludes to her function as protector of the crossroads as in the previous document.

> ὧδε οὐρήσαντι ἡ Ἄρτεμ[ις] / [κε]χόλωται. (*I.Ephes.* 2.569)

> Artemis will be angry against the person who urinates here.

This graffito, inscribed on the wall of the city gate, aims at protecting the public space and evokes the power of the goddesses Artemis, the paramount goddess of Ephesos. Pollution against the city is pollution either against Artemis, because it destroys city's beauty (*kallos, decor*), or against Hecate, who protects the roads and bears the powers of the underworld. The verbal form κεχώλοται is the Greek translation of the Latin expression *habebit deos iratos*. This formula is also found in the Greek East, particularly in Phrygia, in the famous *arai epitymbioi*. These were imprecations against grave desecrators, who committed not only a sacrilege against the gods and the souls of the deceased, but also against the sacred landscape.[53] Moreover, the desecration and opening of a grave could also be the cause of hygienic problems and environmental pollution. These graffiti on the city wall seem to represent a sort of greeting for the visitors coming into the city, but admonished them to respect the cleanness of the public spaces. Moreover, they could testify to the actual problem of passers-by fulfilling their bodily needs and soiling those places.

Not only walls and gates, but also public monuments were not respected in the ancient world, as this inscription, affixed on the monumental arch of Thigibba (Tunisia) testifies:

> *Si qui hic urinam / fecerit // habebit Martem / iratum.* (*AE* 1949.48)

> If someone urinates here, this person will anger Mars.

This monument, which probably commemorated a victory achieved by the favour of the war god, evoked Mars for protecting it against human lack of respect. The monumental style of the letters indicates the public nature of the inscription and that the arch was under the protection of Mars: to disrespect his monument was to disrespect the god.[54]

---

[53] STRUBBE 1997, collected a corpus of these inscriptions from the Greek East.

[54] Sacred and public monuments were often damaged and disrespected. This is evident, for instance, in an inscription from Rome, which invokes the anger of the Roman people's Genius against anyone who does not respect the altar. *CIL* 6.29944: *]ICO / quis hanc ara / laeserit habeat / Genium iratum / populi Romani / et numina divo/rum.*

A similar case can be found in Caracalla' biography, where we read of the curse against all persons who urinated in the place where the images of the emperors were set up, thus disrespecting their holiness.[55]

Altogether these inscriptions show that, notwithstanding – or in addition to – the dispositions of law and fines, civic authorities seem to evoke the power of the gods as a means to guarantee the protection of public spaces (i. e. roads, city gates, walls, and monuments) against human pollution.

5. Pollution of private spaces

Many inscriptions against *cacatores* in Pompeii, mostly employing the formula *cacator cave malu(um,)* (shitter, beware of misfortune) were found in the external wall of a *domus*, testifying that the cleanness of roads and walls must have been an everyday problem in the city.[56] A considerable number of graffiti, written by *cacatores* on the city walls, attests to the vital importance of this problem for public health as well as for the civic *decor*.[57]

One case in point is a graffito, impressed on a fresco that portrays two great snakes, surrounding a *cacator*, near the gigantic figure of the goddess Tyche.[58] The graffito admonishes: *Cacator, / cave malum.* The relationship between image and text is significant. The two snakes are likely the representation of the *genius loci* who protected the *domus,* while the city's *decor* is represented by Tyche.

The habit of erecting figurative panels and reliefs with admonitions against *cacatores* must have been common throughout the Roman Empire. Evidence comes not only from Pompeii, but also, for example, from Aquileia (Northern Italy). In this relief, Zeus is about to strike a man who is in the act of defecating in a private or public space and so is disregarding the prohibition expressed by the inscription and by the relief: Zeus' wrath befalls on anyone who pollutes the space in front of this image. The depicted inscription, which probably contained the admonitions, on the relief is unfortunately disappeared.[59]

The father of the gods is also invoked in another Pompeian graffito:

cacator cave malum / aut si contempseris habeas / Iove(m) iratum. (CIL 4.7716)

Shitter, beware of misfortune, or if you disrespect this admonition, you will anger Jupiter.

Another example has been found in Aquincum (Budapest). The inscription written on the basis of a statue shows: *Minerv[ae] // aliqui<d=T> spur/ci velle fece/rit habeat et / superos et / infer{n}os / deos iratos* (*AE* 1937.197).

[55] HA, *M. Ant.* 5.7: *damnati sunt eo tempore qui urinam in eo loco fecerunt in quo statuae aut imagines erant principis*; see also Iuv. 1.1.131: *cuius ad effigiem non tantum meiere fas est.*

[56] The formula *cacator cave malum* is quite frequent: see for example CIL 4.3832; 5381; 5438; 8870.

[57] CIL 4.3146: *Secundus / hic (c)acat / hi[c cac]at / hic cacat*; CIL 4.5242: *Quodam quidem testis eris quid senserim / ubi cacaturiero veniam / cacatum*; CIL 4.5244: *Marthae hoc tric{hi}linium est nam in tric{hi}linio / cacat*; CIL 4.10070: *Lesbiane cacas scribisque [sa]lute(m)*; CIL 4.10619: *Apollinaris medicus Titi Imp(eratoris) / hic cacavit bene.*

[58] This figure can be interpreted as the city goddess on the basis of its wall-shaped crown.

[59] See Thüry 2001: 19 fig. 21.

The admonition is against all *cacatores* who provoke Jupiter's anger by disrespecting his power (*contempto*). This inscription, placed in a narrow street between two houses, and hence likely often littered by *cacatores*, testifies once again to the seriousness of such a problem in the city, a problem that only the fear of the most potent god of Olympus could resolve. Another Pompeian inscription provides a suggestion to the *cacatores* as to how to avoid divine anger:

> *cacator sig(!) valeas / ut tu hoc locum trasea(s). (CIL 6.6641)*
>
> Shitter, you will remain healthy, if you go past this place.

Also, burial spaces were not respected in the Greek and Roman world and often used as outhouses.[60] This inscription from Rome shows the importance of the belief in the gods in safeguarding the burial places against human filth. This boundary stone was set up opposite the grave of a Roman family:

> *C. Caecilius C. / et C. L. Florus / vixit annos XVI / et mesibus VII qui / hic mixerit aut / cacarit, habeat / deos superos et / inferos iratos.*   (*CIL* 6.13740)
>
> C. Caecilius C. and C. L. Florus lived 16 years and 7 months. Who urinates or shits here, this person will anger both the gods of the heavens and of the underworld.

In this case, the inscription aims at protecting the inviolability of the burial area from pollution, on account that such disrespectful behaviour would make the gods angry.

These two sepulchral inscriptions from Rome contain an admonishment to stop anyone from urinating on the funerary monument:

> *Ad h(oc) m(onumentum) u(rinam) f(acere) n(on) l(icet). (CIL 6.3413)*
>
> It is not permitted to urinate on this sepulchral monument.

> *] / vixit anni[s …] mensibus] / X diebus XXVII [… si quis] / hic urinam fec[erit …]. (CIL 6.29904)*
>
> Lived years… 10 months and 27 days. If anyone urinates (in this place).

Probably, the second inscription ended with a formula as *habeat deos superos et inferos iratos,* invoking the power of the gods of the heavens and of the underworld.

This other epigraphic document, discovered in Pompeii, testifies to the problem of a polluted grave.[61]

> *Hospes adhuc tumuli n<e=I> meias ossa prec[antur] / nam si vis (h)uic gratior esse caca / Urticae monumenta vides discede cacator / non est hic tutum culu(m) aperire tibi.*   (*CIL* 4.8899)
>
> Stranger, the bones ask you not to piss at this tomb, for, if you want to be more agreeable to this man, shit. You see Nettle's tomb; away from here, shitter; it is not safe for you to open your bowels her. (Transl. by E. COURTNEY, Atlanta 1995)

---

[60] An overview of the curses for protecting the graves is in LATTIMORE 1942: 108–26.
[61] COURTNEY 1995: 158–9, n. 168b, 168–9.

Notwithstanding the ironic tone of the text, this epigram testifies to the widespread disrespect for sepulchral monuments in the Roman world. To this point, Trimalchio, the rich *libertus* of Petronius' novel, is going to hire someone to protect his monumental grave against the activity of eventual *cacatores*.[62] This stands as yet more evidence of the diffusion of this problem in the Roman world, as well as of an attempt to adopt efficient measures against such disrespectful acts.

6. Pollution of Sacred spaces

Not only public, private, and funerary places seem to have suffered from the actions of disrespectful passers-by, but also the sacred spaces of temples and sanctuaries. The analysis of this point may begin from this noteworthy graffito from a temple of Pan in Egypt:

Ἱερὸς ὁ τόπος. Ὃς ἐὰν ὦδε ὁ[μ]οῦ [χ]έσῃ / ἢ οὐρήσῃ Πᾶνα ἕξ[ει κεχολωμένον].
(A. BERNARD, *Pan du desert*, Leiden 1977, Nr. 15)

This is a holy place, if someone shits or urinates here, this person will anger Pan.

The text begins with the expression 'this is a holy place', and it seems that many people had forgotten this fact. The warning concerns all persons who defecate or urinate within the bounds of the temple, disrespect the holiness of the space, and of course provoke Pan's anger. This document is not unique; indeed, other sanctuaries also had to tackle with the problem of rubbish and pollution caused by the coming and going of pilgrims. A sacred law of Epidauros prohibited, for instance, littering of the wall of the holy hall (*LSCG* 24.2–3: μὴ λυμαίνειν τὰν / στοὰν τὰν ἱερὰν ἐν τῷ τεμένει), probably because many pilgrims did urinate there. According to the inscription, the trespasser was fined 100 drachms. Furthermore, the inscription expresses the prohibition of dumping excrements and embers into the *temenos* in order to protect the cleanness of the holy place (μηδὲ κόπρον μηδὲ σποδὸν ἐκβάλλειν ἐν τῷ τεμένει). Another interesting testimony comes from a sacred law from Delos, dating to the 3rd century BC, that forbids to dump excrements, embers, or anything else that might dirty the holy place into the sanctuary of Leto (*LSCG* 53.6–9: μηδ' εἰς τὸ τέμενος τὸ τῆς Λητοῦς μήτε κό/προν, μήτε σποδόν, μήτε ἄλλο μη/θέν).[63]

    The wrath of the god is invoked against those pilgrims who heed to their bodily needs, disrespecting the holiness of the place and the power of the god. This inscription shows hygienic problems, due to the frequentation of a great numbers of pilgrims in the sanctuary and how the religious institutions perceived the problem and tried to solve it. This owes to the lack of sanitary facilities in many sanctuaries, on one hand, and to the need to preserve the holiness of the temple from pollution and violation of human excrements, on the other. Furthermore, the epigraphic texts from Epidauros and Delos indicate that the term *kopros*

---

[62] Petr., 71.8: *custodiae causa, ne in monumentum meum popolus cacatum currat.*

[63] Another example is from Thasos (*IG* 12/8.265-3–6: ὁ ἀναι]/ραιρημένος τὸν κῆ[πον τὸ χωρ]ίον καθαρὸν παρέξει [τὸ περὶ τὰς]/πύλας, ὅπου ἡ κόπρος [ἐξεβάλλ]ετο. ἢν δέ τις ἐγβάλλῃ[ι κόπρον ἐς]/τὸ χωρίον, εἶναι τὸ ἄγγος τοῦ ἀναιρερημένου τὸν κῆπο[ν, τὸν δὲ]/δοῦλον μαστιγώσαντα ἀθῷϊον εἶναι). Other epigraphic documents on the prohibition to throw excrements or rubbish into a sanctuary have been collected by DILLON 1997: 125–6.

not only referred to animal excrements, but was also considered a kind of rubbish, and inappropriate for a holy place.[64]

An inscription from Échelle (*Vicus Augusti* in Gallia Narbonensis) shows that Roman temples were not always respected either:

> *Lex rivi Ul[…] / si quis in eo m/ix{s}erit spurcit(iam) / fecerit in temp(lum) / Iovis D(omestici?) |(denarium) d(ato) de/l(atoris) pars dim(idia) esto / n<i=E>si l(ongius) p(assibus) V.*   (*CIL* 12.2426)

> Law of the river Ul…, if anyone urinates or pollutes inside the temple of Jupiter Domesticus (?), may the denouncer receive the half part of the fine, if this is not longer than five miles.

This document shows problems of hygiene within the temple of Jupiter, caused by people who pollute the sacred spaces by urinating. The civic authorities promise a reward for the *delator*.

Another inscription on a *cippus* from the Picenum attests to the attempt of the civic authority to protect the holiness and purity of a sacred water source, the pollution of which could jeopardize pilgrims' health.

> *M(arco) Lepido L(ucio) Arrunti(o) / co(n)s(ulibus) d(ecreto) d(ecurionum) posit(us) / qui intra stercus / fuderit multae a(sses) IIII d(abit).*   (*AE* 1988.486)

> During the consulship of Marcus Lepidus and Lucius Arruntius, according to the decree of the decurions: If anyone drops excrements inside the city, he will pay a fine of four asses.

Another similar document was found in the territory of the Regio X (*Venetia* and *Histria*)

> *Stercus intra / cippos / qui fecerit / aut violarit nei / luminibus fruatur.* (*ILS* 8207b)

> If anyone drops excrements in the vicinity of the cippus or violates (this holy space), may he no longer enjoy eyesight.

These inscriptions, from Rome, Greek East, Egypt, and the Western provinces testify that littering and pollution of sacred spaces were a common problem in the Greek and Roman world. Such a pollution was caused by the great number of pilgrims visiting sanctuaries which were hardly equipped with sanitary facilities, and also by the excrements of the sacrificial victims, as well as simply by human lack of respect. Against such a sacrilege, the inscriptions invoke the wrath of the gods who protected the holiness and purity of their temples. The analysis of these epigraphic documents brings to light a common sacred language that was employed to evoke the gods and their power to the purpose of protecting a private or public place, a grave, a road, and a sanctuary against human fouling and pollution; this is the case, for instance, of the formula *habebit deos iratos*, the Latin translation of the Greek κεχώλωται. Finally, these epigraphic texts prove that the disposal of human litter was a problem of everyday life and that religion could offer a solution to a problem that worldly laws didn't seem able to resolve.

---

[64] Greek attitudes to *kopros* in Hes. *erg.* 726–32, 757–9; *kopros* as *miasma* in PARKER 1983: 293–4; DILLON 1997: 126, n. 123.

7. Protection of rivers

Religion and sacred laws played an important role also in the protection of rivers, which were often polluted by human activities. The sanctity and the sacralisation of rivers could not allow such corruption.

A prime example is a sacred law from Athens, dating to 420 BC, which prohibited the washing of coats of the sacrificed animals in the Hilissos, to the purpose of protecting the purity of this river.[65]

> *LSCG* 4: ἐ[πιμέλ]εσθαι δὲ / τομ βασιλέα γράφσαι δ/ὲ ἐστέλει καὶ /στεσαι ηεκατέροθι με/δὲ δέρματα σέπεν ἐν το/ι Ηιλισοι καθύπερθεν / το τεμένος το Ηρακλέ/[ο]ς, [μ]εδὲ βυρσοδεφσεν, μ/[εδὲ καθά] ρμ[ατ]α (ἐ)ς τὸν π[οταμὸν βάλλεν.

> The archon *basileus* must take care of writing this decree into a stele of stone and of setting it up on both banks of the river. It is forbidden to soak the coats in the Hilissos above the sanctuary of Heracles and to tan them. It is forbidden to throw the residue of the laundering into the river.

Franciszek Sokolowski has interpreted this text as a sacred law since it, involved the city authorities who issued the decree. The mentioned *dermata* are probably the coats of the animals sacrificed in the sanctuary. This act aimed at protecting the river, vital to the city, by employing religious authority. This decree has been related with Plato's bucolic description of the Hilissos area, in his *Phaedrus* (227 a ff., especially 230 b-c): this could be an indication that sanitary conditions were restored to this area in about twenty years.[66]

The waters and rivers of a sanctuary, used for cultic practices, were of course considered sacred and should not be violated. A sacred law from Samos (*LSCG* 81), for example, prohibited drawing water from the river Imbrasos, in order to preserve its purity. According to Thucydides (4.97), the Thebans complained that the Athenian troops occupying the sanctuary at Delion were using water that should have been only employed for purification before sacrifice, for profane purposes. The two sources here discussed reveal the importance of preserving the holy waters of a sanctuary against any type of impurity and pollution.

In Rome, we may refer to the project of modifying the natural course of the Tiber: Tacitus (*Ann.* 1.79) recounts that the Roman Senate did not do anything against the inundations of the Tiber, notably changing the river' course, because they decided to respect what nature had created, the local cults of the allies (*religiones sociorum*), and finally the beauty of the river itself, which was the river of Rome:

> *Actum deinde in senatu ab Arruntio et Ateio an ob moderandas Tiberis exundationes verterentur flumina et lacus, per quos augescit; auditaeque municipiorum et coloniarum legationes, orantibus Florentinis ne Clanis solito alveo demotus in amnem Arnum transferretur idque ipsis perniciem adferret. congruentia his Interamnates disseruere: pessum ituros fecundissimos Italiae campos, si amnis Nar (id enim parabatur) in rivos diductus supersta gnavisset. nec Reatini silebant, Velinum lacum, qua in Narem effunditur, obstrui recusantes, quippe in adiacentia eruptirum; optume rebus mortalium consuluisse naturam, quae sua ora fluminibus, suos cursus utque originem, ita finis dederit; spectandas etiam religiones sociorum, qui sacra et lucos et aras patriis amnibus dicaverint: quin ipsum Tiberim*

---

[65] See Rossetti 2002.
[66] This is the interpretation of Lind 1987, with a reconstruction of the historical context.

nolle prorsus accolis fluviis orbatum minore gloria fluere. seu preces coloniarum seu difficultas operum sive superstitio valuit, ut in sententiam Pisonis concederetur, qui nil mutandum censuerat.

A question was then raised in the Senate by Arruntius and Ateius whether, in order to restrain the inundations of the Tiber, the rivers and lakes which swell its waters should be diverted from their courses. A hearing was given to embassies from the municipal towns and colonies, and the people of Florentia begged that the Clanis might not be turned out of its channel and made to flow into the Arnus, as that would bring ruin on themselves. Similar arguments were used by the inhabitants of Interamna. The most fruitful plains of Italy, they said, would be destroyed if the river Nar (for this was the plan proposed) were to be divided into several streams and overflow the country. Nor did the people of Reate, remain silent. They remonstrated against the closing up of the Veline lake, where it empties itself into the Nar, 'as it would burst in a flood on the entire neighbourhood. Nature had admirably provided for human interests in having assigned to rivers their mouths, their channels, and their limits, as well as their sources. Regard, too, must be paid to the different religions of the allies, who had dedicated sacred rites, groves, and altars to the rivers of their country. Tiber himself would be altogether unwilling to be deprived of his neighbour streams and to flow with less glory.' Either the entreaties of the colonies, or the difficulty of the work or superstitious motives prevailed, and they yielded to Piso's opinion, who declared himself against any change.[67] (Engl. transl. by J. Jackson 1937)

This text reveals the existence of a controversy over the possibility of changing the natural course of the most important river of Roman tradition. We should not forget the existence of the cult to the river god Tiberinus who was worshipped as a healing god, as shown by the archaeological record briefly discussed above. Moreover, this god was worshipped also by the neighbouring Italic peoples, likely with temples and cults. A change of the river's course would have been disrespectful to its holiness as well as to the religiosity of the allies of Rome. In other words, changing the course of the river would have meant altering the natural order, a violation of the *kosmos*. This text reveals a sensibility for environmental issues in the ancient world, in connection to religion and to an outlook on nature as *kosmos*. Finally, this literary source is a precious testimony of how such problems were addressed as public business by the Roman senate.

Conclusions

In the ancient world rivers and waters were considered not only natural elements, but also personified gods, who provided wealth and prosperity to the city and its population. Rivers represented central elements in the sacred landscape of a *chora* where the mythical local traditions of the polis took place. Such a sacralisation allowed the local authorities to use religion and the power of the gods to the purpose of protecting the rivers and waters against human pollution. This was a common problem in the Greek and Roman world, as we have seen by the various examples discussed above. The pollution of a fountain, a source of water, as well as of a public and private space, which was under the protection of a god, was considered a sacrilege and could be punished not only by human justice, but also by divine power.

---

[67] See Cappelletti 2009 and the contribution by Bianchi in this volume.

Bibliography

ALBRECT, N. 2014. *Römerzeitliche Brunnen und Brunnenfunde im rechtsrheinischen Obergermanien und in Rätien.* Inauguraldissertation zur Erlangung der Doktorwürde der Philosophischen Fakultät der Universität Heidelberg. (unpublished)

ASSMANN, J. 1990. *Ma'at. Gerechtigkeit und Unsterblichkeit im alten Ägypten.* München.

BRADLEY, M. and K. STOW (eds.). 2012. *Rome, Pollution and Propriety. Dirt, Disease and Hygiene in the Eternal City from Antiquity to Modernity.* Cambridge.

BRAVI, A. 2012. *Ornamenta Urbis. Opere d'arte greche negli spazi romani.* Bari.

CAMPBELL, B. 2012. *Rivers and the power of ancient Rome.* Chapel Hill.

CAPPELLETTI S., 2009. 'Il progetto sull'esondazione del Tevere. Tacito, Annales 1.76 e 1.79', *Acme* 62, 235–53.

CHANIOTIS, A. 1997. 'Tempeljustiz im kaiserzeitlichen Kleinasien: Rechtliche Aspekte der Beichtinschriften', in G. VELISSAROPOULOS KARAKOSTA (ed.), *Symposion 1995. Vorträge zur griechischen und hellenistischen Rechtsgeschichte.* Böhlau-Köln 1997, 353–84.

CHANIOTIS, A. 2004. 'Under the Watchful Eyes of the Gods: Aspects of Divine Justice in Hellenistic and Roman Asia Minor', in S. COLVIN (ed.), *The Greco-Roman East. Politics, Culture, Society* (Yale Classical Studies 31), 1–43.

CHANIOTIS, A. 2012. 'Constructing the fear of gods. Epigraphic Evidence from Sanctuaries of Greece and Asia Minor', in A. CHANIOTIS (ed.), *Unveiling Emotions. Sources and Methods for the Study of Emotions in the Greek World.* Stuttgart, 205–34.

CHIAI, G. F. 2012. 'Die Götter und ihr Territorium: Münzen als Quellen zur *interpretatio* im kaiserzeitlichen Phrygien', in G. F. CHIAI, R. HÄUSSLER and C. KUNST (eds.), Interpretatio romana / graeca / indigena: *Religiöse Kommunikation zwischen Globalisierung und Partikularisierung* (Monografie di Mediterraneo Antico 15), 51–70.

COURTNEY, E. 1995. *Musa Lapidaria. A selection of Latin verse inscriptions.* Atlanta.

CURBERA, J. 1998. 'Onomastics and river-gods in Sicily', *Philologus* 142, 52–60.

CZYSZ, W. and KELLER, E. 1981. *Bedaium, Seebruck zur Römerzeit.* München.

DILLON, P. J. 1997. 'The ecology of the Greek sanctuaries', *ZPE* 118, 113–27.

DOPPELFELD, O. 1953. 'Hafenfunde vom Altermarkt in Köln', *Bonner Jahrbücher* 153, 102–25.

DOUGLAS, M. 1966. *Purity and Danger: an Analysis of Concepts of Pollution and Taboo.* London.

GÜNTHER, L.-M. 2009. 'Quellen, Bäche, Flüsse und ihre Gottheiten im griechischen Sizilien. Zum Bildtypus "Opfers am Altar"', in E. OLSHAUSEN and V. SAUER (eds.), *Die Landschaft und die Religion. Stuttgarter Kolloquium zur Historischen Geographie des Altertums 9* (Geographica Historica 26). Stuttgart, 81–95.

HOPKINS, J. 2012. 'The "sacred sewer": tradition and religion in the Cloaca Maxima', in M. BRADLEY and K. STOW (eds.), *Rome, Pollution and Propriety. Dirt, Disease and Hygiene in the Eternal City from Antiquity to Modernity.* Cambridge, 81–102.

HUGHES, J. D. 2014[2]. *Environmental Problems of the Greeks and Romans. Ecology in the Ancient Mediterranean.* Baltimore.

HUTTNER, U. R. 1997. 'Wolf und Eber: Die Flüsse von Laodikeia in Phrygien', in J. NOLLÉ, B. OVERBECK and P. WEISS (eds.), *Internationales Kolloquium zur kaiserzeitlichen Münzprägung Kleinasiens. 27.30. April 1994 in der Staatlichen Münzsammlung* (Nomismata. Historisch-numismatische Forschungen 1). Milano, 93–109.

IMHOOF-BLUMER, F. 1923. 'Fluß- und Meergötter auf griechischen und römischen Münzen', *SNR* 23, 173–421.

JAHN, O. 1855: *Über den Aberglauben des bösen Blicks bei den Alten* (Berichte über die Verhandlungen der königlichen sächsischen Gesellschaft der Wissenschaft zu Leipzig, Philologisch-historische Klasse 7). Leipzig.

JES, K. 2007. 'Eine Stadt von edler Abkunft und hohem Alter. Kulturelles Gedächtnis in Aizanoi im 2. Jh. n. Chr.', in O. CORDOVANA and M. GALLI (eds.), *Arte e memoria culturale nell'età della Seconda Sofistica.* Catania, 153–68.

KLAFFENBACH, G. 1954. *Die Astynomeninschrift von Pergamon*. Berlin.

KLINGENBERG, E. 1979. 'La legge platonica sulle fontane pubbliche', in A. BISCARDI, H. J. WOLFF, J. MODRZEJEWSKI und P. DIMAKIS (eds.), *Vorträge zur griechischen und hellenistischen Rechtsgeschichte (Gargnano am Gardasee, 5.–8. Juni 1974)*. Koln-Wien 1979, 283–305.

KNÖRZER, K.-H. 1973. 'Römerzeitliche Pflanzenreste aus einem Brunnen in Butzbach (Hessen)', *Saalburg Jahrbuch 30*, 71–114.

KOERNER, R. 1993. *Inschriftliche Gesetzestexte der frühen griechischen Polis aus dem Nachlaß von Reinhard Koerner herausgegeben von Klaus Hallof*. Köln-Böhlau.

LATTIMORE, R. 1942. *Themes in Greek and Latin epitaphs*. Urbana.

LE GALL, J. 1953, *Recherches sur le culte du Tibre*. Paris.

LIND, H. 1987. 'Sokrates am Ilissos. IG I³ 1 257 und die Eingangsszene des platonisches "Phaidros"', *ZPE* 69, 15–9.

LINDNER, B. R. 1994. *Mythos und Identität: Studien zur Selbstdarstellung kleinasiatischer Städte in der römischen Zeit*. Stuttgart.

LONGO, O. 1991. 'Conciapelli e cultura in Grecia antica', *Lares 57*, 5–24.

LO CASCIO, E. 2001. 'Condizioni igienico-sanitarie e dinamica della popolazione della città di Roma dall'età tardorepubblicana al tardo antico', in J.-N. CORVISIER (ed.), *Thérapies, médecine et démographie antiques*. Arras, 37–70.

MAZZARINO A., 1969–1970. 'Un testo antico sull'inquinamento', *Helikon 9–10*, 643–5.

MORLEY, N. 2005. 'The Salubriousness of the Roman City', in H. KING (ed.), *Health in Antiquity*. London-New York, 192–204.

NOLLÉ, J. 2003. 'Vielerorts war Bethlehem – Göttergeburte im kaiserzeitlichen Kleinasien', *AW 34*, 6, 635–43.

PANCIERA S., 2000. 'Nettezza urbana a Roma organizzazione e responsabili', in J. A. REMOLÀ and X. DUPRÉ RAVENTÓS (eds.), *Sordes urbis: la eliminación de residuos en la ciudad romana, Actas de la Reunión de Roma, 15–16 de noviembre de 1996*. Roma, 95–105.

PANESSA G., 1991. *Fonti greche e latine per la storia dell'ambiente e del clima nel mondo Greco*. Pisa.

PARKER, R. 1983. *Miasma. Pollution and Purification in Early Greek Religion*. Oxford.

PETER, U. 2005. 'Die Bedeutung des Hebros in der Münzprägung von Philippopolis (Thrakien)', in C. ALFARO, C. L. MARCOS y C. L. OFERO (eds.), *XIII Congreso Internacional de Numismática (Madrid 2003), Actas – Proceedings – Actes I*. Madrid, 927–33.

PUCCI, G. 2015. 'Figli improvvidi. Gli Antichi, la Terra, l'ambiente', M. BETTINI e G. PUCCI (eds.), *Terra antica. Volti, miti e immagini della Terra nell'antichità*. Milano, 208–43.

ROBERT, L. 1969. *Laodicée du Lycos. Le Nymphé. Campagnes 1961–63*. Paris.

ROBERT, L. 1981. 'Documents d'Asie Mineure. Fleuves et cultes d'Aizanoi', *BCH 105*, 331–60 (= *Documents d'Asie Mineure*. Athen 1987, 241–70).

ROSSETTI, L. 2002. 'Il più antico decreto ecologico a noi noto e il suo contesto', in T. M. ROBINSON and L. WESTRA (eds.), *Thinking about the Environment. Our Debt to the Classical and Medieval Past*. Lanham MD, 44–57.

SCOBIE, A. 1986. 'Slums, sanitation, and mortality in the Roman world', *Klio 68*, 399–433.

STRUBBE, J. 1997. ΑΡΑΙ ΕΠΙΤΥΜΒΙΟΙ. *Imprecations against Desecrators of the Grave in Greek Epitaphs of Asia Minor. A Catalogue (IGSK 52)*. Bonn.

THOMMEN, L. 2009. *Umweltgeschichte der Antike*. München.

TORBRÜGGE, W. 1970/1971. 'Vor- und frühgeschichtliche Flußfunde. Zur Ordnung und Bestimmung einer Denkmälergruppe'. *Ber. RGK 50–1*, 1–146.

THÜRY, G. E. 1995. *Die Wurzeln unserer Umweltkrise und die griechisch-römische Antike*. Salzburg.

THÜRY, G. E. 2001. *Müll und Marmorsäulen. Siedlungshygiene in der römischen Antike*. Mainz.

VERSNEL, H. S. 1991. 'Beyond the Cursing: The Appeal to Justice in Judicial Prayers', in C. A. FARAONE (ed.), *Magika Hiera. Ancient Greek Magic and Religion*. Oxford, 60–106.

VERSNEL, H. S. 2009. *Fluch und Gebet: Magische Manipulation versus religiöses Flehen? Religionsgeschichtliche und hermeneutische Betrachtungen über antike Fluchtafeln*. Berlin.

von Haulock, H. 1979. 'Zur Münzprägung von Aizanoi', in R. Naumann (ed.), *Der Zeustempel von Aizanoi*. Berlin, 82–94.

Weeber, K.-W. 1990. *Smog über Attika. Umweltverhalten im Altertum*. Zürich-München.

Weiss, C. 1984. *Griechische Flussgottheiten in vorhellenistischer Zeit*. Würzburg.

Wilmott, T. 1982. 'Excavations at Queen Street, City of London 1953 and 1960, and Roman Timber-Lined Wells in London', *Transactions of the London and Middlesex Archaeological Society* 33, 1–78.

Arnaldo Marcone

L'evoluzione della sensibilità ambientale a Roma all'inizio del Principato

1. Il concetto romano di inquinamento

Non è evidente il concetto romano di 'inquinamento', quanto meno di 'inquinamento ambientale'. Se si accettano le idee sviluppate dall'antropologa MARY DOUGLAS proprio il concetto che una società ha di inquinamento rappresenta uno degli indicatori dell'ordine sociale e dei sistemi culturali sviluppati da una determinata società.[1] Come è emerso, tra l'altro, dalle giornate di studio che hanno avuto luogo alla British School di Roma nel giugno del 2007,[2] proprio la storia di Roma rappresenta una verifica importante per la messa a fuoco di questi concetti.[3] Roma antica, in particolare la megalopoli che si viene costituendo e organizzando a partire dal secondo secolo a. C., anche per i valori connessi a quella che oggi siamo soliti chiamare 'tutela ambientale' sembra avere le caratteristiche, almeno apparentemente contraddittorie, di una società premoderna.

Si sono ricercati, come vedremo, con determinazione i precedenti di tutela ambientale a Roma a livello giuridico. Va premesso che la letteratura medica, per quanto per altri aspetti sofisticata e ricca di elementi scientificamente significativi, sotto questo aspetto è deludente. In generale, a essere oggetto di valutazione come causa di malattia di tipo epidemico sono la capacità dell'essere umano di adattarsi a determinate condizioni di vita e la tipologia dell'alimentazione. È interessante ad esempio, come Cornelio Celso, attivo nel primo secolo d. C., riconosca che gli abitanti della città abbiano un'aspettativa di vita inferiore a quella di chi abita in campagna.[4] Ma non si discosta da valutazioni genericamente moralistiche: la ragione di tale precoce mortalità è attribuita agli effetti della vita cittadina in cui un'alimentazione poco sana si accompagna a abitudini in cui il lusso è prevalente.[5] Soprattutto sembra assente ogni suggerimento concreto per far fronte a queste emergenze.

Quando si parla di condizioni ambientali a Roma nei primi decenni del Principato è d'obbligo un riferimento a un episodio assai noto. All'inizio del suo racconto del regno di Vespasiano, Svetonio elenca i presagi che annunciano il regno del futuro imperatore. Tra questi c'è la storia di come una volta un cane randagio gettasse sotto il tavolo di Vespasiano,

---

[1] DOUGLAS 1966.
[2] BRADLEY 2012.
[3] L'insalubrità delle condizioni di vita delle città antiche è talvolta particolarmente enfatizzata: cfr. SCOBIE 1986.
[4] Cel., *de med.* 1.2.
[5] NUTTON 2000: 66.

che stava facendo colazione, una mano appartenente al cadavere di un uomo che giaceva a un incrocio.[6] Questo luogo è citato immancabilmente in tutti gli studi dedicati all'esame delle condizioni di vita a Roma in età imperiale dove trovare una mano umana in mezzo a una strada era all'ordine del giorno. In effetti *Dealing with the dead in ancient Rome*, per riprendere il titolo di un lavoro di John Bodel, era una questione seria.[7] Si tenga presente che Cicerone, nel riprendere *Tab.* 10.1 in merito al divieto di seppellire morti in città e di bruciarli (*hominem mortuum in urbe ne sepelito neve urito*) di fatto non si poneva un problema di norma igienica, ma si preoccupava del possibile sviluppo di incendi.[8]

Naturalmente al di là di questo episodio la discussione è aperta.

2. Gli ultimi due secoli della Repubblica

Malgrado la mancanza di dati archeologici assolutamente sicuri, l'evoluzione conosciuta dalla città di Roma nei due secoli finali della Repubblica è nota almeno nelle sue linee generali. Quando la popolazione cominciò a crescere – fino forse a mezzo milione di persone già verso la fine del secondo secolo a. C. – dovette crescere in parallelo la concentrazione delle case, vicinissime une alle altre e sovraffollate. I siti in collina, ben drenati e ventilati, erano naturalmente molto ricercati, ma i membri dell'élite così come anche gli appartenenti ad altri ceti sociali favorivano con i loro insediamenti lo sviluppo di aree meno salubri, come le valli e le aree soggette ad inondazione. Le strade dovevano essere terribilmente affollate e rumorose, sporche di ogni genere di rifiuto della vita romana di ogni giorno, ivi compresi escrementi animali e umani, cibo scartato e addirittura cadaveri dei miserabili. Cumuli di immondizie dovevano essere dovunque.[9]

Andreas Wacke ha enucleato in modo assai persuasivo i problemi fondamentali in numerosi suoi interventi.[10] Avendo ben presente i suoi argomenti mi sembra utile in questa sede far riferimento alle questioni affrontate da alcuni giuristi in un dibattito in cui gli studiosi italiani – in particolare quelli di diritto romano – negli ultimi anni sono stati particolarmente attivi. Le opinioni in merito sono sorprendentemente differenziate. Per taluni studiosi Roma aveva condizioni sanitarie relativamente avanzate rispetto ai suoi tempi mentre per altri era un luogo inospitale e pericoloso. E, secondo i primi, ci sarebbero prove di una legislazione ambientale sufficientemente articolata: Andrea Di Porto, per esempio, ha una visione ottimistica della sensibilità ecologica romana.[11]

I primi interventi giurisdizionali in materia ambientale sembrano essere invero di ambito limitato, quali le controversie cui accenna Cicerone nel ricordare l'attività di Aquilio

---

[6] Suet., *Vesp.* 5.4. Sempre Svetonio nella vita di Nerone racconta (48) come il cavallo dell'imperatore in fuga si impennò per il fetore di un cadavere abbandonato. Cfr. Panciera 2000: 96.

[7] Bodel 2000.

[8] Cic., *de leg.* 2.5.8.

[9] 'L'aspetto di Roma antica non doveva essere quello di un ridente villaggio svizzero' (Panciera 2000: 97).

[10] Cfr. in particolare Wacke 2002.

[11] Di Porto 1990; Di Porto-Gagliardi 1999. Si veda da ultimo ora anche Di Porto 2015.

Gallo nel campo della definizione del *litus* (*cum de litoribus ageretur, quae omnia publica esse vultis*), tra *res nullius* e i *loca publica in publico usu*.[12]

Come cercherò di suggerire rapidamente, se attenzione per problemi ambientali a Roma in termini generali ci fu, questa riguarda essenzialmente la qualità dell'acqua e la sua tutela. Quando si parla di ambiente a Roma il bene che suscita il maggiore interesse è l'acqua in tutte le sue possibili modalità d'uso pubblico e privato. L'acqua, come bene pubblico, era comprensibilmente tutelata da specifiche norme di legge. A livello di legislazione generale è sufficiente ricordare la *Lex Quinctia de aquaeductibus* (9 a.C.), una legge promulgata a Roma al tempo di Augusto, dal console Tito Quinzio Crispino Sulpiciano per la regolamentazione degli acquedotti il cui testo ci è noto attraverso Frontino.[13]

La legge riepilogava le norme a tutela degli acquedotti e stabiliva pesanti pene pecuniarie per chi le contravveniva. In particolare era vietato danneggiare gli acquedotti, ma anche captarne l'acqua per uso agricolo o di altra natura senza averne avuto il permesso. La legge vietava inoltre qualsiasi attività edilizia, agricola, o pastorale presso gli acquedotti, che non fosse strettamente necessaria per la manutenzione dei luoghi o delle opere preesistenti.

Per dispute riguardanti privati è indubbiamente significativo il ricorso all'interdetto *quod vi aut clam,* per il quale abbiamo alcuni riscontri, che riguarda specificamente la tutela delle acque di superficie che erano soggette al rischio di contaminazione. Si tratta di un'importante misura pretoria di tipo restitutivo nel senso che stabiliva l'immediato ripristino della situazione precedente al guasto prodotto tramite l'eliminazione del guasto stesso.[14] Tuttavia la tutela interdittale veniva concessa solo per garantire la conduzione di acque perenni, forse anche in una forma di ricezione delle concezioni correnti presso i naturalisti romani.[15] In realtà le sorgenti potevano essere sfruttate dal proprietario senza che ci fosse alcuna valutazione preventiva dell'impatto ambientale che poteva determinare, ad esempio, l'impianto di una *fullonica*.[16] I principi dell'*actio aquae pluviae arcendae* sembrano scaturire dalla volontà di garantire di fatto la sola tutela delle persone private.

3. Labeone

In proposito è certamente da considerare la precoce attenzione mostrata da Labeone, già in età augustea, per i problemi relativi all'inquinamento idrico sia pure a livello di diritto privato, per i problemi della *salubritas,* dunque per un ambiente non contaminato. Ed è comunque notevole come Labeone consideri il danno derivante dall'inquinamento dell'acqua 'viva' e 'non viva' in città come in campagna (nel senso di *aliquid effundere* in una peschiera o in una cisterna): *Is qui in puteum vicini aliquid effuderit, ut hoc facto aquam corrumperet, ait Labeo interdicto quod vi aut clam eum teneri: portio enim agri videtur aqua viva, quemad-*

---

[12] Cic., *top.* 7.32.

[13] Cfr. Maganzani 2012.

[14] Fargnoli 2012. Di Porto, in particolare, pone con forza il problema dei limiti entro i quali Labeone e Ulpiano affermano l'impiego dell'interdetto *quod vi aut clam* nel campo dell'inquinamento idrico (Di Porto 1990: 5).

[15] In particolare Sen., *quaest. nat.* 3.23. Cfr. Capogrossi Colognesi 1966: 5.

[16] Cfr. Fiorentini 2006, critico della valutazione ottimistica di Zamora Manzano 2003: 19–67, in merito al presunto interventismo romano in materia di tutela delle acque.

modum si quid operis in aqua fecisset.[17] Qualcuno, versando alcunché nel pozzo del vicino, ne ha inquinato l'acqua; Ulpiano, aderendo all'opinione di Labeone, ritiene utilizzabile in siffatta ipotesi l'*interdictum quod vi aut clam*. Per giustificare tale applicazione, Labeone (ma forse l'illustrazione della *ratio* va ricondotta ad Ulpiano oppure ad entrambi) sostiene che vi è una equivalenza di fatto tra acqua viva e *portio agri*, ragione per cui è come se un *opus* fosse stato realizzato nell'acqua.

Sempre Labeone dà un'interpretazione estensiva dell'editto pretorio *de aqua cottidiana et aestiva*:[18] l'interdetto si indirizza non solo contro coloro che precludono l'adduzione di acqua da parte di terzi ma, più in generale, contro tutti coloro che pregiudicano la salubrità dell'acqua di un acquedotto.[19]

Tuttavia presupporre elaborazioni compiute di una 'autentica coscienza ambientalistica' nel mondo romano, secondo quanto suggerito in un breve ma puntuale saggio di recente pubblicazione di LUIGI LABRUNA, è impegnativo.[20] Rispetto a troppo ottimistiche valutazioni di elementi embrionali di tutela ambientale a Roma è giusto osservare una certa prudenza.[21] Così come non è evidente invocare puntuali precedenti romani nella gestione delle acque.[22] In proposito sembra giusto ritornare a una considerazione premessa alla riflessione svolta da MARIO FIORENTINI in uno studio di una decina di anni fa, che metteva in chiaro come solo la fine della fiducia nella crescita illimitata e, quindi, dello sviluppo abbia determinato una riflessione critica nella coscienza collettiva contemporanea, altrimenti solo embrionale o episodica.[23] Appare quindi eccessivo ricercare puntuali suggestioni di questo genere nell'antica Roma.

4. Frontino

È comunque da verificare oltre alla percezione che i Romani avevano del degrado ambientale e alle politiche da loro adottate per gestirlo, se tale percezione sia mutata nel tempo. E come noi siamo in grado di cogliere questo eventuale mutamento. Una fonte, a mio modo di vedere, deve essere considerata in via preliminare. Frontino, *curator aquarum*, dopo aver ricordato come le acque di scolo inquinino l'aria e la rendano irrespirabile, si rallegra per il fatto che problemi di questo genere non esistano più a Roma, mentre erano esistiti in passato contribuendo alla cattiva fama della città.[24] Al di là della componente propagandistica, di celebrazione del *beatissimum saeculum* dell'imperatore regnante (Nerva) e di sé medesimo presente nello scritto di Frontino, il giudizio su una situazione negativa del passato, rispetto a quella presente, merita attenzione anche perché doveva essere evidentemente un tema sentito a livello di amministrazione imperiale (*alia munditiarum facies, purior spiritus, et causae gravioris caeli quibus apud veteres urbis infamis aer fuit, sunt remotae*). Si segnala dunque un

---

[17] Ulp., *l.* 71 *ad ed.*: D. 43.24.11 *pr*. Cfr. DI PORTO 1990: 56–7.

[18] Ulp., *l.* 70 *ad ed.*: D. 43.20.1.27.

[19] Cfr. CAPOGROSSI COLOGNESI 1976: 12.

[20] LABRUNA 2008.

[21] In particolare ZAMORA MANZANO 2003.

[22] Cfr. WESTCOAT 1997.

[23] FIORENTINI 2006.

[24] Front., *de aq. u. R.* 88.

prima e un poi, una situazione critica cui solo da ultimo si è posto rimedio. Se poi questo rimedio sia stato davvero risolutivo è questione aperta.

Il passo di Frontino è significativo perché mette in evidenza come ci sia un momento di svolta nell'attenzione per l'ambiente a Roma con un esito specifico verso la fine del primo secolo d. C.

Che ci sia stato un aggravamento nelle condizioni ambientali di Roma sembra sicuro.

Forme di sensibilità ambientale possono peraltro avere manifestazioni diverse. È senz'altro legittimo porsi il problema se l'esperienza giuridica romana abbia reagito o meno a quelli che oggi chiameremmo problemi di inquinamento, ovvero se abbia risposto a esigenze sentite all'interno della società.[25] Il periodo decisivo per un possibile mutamento nella percezione di questi problemi è senz'altro quello tra Tarda Repubblica e primo Impero quando si vengono a profilare le prime risposte a problemi concreti. È l'epoca del grande sviluppo urbanistico di Roma, soprattutto in alcuni quartieri, con le costruzioni ad *insulae* e un generale sovraffollamento.

5. Attività inquinanti

A essere considerati – a livello di intervento giuridico circostanziato – risultano essere stati prevalentemente i problemi legati a determinate forme di lavorazione. È soprattutto il problema dell'inquinamento dell'acqua prodotto dalle *fullonicae* che risulta al centro del dibattito giurisprudenziale tra Tarda Repubblica a Alto Impero. Questo è spiegabile anche in relazione alla molteplicità delle operazioni di queste botteghe artigianali, che comprendevano il trattamento delle vesti e dei tessuti con acqua mista a sostanze depuranti come il *nitrium* e, soprattutto, urina umana oltre alla *creta fullonia,* cioè argille dal forte potere assorbente.[26] Tra l'altro, oltre che in città, le *fullonicae* erano collegabili anche all'economia della villa. Dato il gran bisogno di acqua delle *fullonicae* si capiscono le ragioni di questa attenzione da parte dei giurisperiti e si possono anche cogliere le premesse di un'embrionale forma di sensibilità ambientale.

Quando parliamo di Roma in età tardorepubblicana o altoimperiale è abbastanza evidente la presenza in città di attività di per sé stesse inquinanti con immissioni nell'ambiente di sostanze banalmente sgradevoli all'olfatto, oltre che potenzialmente dannose alla salute. Non è sorprendente che l'élite politica e sociale sviluppasse una propria strategia di autotutela a prescindere dall'elaborazione di motivazioni strettamente ambientalistiche. È abbastanza evidente che, come accade, la linea di confine tra attività produttiva e lusso dannoso e riprovevole si fa progressivamente incerta (nella letteratura di età imperiale prevale il secondo aspetto). Un caso significativo è rappresentato dal compromesso tra l'appropriabilità delle coste, mai messa in discussione, e l'accesso al mare e alla costa tra proprietari di ville

---

[25] Cfr. Dɪ Pᴏʀᴛᴏ 1990 e 2015; Fᴇᴅᴇʟɪ 1990.

[26] Cfr. Mᴏᴇʟʟᴇʀ 1976: 20; Cᴜᴏᴍᴏ ᴅɪ Cᴀᴘʀɪᴏ 2007: 695 (i riferimenti sono a Plin., *nat. hist.* 35.57.196–7; 17.4.46). L'asciugatura dei panni era uno dei problemi che si palesavano nelle città antiche soprattutto in quelle affollate tanto è vero che i lavatori di panni ottenevano il diritto di usufruire degli spazi davanti alle loro botteghe per far asciugare i panni. Va segnalato come Marziale (7.61) lodi Domiziano per aver disciplinato le botteghe.

marittime e terzi.[27] Labeone introduce un interdetto specifico contro opere sottocosta che potevano recare pregiudizio alla navigazione.[28]

A Roma le concerie e le lavanderie dovevano produrre miasmi particolarmente pesanti da sopportare con conseguenze serie anche sulle acque di scolo. Un'allusione di Marziale può essere esemplificativa: 'Come Taide non puzza nemmeno il vecchio vaso di un lavandaio che s'è appena rotto in mezzo alla strada'.[29] Si capisce che per questo tipo di attività si fosse pensato a un'area specifica della *regio XIV trans Tiberim*. D'altra parte, abbiamo notizia di specifici interventi legali a tutela di un fondo che si fosse visto interessato dalle acque di risulta di lavorazione di una *fullonica*.[30] Il cattivo odore che si sprigionava dalle tintorie è ben comprensibile se si tiene conto del fatto che i tintori lavoravano con l'urina che si procuravano dalle latrine pubbliche, un uso che non a caso Vespasiano volle tassare.

6. Vitruvio

Ci si è chiesto come mai i Romani abbiano fatto fondamentalmente poco per migliorare la qualità di vita della loro capitale.[31] È una domanda cui è difficile dare una risposta sicura. È certo facile e un po' scontato richiamarsi alla scarsa coscienza ambientalistica dei Romani anche se non mancavano richiami a forme di attenzione ecologica. Vitruvio, per il quale la componente teorica gioca un ruolo importante, aveva anche individuato dei metodi per verificare la purezza dell'acqua e aveva anche sottolineato le implicazioni che se ne potevano trarre a livello di salute pubblica. Vitruvio riconosceva l'effetto inquinante derivante dall'utilizzazione di determinate sostanze su coloro che abitavano nelle zone vicine, in particolare riguardo ai modi di trasmissione dell'acqua:

> *Exemplar autem ab artificibus plumbariis possumus accipere, quod palloribus occupatos habent corporis colores. Namque cum fundendo plumbum flatur, vapor ex eo insidens corporis artus et inde exurens eripit ex membris eorum sanguinis virtutes. Itaque minime fistulis plumbeis aqua duci videtur, si volumus eam habere salubrem. Saporemque meliorem ex tubulis esse cotidianus potest indicare victus, quod omnes, et structas cum habeant vasorum argenteorum mensas, tamen propter saporis integritatem fictilibus utuntur.*[32]

Si tratta di dubbi condivisi sui pericoli per la salute rappresentati dalle condutture plumbee: al loro posto erano raccomandate condutture in ceramica (Columella e Plinio il Vecchio e, qualche secolo dopo, Palladio sembrano concordare con Vitruvio).[33] Tuttavia sembra che malgrado queste prese di posizione l'acqua continuò ad essere ditribuita tramite condutture

---

[27] Ner., *l. 5 membr.*: D. 41.1.14 *pr.* […] *quod in litore quis aedificaverit eius erit.* Cfr. Plin., *nat. hist.* 17.9.59.

[28] Ulp., *l. 68 ad ed.*: D. 43.12.1.17.

[29] Mar., *ep.* 6.93.

[30] Ulp. *l. 53 ad ed.*: D. 39.3.3 *pr.* Secondo Di Porto 1990: 77–9, Trebazio, il giurista di tarda età repubblicana, cui il frammento del Digesto fa riferimento, avrebbe avuto in mente un 'modello' di *fullonica* particolarmente complesso (ricollegabile anche a Varr., *de re rus.*, 1.16.3–4 e 1.2.21).

[31] Davies 2012: 70.

[32] Vitr., *de arch.* 8.3.28; 8.4.1.

[33] Cfr. Fedeli 1990: 71–2.

in piombo. I rischi di inquinamento risultano in realtà davvero minimali in considerazione della brevità del contatto tra il piombo e l'acqua.[34]

Merita sottolineare la conclusione di Vitruvio: *itaque minime fistulis plunbeis aqua duci videtur, si volumus eam habere salubrem*.[35] E valorizzava la costruzione di parchi, a scopo di passeggiate e a beneficio della salute oculare, a vantaggio generale della *salubritas* degli abitanti con riferimento alla pianificazione urbana.[36] Vitruvio aveva anche elaborato un metodo per testare la purezza dell'acqua delle fonti che si basava essenzialmente sull'aspetto delle persone che abitavano nelle vicinanze:

> *Quare magna diligentia industriaque quaerendi sunt et eligendi fontes ad humanae vitae salubritatem. Expertiones autem et probationes eorum sic sunt providendae. Si erunt profluentes et aperti, antequam duci incipiantur, aspiciantur animoque advertantur, qua membratura sint qui circa eos fontes habitant homines; et si erunt corporibus valentibus, coloribus nitidis, cruribus non vitiosis, non lippis oculis, erunt probatissimi.*[37]

Tuttavia è possibile tentare un discorso articolato, che introduca alcune distinzioni e lasci intavedere un'evoluzione quanto meno nei comportamenti dei ceti alti e, quindi, un cambiamento di ordine culturale.

In proposito merita considerazione un passo di Vitruvio in cui si può trovare una esauriente descrizione degli espedienti tecnici atti a rendere solide le costruzioni su pendio o, comunque, su terreni non coesi, creando quelle *substructiones* che potevano essere poi utilizzate nei modi più diversi.[38] È interessante come il ricorso al sotterraneo in contesto privato, a scopo abitativo o funzionale, percorra senza soluzione di continuità la storia dell'edilizia privata romana a partire dalla fine della Repubblica. Certamente in parte si tratterà di semplice 'moda', ma certo è che dalla fine del secondo secolo a. C. in poi si inizia a sfruttare il sottosuolo, come emerge dalle case sul Palatino e dalle ricche ville marittime della costa.[39] Appare evidente che a questa soluzione si sia arrivati in primo luogo per esigenze pratiche dovute alla necessità di ampliare il piano d'uso della casa creando ulteriori spazi anche ad uso abitativo. Sembra peraltro accertato che alle necessità di ordine pratico iniziali si accompagnarono presto motivazioni forti di ordine ideologico che indussero gli architetti a creare anche dei sotterranei artificiali. Abbiamo a che fare indubbiamente con manifestazioni di lusso, contro cui era facile per i moralisti scagliare le loro invettive: *quo longiores porticus expedierint, quo altius turres sustulerint, quo latius vicos porrexerint, quo depressius aestivos specus foderint, hoc plus erit quod illis caelum abscondat.*[40]

---

[34] Cfr. Vuorinen, Juuti e Katko 2013.

[35] Vitr., *de arch.* 8.6.1.

[36] Vitr., *de arch.* 1.4; 1.6. Cfr. Davies 2012: 75.

[37] Vitr., *de arch.* 8.4.

[38] Vitr., *de arch.* 6.8.1. Cfr. Basso, Bonetto e Ghedini 2001.

[39] Cfr. Papi 1998.

[40] Sen., *Cons. ad Helviam matrem* 9.2 ('quanto più alte innalzeranno le loro torri, quanto più vasti edificheranno i loro caseggiati, quanto più profonde scaveranno le loro grotte per l'estate, quanto più sovraccarichi saranno i soffitti delle loro sale da pranzo, tanto più, tutto questo, nasconderà loro il cielo').

7. Le ville di lusso

È lecito tuttavia immaginare un ordine di motivazioni più articolato. Un accenno contenuto nel *De re Rustica* di Varrone merita attenzione: emerge un nuovo tipo di *villa*, detta *inutilis* quella di Appius Claudius Pulcher, dal prezioso arredamento, quindi un segno di raffinatezza culturale ed estetica, è contrapposta alla semplicità della villa rustica, volta all'esclusiva gestione della produzione agricola.[41] Un altro interlocutore del dialogo, Axius, afferma che la sua villa, pure lussuosa, ha un superiore valore culturale, oltre che economico, perché è economicamente produttiva.[42] Siamo all'inizio di un processo: alla fine del primo secolo d. C. abbiamo vari riscontri di come la componente estetica di una villa abbia ormai acquisito la preponderanza sul suo valore economico.

La villa di Manlius Vopiscus, di cui Stazio ci ha lasciato un'ampia presentazione, ne è un buon esempio: i primi 34 versi sono dedicati a una descrizione del paesaggio che caratterizza il sito in cui è situata la villa di Tivoli.[43] Poi prosegue per altri undici versi a descrivere la villa dall'interno. L'importanza della prospettiva dall'interno serve come argomento per valorizzare l'*amoenitas* della villa. Le ville di lusso di età imperiale sono valorizzate per questa doppia considerazione, dell'*amoenitas* (quindi la collocazione della villa rispetto all'ambiente esterno) e del *prospectus*, la vista dall'interno della casa. Di un caso (fittizio) di dispute tra vicini per ragioni di *prospectus* tratta Seneca il Vecchio in una *controversia*:[44] dovevano essere non rare le questioni legali relative alle tutele del *prospectus* da parte dei ricchi proprietari che tenevano a tutelare il *prospectus* della loro villa per accrescerne il valore.

In questo cambiamento di stile di vita sarà entrato in gioco in primo luogo, l'interesse economico di acquisire spazi abitativi senza l'acquisto di ulteriori lotti di terreno. Dall'altra è possibile anche che giocasse un ruolo anche il desiderio di creare uno spazio il più possibile protetto all'interno della propria dimora, immune in qualche misura dal contagio esterno. Un luogo della *Elegia in Maecenatem*, databile probabilmente poco dopo la sua morte, pur nella sua convenzionalità, può meritare interesse: il ministro di Augusto appare un perfetto epicureo nella deliziosa pace dell'ombra del suo giardino, che gode dei *pauca iugera certa pomosi soli* (è evidente il richiamo al *senex Corycius* virgiliano *cui pauca relicti iugeris iugera ruris erant*)[45] mentre la poesia gli fluisce abbondante senza sforzo alcuno.[46]

Considerazione merita il ricorso ai cosiddetti criptoportici, un termine che appare come un neologismo nell'epistolario di Plinio il Giovane. Si deve tener presente che si tratta di strutture che potevano essere poste sia al di sotto della superficie del terreno ma anche al di sopra di questa (*aestiva cryptoporticus in edito posita*)[47]: Plinio ci dà una descrizione di un criptoportico della sua villa laurentina, aperto su due lati con finestre che affacciavano ri-

---

[41] Vitr., *de arch.* 3.17.1.

[42] ZEINER 2005: 78.

[43] St., *sil.* 1.3.

[44] Plin. Iun., *ep.*, 5.5.

[45] Verg., *georg.* 4.127–8.

[46] *El. Maec.*, 34.

[47] Plin. Iun., *ep.* 5.6.29. A proposito del termine 'criptoportico' è stata giustamente sottolineata l'opportunità di un suo uso mirato e non generico: altri sono infatti i termini più appropriati riferiti a complessi pubblici a funzionalità differenziata. Sembra preferibile utilizzare il termine 'criptoportico' specificamente per definire ambienti che siano compresi in spazi residenziali privati.

spettivamente su un giardino e su una terrazza.[48] Evidentemente si tratta di ambienti concepiti esclusivamente per la piacevolezza del soggiorno riservati al passeggio e in funzione del godimento del paesaggio: la loro utilità era arricchita dai vantaggi termici, igienici e anche terapeutici che potevano fornire.

È da prendere in considerazione la nota lettera di Plinio il Giovane indirizzata a Domizio Apollinare, che fa da pendant alla precedente in cui presenta la sua villa *in Tuscis* come un vero *locus amoenus*:

> L'aspetto del paese è bellissimo: immagina un anfiteatro immenso e quale soltanto la natura può crearlo. Una vasta e aperta piana è cinta dai monti, e le cime dei monti hanno boschi imponenti ed antichi. [...] Eppure, benché vi sia abbondanza di acque, non vi sono paludi, perché la terra in pendio scarica nel Tevere l'acqua che ha ricevuto e non assorbito.[49]

Il clima della zona, specie in estate, risulta particolarmente benigno e propizio, tanto è vero che Plinio segnala un'anomala presenza di anziani, con nonni e bisnonni insieme a giovani nel fiore dell'età.[50] Ma c'è un luogo, in particolare, che merita segnalare. Plinio sottolinea come, anche in una sorta di valutazione comparativa con altre sue ville di sua proprietà più vicine a Roma (Tusculum, Tivoli, Preneste), in questa *in Tuscis purius est caelum* e *l'aer liquidior*.[51] Dunque l'aspirazione – si direbbe di chi vive normalmente in una città maleodorante – è per un ambiente salubre, dove si possa respirare a pieni polmoni.[52]

È notevole come Plinio evidenzi che il luogo è salubre anche per i suoi dipendenti (in una considerazione che sembra voler suggerire che trae da questo un profitto aggiuntivo): *mei quoque nusquam salubrius degunt; usque adhuc certe neminem ex iis, quos eduxeram mecum, ibi amisi.*[53]

8. La villa 'letteraria'

È suggestivo immaginare che il motivo, ben noto per Roma e la Campania, della villa 'letteraria', della *domus* immaginata attraverso la descrizione delle fonti, possa essere una diversione dai problemi di inquinamento in senso lato dell'ambiente urbano che le élites cercavano in vario modo di fronteggiare. È comunque interessante come nella ricerca odierna si riconosca il progressivo affermarsi del modello della tipologia della villa signorile di campagna su quello della *domus* cittadina tradizionale anche all'interno dello spazio urbano. Si può ragionevolmente ipotizzare che sia maturata una nuova sensibilità culturale anche per quel che riguarda la valorizzazione dell'ambiente per la valenza che sembrano assumere gli

---

[48] Plin. Iun., *ep.* 2.17.16–7.

[49] Plin. Iun., *ep.* 5.6.7 e 11: *Regionis forma pulcherrima. Imaginare amphitheatrum aliquod immensum, et quale sola rerum natura possit effingere. Lata et diffusa planities montibus cingitur, montes summa sui parte procera nemora et antiqua habent* [...] *sed ubi aquae plurimum, palus nulla, quia devexa terra, quidquid liquoris accepit nec absorbuit, effundit in Tiberim.* Cfr. Lefèvre 1977.

[50] Plin. Iun., *ep.* 5.6.6.

[51] Plin. Iun., *ep.* 5.6.45.

[52] Cfr. Galimberti Biffino 2014.

[53] Plin. Iun., *ep.* 5.6.46.

interventi di riqualificazione architettonica, che si possono leggere come manifestazioni del gusto dei committenti, oltre che del loro *status*. La ricerca di località ben aerate, almeno da parte dei ceti agiati, dovette essere abbastanza precoce e, comunque, contestuale al rapido sovraffollamento della città. I bagni posizionati quasi dovunque dovevano servire come una sorta di via di fuga a un ambiente altrimenti poco tollerabile.[54]

Almeno una rapida considerazione merita ancora Stazio che nelle *Silvae*, in una poesia pur molto raffinata, ci dà qualche indicazione importante. In particolare *Silvae* 2.2 può valere come una meditazione attenta sull'importanza della villa e delle sue valenze come *locus amoenus*. È stato giustamente riconosciuto come la topografia della proprietà di Pollione esprima metaforicamente e visivamente il valore della vita intellettuale, dedita al perseguimento della virtù, dell'epicureismo lontano dalla vita della capitale. Le implicazioni di questo rifugio ideale, anche rispetto all'autorità culturale di Roma sono difficili da sottovalutare. L'*amoenitas* assurge, quindi, al valore di criterio distintivo e simbolico, al punto che il proprietario si preoccupa di trovare forme di tutela legale per il *prospectus*.[55] L'*amoenitas* diventa un valore da perseguire in quanto tale, dove fioriscono amicizia e poesia, come risulta anche da *Silvae* 1.5 in cui l'ambiente propizio è costituito dai bagni di Etruscus. Stazio nella sua lode di Napoli e del suo golfo fa non a caso riferimento al vantaggio che questi hanno di essere liberi dal rumore e dai fastidi della capitale.[56]

9. *Amoenitas* e *salubritas*

Alla fine, quanto meno per i romani ricchi, una proprietà in campagna non era solo un affare economico, ma anche un luogo di relax e di ristoro ed era ricercata per i suoi pregi estetici e per quanto poteva offrire per il benessere fisico.[57] E tali caratteristiche vengono progressivamente tutelate a livello giuridico. Il Digesto lascia percepire elementi significativi, per quanto limitati, di questa evoluzione. Risulta che un proprietario poteva attingere l'acqua non solo a scopi di irrigazione ma anche *amoenitatis causa*, in buona sostanza per alimentare un lago artificiale con fontane e giochi d'acqua.[58] Siamo evidentemente in una fase evoluta delle relazioni sociali e di evoluzione economica.[59]

I concetti di *amoenitas* e *salubritas* giocano un ruolo caso peculiare di cui si è conservato riscontro nel Digesto.[60] Il proprietario di una casa aveva acquistato un giardino adiacente per abbellire la propria proprietà. Aveva anche fatto costruire una via di comunicazione tra le due parti. In questo modo il giardino figurava come comunicante con la proprietà originaria. Ne risultava che il proprietario aveva creato una sorta di annesso: come bene ereditario

---

[54] Cfr. Guilhembet 2001 e 2008.

[55] Zeiner 2005: 78.

[56] Cfr. Delarue 2014.

[57] Capogrossi Colognesi 2008: 8, parla di 'micropaesaggi integrati all'interno delle ville' della prima età imperiale.

[58] Pomp., *l*. 34 *ad Sab.*: D. 43.20.3 *pr*.

[59] Cfr. Capogrossi Colognesi 2008: 19, che sottolinea come si abbia riscontro di una concezione 'più elastica e comprensiva dell'*utilitas fundi*' che ricomprende non solo gli effettivi bisogni agrari ma anche la semplice *amoenitas fundi*'.

[60] Pap., *l*. 7 *resp.*: D. 32.91.5.

entrambe le parti figuravano come costitutive del medesimo bene. *Amoenitas* e *salubritas* sono considerate i valori decisivi nella considerazione dell'eredità (*ut amoeniorem domum ac salubriorem possideret*). Quindi la *salubritas* e l'*amoenitas* entrano in qualche modo nella considerazione giuridica cosa che si può considerare un esito dell'evoluzione sociale.

In proposito va segnalata una considerazione del giurista Paolo (terzo sec. d.C.), in un idealistico apprezzamento della fruizione del paesaggio che fa sì che anche l'ombra di un cipresso debba essere oggetto di tutela a favore dell'usufruttuario. È notevole come al concetto di *amoenitas* sia riconosciuto un valore meritevole di tutela in quanto tale, per cui oggetto di interdetto sono anche le *arbores non frugiferae* come i *cupressi*: *sed si amoenitas quaedam ex huiusmodi arboribus praestetur, potest dici ex fructuarii interesse propter voluntatem et gestationem et esse huic interdicto locum.*[61]

Le testimonianze sono diverse, perché diverse sono le situazioni che danno origine alle prese di posizione dei giuristi. Ma nell'insieme ci danno la percezione del cambiamento delle aspettative e della sensibilità sociale. Minturnae, una località portuale del Lazio meridionale sulla via Appia (più o meno a metà strada tra Roma e Napoli), alla foce del fiume Liri si trovava in un ambiente paludoso e malsano. Sappiamo ancora dal Digesto che un certo Cerellius Vitalis pretendeva di veder riconosciuto il suo diritto di far causa nei confronti del proprietario di un negozio di formaggi che lavorava nei locali al di sotto dei suoi a causa del fumo che la sua attività comportava e che invadeva la sua proprietà.[62] In realtà il parere di Aristone sembra non aver dato soddisfazione al querelante, che si sarà dovuto rassegnare al cattivo odore per godersi il panorama marino del golfo di Gaeta, almeno se il proprietario in questione aveva ottenuto un diritto di servitù in favore del caseificio.

Aria malsana poteva provenire da certe zone di determinate proprietà. Su un piano strettamente giuridico un terreno insalubre, un *fundus pestilens* non era incedibile. Ma sappiamo come l'acquirente potesse annullare il contratto di acquisto attraverso un'*actio redhibitoria*, se fosse stato in grado di dimostrare di essere stato all'oscuro di queste caratteristiche.[63] La *salubritas* era valutata tra le qualità di un appezzamento insieme all'estensione, l'*amplitudo* e la qualità del terreno (la *bonitas*). Va notato peraltro che la locuzione *caelum salubre* non compare nelle fonti giuridiche.

Abbiamo dunque indicazioni sufficienti nel Digesto per poter sostenere che ci fossero forme di tutela di proprietari privati rispetto a emissioni insalubri di tipo industriale. Si tratta certamente di disposizioni occasionate da circostanze occasionali e dal valore limitato. Ma sono indicative di una tendenza e di un mutamento di sensibilità. Ancora una volta però va osservato come sia la gestione dell'acqua a richiamare su di sé l'attenzione maggiore. Solo in secondo piano figura la considerazione per la qualità dell'aria e, comunque, in riferimento non a problematiche di ordine generale ma a casi specifici.

Vero è che, non so quanto possa consolare – soprattutto per quanti cercano precedenti romani di situazioni contemporanee, constatare che il fenomeno delle costruzioni indiscriminate, dell'edilizia aggressiva sembra aver preso piede già a Roma nella tarda Repubblica. Così almeno a credere almeno al grido di dolore di Orazio contro i palazzinari che costrui-

---

[61] Paul., *l.* 67 *ad ed.*: D. 43.24.16.1.
[62] Ulp., *l.* 17 *ad ed.*: D. 8.5.8.5.
[63] Ulp., *l.* 8 *disp.*: D. 21.1.49.

vano anche nel mare, togliendo ai pesci il loro spazio vitale: 'erigi residenze in quel di Baia, affannandoti ad interrare la battigia dove scrosciano i marosi' (*immemor struis domos/marisque Bais ostrepentis urges/submovere litora*).[64]

Bibliografia

BASILE, A. 2014. 'Alcune riflessioni sulla rappresentazione letteraria delle ville campane in età flavia', in O. DEVILLERS (ed.), Neronia IX. *La villégiature dans le monde romain de Tibère à Hadrien*, Actes du IXᵉ Congrès de la SIEN. Paris, 79–88.

BASSO, P., J. BONETTO, e F. GHEDINI. 2001. 'L'uso del sottosuolo nell'edilizia privata della Cisalpina romana', in M. VERZÀR BAS (ed.), *Abitare in Cisalpina. L'edilizia privata nelle città e nel territorio in età preromana* (Antichità Alto Adriatiche 49). Trieste, 141–93.

BELL, M. 1998. 'Le stele greche dell'Esquilino e il cimitero di Mecenate', in M. CIMA e E. LA ROCCA (ed.), *Horti romani*, Atti del convegno internazionale Roma 4–6 Maggio 1995. Roma, 295–314.

BODEL, J. 2000. 'Dealing with the dead in ancient Rome. Undertakers, Executioners, and Potter's Fields in Ancient Rome', in M. HOPE e E. V. MARSHALL (eds.), *Death and Disease in the Ancient City*. London-New York 2000, 128–51.

CAPOGROSSI COLOGNESI, L. 1966. *Ricerche sulla struttura delle servitù d'acqua in diritto romano*. Milano.

CAPOGROSSI COLOGNESI, L. 2008. 'Préface. Acque, terre e paesaggi nella storia di Roma' in E. HERMON (ed.), *Vers une gestion intégrée de l'eau dans l'Empire romain*. Roma, 13–20.

CUOMO DI CAPRIO, N. 2007. *La ceramica in archeologia, 2: antiche tecniche di lavorazione e moderni metodi di indagine*. Roma.

DAVIES, P. 2012. 'Pollution, propriety and urbanism in Republican Rome', in M. BRADLEY e K. STOW (eds.). 2012. *Rome, Pollution and Propriety. Dirt, Disease and Hygiene in the Eternal City from Antiquity to Modernity*. Cambridge 67–80.

DELARUE, F. 2014. 'L'eau et l'imaginaire. Les villas des Silvae de Stace', in O. DEVILLERS (ed.), Neronia IX. *La villégiature dans le monde romain de Tibère à Hadrien*, Actes du IXᵉ Congrès de la SIEN. Paris, 89–98.

DI PORTO, A. 1990. *La tutela della 'salubritas' fra editto e giurisprudenza: Il ruolo di Labeone. I*. Milano.

DI PORTO, A. 2015. *Salubritas e forme di tutela in età romana. Il ruolo del civis*. Torino.

DI PORTO, A. e L. GAGLIARDI. 1999. 'Prohibitions Concerning Polluting Discharges in Roman Law', in A. GRIECO, S. IAVICOLI, e G. BERLINGUER (eds.), *Contributions to the History of Occupational and Environmental Prevention*. Amsterdam, 121–34.

DOUGLAS, M. 1966. *Purity and Danger. An Analysis of Concepts of Pollution and Taboo*. London- New York.

FARGNOLI, I. 2012. 'Umweltschutz und römisches Recht?', in I. FARGNOLI e ST. REBENICH (eds.), *Das Vermächtniss der Römer. Römisches Recht und Europa*. Bern, 151–75.

FEDELI, P. 1990. *La natura violata. Ecologia e mondo romano*. Palermo.

FIORENTINI, M. 2006. 'Precedenti di diritto ambientale a Roma?', *Index* 11, 353–400.

GALIMBERTI BIFFINO, G. 2014. 'Amoenitas, utilitas e voluptas in ville e villeggiatura: testimonianze di Vitruvio e di Plinio il Giovane', in O. DEVILLERS (ed.), Neronia IX. *La villégiature dans le monde romain de Tibère à Hadrien*, Actes du IXᵉ Congrès de la SIEN. Paris, 37–46.

GUILHEMBET, J.-P. 2001. 'Les résidences aristocratiques de Rome, du milieu du Ier siècle avant n. è. à la fin des Antonins', *Pallas* 55, 215–41.

GUILHEMBET, J.-P. 2008. 'L'aristocratie en ses quartiers (IIᵉ s. av. J.-C. – IIᵉ s. ap. J.-C.)', in M. ROYO, É. HUBERT e A. BÉRENGER (éds.), *Rome des quartiers: des vici aux rioni. Cadres institutionnels, pratiques sociales et requalification entre Antiquité et époque moderne*. Paris, 193–227.

---

[64] Hor., *car.* 2.18.19–22. Cfr. LABRUNA 2008: 277–80.

LABRUNA, L. 2008. 'Rome et le droit de l'environnement', in E. HERMON (ed.), *Vers une gestion intégrée de l'eau dans l'Empire romain*. Roma, 277–80.

LEFÈVRE, E. 1977. 'Plinius-Studien I. Römische Baugesinnung und Landschaftsauffassung in den Villen-briefen (2,17; 5,6)', *Gymnasium* 84, 519–41.

MAGANZANI, L. 2012. 'Senatusconsulta de aquis e lex Quinctia de aquaeductibus', in G. PURPURA (ed.), *Revisione e integrazione dei* Fontes Iuris Romani Anteiustiniani (FIRA). *Studi preparatori I Leges*. Torino, 135–51.

MOELLER, W. 1976. *The Wool Trade of Ancient Pompei*. Leiden.

NUTTON, V. 2000. 'Medical thoughts on urban Pollution', in V. M. HOPE e E. MARSHALL, (eds.), *Death and disease in the Ancient city*. London, 65–73.

PANCIERA, S. 2000. 'Nettezza urbana a Roma. Organizzazione e responsabili', in J. A. REMOLÀ e X. DUPRÉ RAVENTÓS (eds.), *Sordes urbis : la eliminación de residuos en la ciudad romana*, Actas de la Reunión de Roma, 15–16 de noviembre de 1996. Roma, 95–105.

PAPI, E. 1998. '*Domus est quae nulli villarum mearum cedat. (Cic., Fam. 6, 18, 5)*. Osservazioni sulle residenze del Palatino alla metà del I secolo a. C.', in M. CIMA e E. LA ROCCA (eds.), *Horti romani*. Roma, 45–67.

SCOBIE, A. 1986. 'Slums, Sanitation and Mortality in the Roman World', *Klio* 68, 407–22.

VUORINEN, H. S., P. JUUTI, e T. S. KATKO. 2013. 'Safety and water pipes: history and present', in T. S. KATKO, P. JUUTI, e KL. SCHWARTZ (eds.), *Water Services Management and Governance: Lessons for a Sustainable Future*. London, 89–96.

WACKE, A. 2002. 'Umweltschutz im römischen Recht?', in *Orbis Iuris Romani* 7, 101–39.

WESCOAT, J. L. JR. 1997. 'Toward a modern map of Roman water law', *Urban Geography* 18, 100–5.

ZAMORA MANZANO, J. L. 2003. *Precedentes romanos sobre el derecho ambiental: la contaminación de aguas, canalización de las aguas fecales y tala ilícita forestal*. Madrid.

ZEINER, N. K. 2005. *Nothing Ordinary Here: Statius as Creator of Distinction in the Silvae*. New York-London.

Edoardo Bianchi

Floods of the Tiber in Rome under the Julio-Claudians

Nullique fluviorum minus licet inclusis utrimque lateribus, nec tamen ipse depugnat, quamquam creber ac subitus incrementis est, nusquam magis aquis quam in ipsa Urbe stagnantibus. Quin immo vates intelligitur potius ac monitor, auctu semper religiosus verius quam saevus.

(Plin., *nat. hist.* 3.55)

Introduction

Thanks to the growing interest in environmental problems of the Ancient World, floods of the Tiber in Rome have caught the attention of many scholars over the last few years. For example, at the Laval University Conference (2006), PHILIPPE LEVEAU gave a paper on the related effects of public policy and climatic changes on the hydrographical phenomena attested in Imperial Rome. Furthermore, in 2009, SILVIA CAPPELLETTI dedicated an article to the severe flood of 15 AD and its political consequences, of which Tacitus offers a detailed testimony.[1] In 2007, the monograph by GREGORY S. ALDRETE, *Floods of the Tiber in Ancient Rome*, also appeared. This book gathers all the evidence about the thirty flood episodes documented in Rome from 414 BC to 398 AD, and usefully supplements the classic work by JOËL LE GALL, *Le Tibre, fleuve de Rome dans l'antiquité*, dedicated to the cohabitation of the Romans with their river from the Early Republic to Late Antiquity.[2]

In the light of these rich contributions, my purpose here is not to offer a new assessment of the Tiber hydrographical problems throughout the history of Rome. On the contrary, I will focus on the flood emergencies in Rome as they appear under the Julio-Claudian emperors, especially Augustus and Tiberius, since the ancient sources (above all Cassius Dio) even attest eight overflows of the Tiber concentrated in a period of less than seventy years, between 27 BC and 36 AD. Such a high frequency of floods cannot be disputed because of the documentary gaps in other historical periods.[3] Therefore, it seems necessary to explain: 1) what the consequences of flooding in Rome during the Julio-Claudian era were; 2) how the Roman rulers reacted to the problem, in practical and ideological terms; 3) whether their solutions were effective or not.

---

[1] LEVEAU 2008 (with a comparative approach); CAPPELLETTI 2009 (taking a juridical point of view).

[2] ALDRETE 2007, with complete list of known floods on p. 15; LE GALL 1953 (I quote here from the revised edition by MOCCHEGIANI CARPANO and PISANI SARTORIO: LE GALL 2005, with partial list of known floods on pp. 35–6).

[3] See FEDELI 1990: 82; GIARDINA 1997: 165–6; LEVEAU 2008: 138 and 144.

Eight episodes of flooding between 27 BC and 36 AD

After a dubious episode in 32 BC, the first flood emergency in the Augustan age happened with certainty in 27 BC, on the night between the 16th and 17th of January, when the Tiber suddenly overtopped its banks and inundated the city.[4] Cassius Dio, the only source on the subject, does not explain what the material consequences for the Romans were, nor does he say how long the flooding persisted.[5] It is only clear from his account that floodwaters covered all the low-lying (and highly populated) areas of Rome, rendering them navigable for boats. But apart from the extent of the flood, the most important aspect is that the city was inundated just on the night after Octavian was granted the name of Augustus by the senate, and it was not difficult to see a link between the two events.[6] In other words, the coincidence did not escape the religious sensitivity of the Romans and we have reason to believe that it was given an official interpretation, endorsed by the senate. The flood was seen, in fact, as a good prodigy, since, according to Cassius Dio, the μάντεις, probably the *haruspices*, said that the rising of the Tiber meant that Augustus would rise to great heights and hold the entire world under his sway.[7]

Such an official interpretation was exceptional, because floods (*aquae magnae* or *ingentes* in Latin) generally assumed importance (and therefore were recorded in the ancient sources) when they were destructive enough to be seen as signs or portents that either foretold or accompanied negative events for the Roman community, due to the *ira deorum*. This is proved, for example, by the flood of 193 BC and by the consequent collapse of buildings at the *porta Flumentana*, which were placed on the same level of the contemporaneous lightning that struck the *porta Caelimontana*.[8] Furthermore, the expiation of similar prodigies

---

4 On the episode of 32 BC see Cass. Dio 50.8.3 (very vague): καὶ συχνὰ μὲν ὑπὸ χειμῶνος ἐπόνησεν, ὥστε καὶ τρόπαιόν τι ἐν τῷ Ἀουεντίνῳ ἑστὸς καὶ νίκης ἄγαλμα ἀπὸ τῆς τοῦ θεάτρου σκηνῆς πεσεῖν, τήν τε γέφυραν τὴν ξυλίνην πᾶσαν καταρραγῆναι. ALDRETE 2007: 23–4, considers the tempest (χειμών) and the following destruction of the *pons Sublicius* (ἡ ξυλίνη γέφυρα, the wooden bridge) as evidence of flooding. See also LEVEAU 2008: 138; LONARDI 2013: 12; but LE GALL 2005: 36, does not consider this episode in his list of floods.

5 See Cass. Dio 53.20.1–2: Αὔγουστος μὲν δὴ ὁ Καῖσαρ, ὥσπερ εἶπον, ἐπωνομάσθη, καὶ αὐτῷ σημεῖον οὐ σμικρὸν εὐθὺς τότε τῆς νυκτὸς ἐπεγένετο· ὁ γὰρ Τίβερις πελαγίσας πᾶσαν τὴν ἐν τοῖς πεδίοις Ῥώμην κατέλαβεν ὥστε πλεῖσθαι, καὶ ἀπ᾽ αὐτοῦ οἱ μάντεις ὅτι τε ἐπὶ μέγα αὐξήσοι καὶ ὅτι πᾶσαν τὴν πόλιν ὑποχειρίαν ἕξοι προέγνωσαν. According to MAZZARINO 1966: 624, this flood lasted only one night and caused no damage. See also BECHER 1985: 474–6.

6 In the *Res gestae* Augustus prides himself on having repaired the *via Flaminia* and its bridges, with the exception of *pons Mulvius* and *pons Minucius* (20.5): Con[s]ul septimum viam Flaminiam a[b urbe] Ari[minum refeci pontesque] omnes praeter Mulvium et Minucium. This intervention was realized in 27 BC (when Augustus was in his seventh consulate) and was possibly due to the damage caused by the flood of the same year.

7 On μάντεις = *haruspices* see SCHMITTHENNER 1962: 38; FREYBURGER 1999: 24–9. *Contra* BECHER 1985: 475, and RICH 1990: 153, according to whom the μάντεις mentioned by Cassius Dio were 'private' soothsayers. Recent discussion of the problem in MONTERO 2012: 276–9, who does not exclude that the soothsayers were actually astrologers.

8 According to Livy 35.9.2–5, the *Xviri sacris faciundis* consulted the Sibylline Books in order to understand the origin of these *prodigia*: Aquae ingentes eo anno fuerunt et Tiberis loca plana urbis inundavit; circa portam Flumentanam etiam conlapsa quaedam ruinis sunt. Et porta Caelimontana fulmine icta est murusque circa multis locis de caelo tactus; [...]. Horum prodigiorum causa decemviri

was usually demanded by the senate to the priestly college of the (*quin*)*decemviri sacris faciundis*, in charge of consulting the Sibylline Books. This religious procedure could have political effects, especially negative for magistrates in office during the last turbulent decades of the Republic. The best evidence comes from the severe flood of 54 BC that was interpreted as a consequence of the anger of the gods over the illegal conduct of the legate A. Gabinius and was finally expiated through his banishment.[9] This implies that the overflow of the Tiber could take on the political meaning of a personal presage, denouncing a situation of 'tyrannical' power, which had to be removed.[10]

Nothing like that happened in 27 BC: as a unique example, the flood of this year was seen as a good omen with reference to the assumption of the title Augustus by Octavian, symbolizing the beginning of a new and prosperous age for Rome.[11] Evidently the interpretation of the μάντεις persuaded the senate at this time and made it unnecessary to entrust the *quindecemviri* with the consultation of the Sibylline Books. The problem arose when the populace of Rome had to face other floods of the Tiber, whose material damage could not be positively interpreted any more; hence, the absolute necessity for Augustus to prevent the new emergencies from having repercussions of a political nature.[12]

Two new floods occurred, in fact, a few years later, in 23 and 22 BC, at a short distance from one another. In the first case, Cassius Dio says that Rome was submerged by the Tiber for a period of three days and that the *pons Sublicius*, the wooden bridge, was destroyed.[13] In addition, if we accept seeing allusion to the flooding of this year in Horace's *Carm.* 1.2, we can point out that the overflow reached the *Forum Romanum*, including the *Regia* and the *aedes Vestae*.[14] As for the second case, we can rely on no other literary evidence except for a brief passage of Cassius Dio, who limits himself to registering that the flood of 22 made the

---

     libros adire iussi, et novemdiale sacrum factum et supplicatio indicta est atque urbs lustrata. For a comment see FEDELI 1990: 84.

[9] Cass. Dio 39.61.1–3 speaks of flood ἐκ παρασκευῆς δαιμονίου τινός, due to the fact that A. Gabinius, legate of Syria, had returned the reign of Egypt to Ptolemaeus XII Auletes, despite the opposition of the senate (Cic., *ad Quint.* 3.7.1; App., *Syr.* 51). See MAZZARINO 1966: 622; BECHER 1985: 473–4; ALDRETE 2007: 220–1; CAPPELLETTI 2009: 237; MONTERO 2012: 273–5.

[10] See MONTERO 2012: 272–3, according to whom the flood of 54 BC was the first one to be interpreted as a personal presage.

[11] MAZZARINO 1966: 622–3; BECHER 1985: 475.

[12] Note that, at this time, Augustus had already been co-opted into the college of *quindecemviri*, but probably he was not *magister* of the college and did not have control over it: see RÜPKE and GLOCK 2005: 138 and 839.

[13] Cass. Dio 53.33.7: ὅ τε Τίβερις αὐξηθεὶς τήν τε γέφυραν τὴν ξυλίνην κατέσυρε καὶ τὴν πόλιν πλωτὴν ἐπὶ τρεῖς ἡμέρας ἐποίησε. It was usual that, in case of severe flooding, the wooden bridge was destroyed: see *supra*, n. 4; further discussion in DESNIER 1998: 520; ALDRETE 2007: 23; GRIFFITH 2009: 316–8; MONTERO 2012: 264.

[14] Hor., *Carm.* 1.2, esp. ll. 13–6: *vidimus flavum Tiberim, retortis litore Etrusco violenter undis, ire deiectum monumenta regis templaque Vestae.* I follow here MAZZARINO 1966: 624: according to him, Horace wrote the ode in 23 BC, and alluded to the flood of the same year. Identical opinion can be found in GIARDINA 1997: 188 n. 179. For different points of view, due to the uncertain date of the poem, see LE GALL 2005: 36 n. 56 (the flood Horace alludes to could be one not recorded in available documentation) and ALDRETE 2007: 21–3 (the flood Horace alludes to should be dated to 44, when many portents accompanied Julius Caesar's death). Further bibliography in BECHER 1985: 475–6 n. 33; LONARDI 2013: 12.

city navigable by boat again, without offering any details about its duration and magnitude.[15] At any rate, this was probably the time when the _pons Fabricius_ – the stone bridge built in 62 between the left bank and Tiber Island – was damaged by the fury of the water, since a double inscription of 21 testifies to the restoration of the bridge and its subsequent new test (_probatio_) by the consuls in office, Q. Lepidus and M. Lollius.[16]

It is quite clear that both the floods of 23 and 22 BC were traumatic experiences for the population of Rome and could be seen as negative prodigies, but Augustus avoided resorting to the Sibylline Books in order to find a religious solution to the crisis.[17] On the contrary, we can be sure that the emperor, just after these floods, decided to carry out a series of tangible measures aiming not only at the restoration of damaged buildings, but also at the prevention of future injuries.

Suetonius must allude particularly to this case when, also without giving precise chronological references, he confirms that the emperor _ad coercendas inundationes alveum Tiberis laxavit ac repurgavit completum olim ruderibus et aedificiorum prolapsionibus coartatum._[18] In other words, Augustus tried to intervene decisively, doing the same things modern-day authorities still do after a flood with devastating effects: he widened the bed of the river and cleaned it up, removing not only the discarded materials accumulated over time but also the rubble of the buildings just collapsed.[19] Similar interventions required some kind of specific supervision, and Suetonius is therefore right when ascribing to Augustus the institution of the senatorial board of the _curatores alvei Tiberis_, later turned into _curatores riparum et alvei Tiberis._[20] On this point, scholars have often given credit to Cassius Dio's account in

---

[15] Cass. Dio 54.1: τῷ δ᾽ ἐπιγιγνομένῳ ἔτει, ἐν ᾧ Μᾶρκος τε Μάρκελλος καὶ Λούκιος Ἀρρούντιος ὑπά τευσαν, ἥ τε πόλις πελαγίσαντος αὖθις τοῦ ποταμοῦ ἐπλεύσθη. The historian goes on to recount that the flood was accompanied by other natural calamities, especially a pestilence which was followed by famine; the crisis in the food supply prompted Augustus to appoint the first _curatores frumenti_ (see Suet., _Aug._ 37 and Daguet-Gagey 2011: 355).

[16] _CIL_ 6.1305 (= 31594) = _ILS_ 5892. See Orlandi 2008; _EDR_102346 (V. Gorla).

[17] The crisis coincided with a period of political tensions (begun with the conspiracy of the consul Licinius Murena in 23), that Augustus did not want to aggravate. In any case, interpretations of these floods as prodigies could circulate unofficially: see Montero 2012: 279–80.

[18] Suet., _Aug._ 30: for a comment, see Masi Doria and Cascione 2010: 289. The passage raises a philological question, since the majority of manuscripts show the word _prolationibus_ instead of _prolapsionibus_; but I would keep the variant _prolapsionibus_ because it is well suited to the context, possibly alluding to the collapse of buildings after flooding. On the two variants, Le Gall 2005: 134 n. 24; however, I cannot follow this scholar – _contra_ Gascou 1984: 642–3, and Montero 2012: 301 – when he suggests that the measures Suetonius refers to were contemporary to the delimitation of riverbanks (_terminatio riparum_) made in 8–6 BC (see _infra_, n. 25).

[19] An allusion to these works can be found in the late _Historia Pseudo-Isidoriana_ 5, according to which Augustus intervened in the course of the Tiber not only at Rome, but also for twenty miles upstream and downstream from the city. On the reliability of this source, Le Gall 2005: 134–5.

[20] Suet., _Aug._ 37: _Quoque plures_ [sc. _senatores_] _partem administrandae rei publicae caperent, nova officia_ [sc. _Augustus_] _excogitavit: curam operum publicorum, viarum, aquarum, alvei Tiberis, frumenti populo dividundi, praefecturam urbis, triumviratum legendi senatus et alterum recognoscendi turmas equitum, quotiensque opus esset._ From the time of Tiberius onwards, inscriptions usually attest _curator(es) riparum et alvei Tiberis_ or _curator(es) alvei Tiberis et riparum_: see Le Gall 2005: 170–1; Lonardi 2013: 25–7 (analysing all the inscriptions); Daguet-Gagey 2011: 352–4 (giving credit to Suetonius' account).

claiming that Augustus took care of the Tiber directly and that it was Tiberius who instituted the *curatores*, but, as we shall see below, it is more likely that the second emperor of Rome limited himself to defining (and making permanent) a (temporary) office already set up by Augustus, probably after the floods of 23 and 22 BC.[21]

Moreover, at the beginning of July 13 BC, a new flood affected Rome, while the emperor was returning to the city from his stay in the Western provinces and L. Cornelius Balbus was dedicating his new stone theatre in the *Campus Martius*. For this reason, according to Cassius Dio, Balbus 'began to put on airs, as if it were himself who was going to bring Augustus back, although he was unable even to enter his theatre, except by boat, on account of the flood of water caused by the Tiber'.[22] The piece of information is too brief to make us understand the real effects of the flood on the Roman people, but we can believe that it was not long nor very damaging, since Augustus, once arrived in the city, was able to attend the *constitutio* of the *Ara pacis* right in the *Campus Martius*, on the 4th of July.[23] A very damaging flood happened, on the other hand, some years later, in 5 AD, when the fury of the Tiber killed men and destroyed houses for a period of eight days, according to Cassiodorus, and made the city navigable by boat for a period of seven days, according to Cassius Dio.[24] Evidently it was not enough that, over the previous years, Augustus had continued to monitor the situation of the Tiber, as can be confirmed by the delimitation of riverbanks (*terminatio riparum*) that was committed to the consuls of 8 BC and was completed under his supervision in 7–6 BC.[25] Furthermore the flood of 5 AD was not the last one of the Augustan age: another emergency (probably not a major one) occurred in May 12 AD, when the Tiber inundated the *Circus Maximus* and therefore the *Ludi Martiales*, planned for those days,

---

[21] Cass. Dio 57.14.8 (quoted in n. 40). Note that Augustus was away from Rome between 22 and 19 BC and so could not devote himself to the problems of the Tiber directly (Aug., *RG* 11). On the maintenance of the Tiber before the Augustan age, see Viganò 1972: 805; Masi Doria and Cascione 2010: 291; Maganzani 2012: 96–7.

[22] Cass. Dio 54.25.2: καὶ ἔτυχε γὰρ ἡ ἀγγελία τῆς ἀφίξεως αὐτοῦ ἐν ἐκείναις ταῖς ἡμέραις ἐς τὸ ἄστυ ἐλθοῦσα ἐν αἷς Κορνήλιος Βάλβος τὸ θέατρον τὸ καὶ νῦν ἐπ' αὐτοῦ καλούμενον καθιερώσας θέας ἐπετέλει, ἐπί τε τούτῳ ὡς καὶ αὐτὸς τὸν Αὔγουστον ἐπανάξων ἐσεμνύνετο, καίτοι ὑπὸ τοῦ πλήθους τοῦ ὕδατος, ὅπερ ὁ Τίβερις πλεονάσας ἐπεποιήκει, μηδὲ ἐσελθεῖν ἐς τὸ θέατρον εἰ μὴ πλοίῳ δυνηθείς. (Engl. transl. by E. Cary, Cambridge MA 1917). For a comment, Mazzarino 1966: 622.

[23] Aug., *RG* 12.2. The *constitutio* of the *Ara pacis* seems to have happened on the same day as the arrival of Augustus in the city: Cass. Dio 54.25.3.

[24] Cassiod., *Chron.* 604: *per dies octo Tiberis impetu miseranda clades hominum domorumque fuit*; Cass. Dio 55.22.3: τότε δ' οὖν ἐπί τε τοῦ Κορνηλίου καὶ ἐπὶ Οὐαλερίου Μεσσάλου ὑπάτων σεισμοί τε ἐξαίσιοι συνέβησαν, καὶ ὁ Τίβερις τήν τε γέφυραν [sc. τὴν ξυλίνην] κατέσυρε καὶ πλωτὴν τὴν πόλιν ἐπὶ ἑπτὰ ἡμέρας ἐποίησε, τοῦ τε ἡλίου τι ἐκλιπὲς ἐγένετο, καὶ λιμὸς συνηνέχθη. Apart from the destruction of the *pons Sublicius* (*supra*, nn. 4 and 13), note the exceptional duration of the flood (the longest ever attested: see Aldrete 2007: 63).

[25] The delimitation of the Tiber banks is epigraphically attested: we have a series of *cippi* inscribed *ex senatus consulto* by the consuls in office C. Marcius Censorinus and C. Asinius Gallus (*CIL* 6.31541; Le Gall 2005: 176–8) in 8 BC; another series of *cippi* inscribed *ex senatus consulto* by Augustus in 7–6 BC (*CIL* 6.31542; Le Gall 2005: 178–80). Note that the year 7 BC was decisive for the administrative reorganization of the city, including the creation of the fourteen regions (Cass. Dio 55.8): thus, in spite of Montero 2012: 301 and 304, it is unnecessary to see in the completion of the *terminatio riparum* by Augustus the consequence of supposed conflict between the emperor and the consuls of the previous year.

were moved to the *Forum Augusti*.[26] In any event, it is important to highlight that none of these floods was officially interpreted as a bad prodigy by the senate, since the sources do not mention any consultation of the Sibylline Books in this regard.

We can now summarize by saying that, under Augustus' principate, Rome suffered up to six episodes of flooding, of variable magnitude.[27] On the contrary, we are going to see that under Tiberius' principate, which lasted twenty-three years, the frequency of such calamities was more limited: only two episodes of flooding are known, in fact, respectively in 15 AD and 36 AD, even if the former was possibly the worst inundation that Rome suffered in the entire Julio-Claudian age.[28]

Beyond a reference in Cassius Dio, we can rely on Tacitus, who presents an interesting account of the dangerous situation caused by the flood of the Tiber at the beginning of 15: *Eodem anno, continuis imbribus auctus Tiberis plana urbis stagnaverat; relabentem secuta est aedificiorum et hominum strages*.[29] First, the duration of the flood is uncertain, but Tacitus emphasizes that there were both human loss and material damage not only at the time of the influx but especially at the reflux of the water.[30] Secondly, the historian goes further, telling us what the reaction of the emperor and senators was in the aftermath of the crisis: *Igitur censuit Asinius Gallus ut libri Sibyllini adirentur. Renuit Tiberius, perinde divina humanaque obtegens; sed remedium coercendi fluminis Ateio Capitoni et L. Arruntio mandatum.* This means that, in a meeting of the senate convened and presided over by Tiberius, C. Asinius Gallus proposed seeking guidance from the Sibylline Books,[31] evidently because he considered the flood as a bad portent, which needed to be expiated.[32] But the emperor did not hesitate in rejecting the proposal and decided, on the contrary, to entrust the senators C. Ateius Capito and L. Arruntius with finding technical solutions to the overflows of the Tiber.[33]

To understand such a decision, it is worth remembering that the personal relationship between Tiberius and Asinius Gallus was not a good one; nevertheless, it was again a matter of political balance with the senate and, in practical terms, of good governance.[34] The evi-

---

[26] Cass. Dio 56.27.4: τά τε Ἄρεια τότε μέν, ἐπειδὴ ὁ Τίβερις τὸν ἱππόδρομον προκατέσχεν, ἐν τῇ τοῦ Αὐγούστου ἀγορᾷ καὶ ἵππων δρόμῳ τρόπον τινὰ καὶ θηρίων σφαγῇ ἐτιμήθη. The *Ludi Martiales*, instituted by Augustus, were usually held on the 12th of May: Aug., *RG* 22.2.

[27] On average, the Tiber flooded once every five to ten years. See statistics in Leveau 2008: 138–9.

[28] The flood of 15 was really a 'catastrophic' inundation according to Cappelletti 2009: 237.

[29] Tac., *ann.* 1.76.1; Cass. Dio 57.14.7, who points out that a major part of the city became navigable for boats: τοῦ τε ποταμοῦ τοῦ Τιβέριδος πολλὰ τῆς πόλεως κατασχόντος ὥστε πλευσθῆναι.

[30] We cannot say whether this flood was longer than that of 5 AD, which remains the longest inundation ever attested throughout the history of ancient Rome: see Aldrete 2007: 63.

[31] C. Asinius Gallus was then *XVvir sacris faciundis* and was well acquainted with the problems of the Tiber, as in 8 BC he had proceeded to the delimitation of riverbanks: see Cappelletti 2009: 237; *supra*, n. 25. On the contrary, it is uncertain whether Tiberius was *XVvir* at this time: see Rüpke and Glock 2005: 886.

[32] The general belief of the Romans is mentioned by Cass. Dio 57.14.7: οἱ μὲν ἄλλοι ἐν τέρατος λόγῳ καὶ τοῦτο, ὥσπερ που τό τε μέγεθος τῶν σεισμῶν ὑφ᾽ ὧν καὶ μέρος τι τοῦ τείχους ἔπεσε, καὶ τὸ πλῆθος τῶν κεραυνῶν ὑφ᾽ ὧν καὶ οἶνος ἐξ ἀγγείων ἀθραύστων ἐξετάκη, ἐλάμβανον.

[33] The last consultation of the Sibylline Books, following a flood, had taken place in 54 BC. Cappelletti 2009: 237, is right when saying that 'la proposta stessa [di Asinio Gallo] di adire ai Libri Sibillini non era consueta nel dibattito politico coevo'. See also Levick 1976: 105–6.

[34] On the difficult relationship between Tiberius and Gallus, exacerbated by the marriage of Gallus with Vipsania, former wife of Tiberius, see Levick 1976: 43 and 114.

dence comes from the fact that Tiberius tackled the problem of flooding by the creation of a special commission with technical expertise, since Ateius Capito had been *curator aquarum* at least for two years.[35] In this way, the new emperor clearly showed his intention to follow and develop the same policy adopted in this regard by Augustus throughout his principate.

The consequence was, according to Tacitus, that some months later a new senate meeting was held, in which Ateius Capito and Arruntius presented a complex project, concerning specific interventions for the upper course of the Tiber and that of its tributaries. The first part of the project was to divert the initial stretch of the river Clanis into the river Arnus. The second one (more difficult to understand) was probably to occlude the Lake Velinus, so that its water could not pour into the river Nar (through the Marmore Falls), and finally to divide up the decreased water of the Nar in irrigation canals in the territory of Interamna Nahartium.[36] Tacitus does not comment on the value of the project, but reports that, during the meeting of the senate, a strong opposition to it was raised by the legacies of communities that would be affected by its realization. The inhabitants of Florentia opposed to the diversion of the Clanis, because they feared that the increased inflow of water in the Arno would make it overflow. Likewise, the inhabitants of Reate and Intermna opposed to the occlusion of the Lake Velinus and to dividing up the Nar, because they feared that their land would become marshy and less fertile.[37] Furthermore, alongside concrete interests, there were also 'naturalistic' and religious reasons to oppose the project, for the inhabitants of Reate argued that it was necessary to respect what had been arranged by nature, and also reminded the senate that an overall intervention in the upper Tiber valley would have touched the religious sensitivity of local communities, which had dedicated cults to their rivers and especially to the Tiber.[38]

In the end, the opposition was so strong that the senate decided to reject the entire project and to adopt the proposal of Cn. Calpurnius Piso to leave everything unchanged. Tacitus, indeed, does not even feel the need to specify whether the prayers of the local communities were more decisive, or the difficulty in carrying out the work, or the religious scruples

---

[35] Front., *aq.* 102 (Ateius Capito *curator aquarum* since 13 AD); Rogers 1982: 171–2; Cappelletti 2009: 238–9. Capito was possibly the author of a treatise on the administration of water (Front., *aq.* 97); furthermore, he and Arruntius had practical experience of the problem, also because they had been *consules* after the flood of 5 AD (Capito was *cos. suff.* in 5; Arruntius was *cos. ord.* in 6). See Le Gall 2005: 156 n. 5.

[36] Tac., *ann.* 1.79.1–3: *Actum deinde in senatu ab Arruntio et Ateio an ob moderandas Tiberis exundationes verterentur flumina et lacus, per quos augescit; auditaeque municipiorum et coloniarum legationes, orantibus Florentinis ne Clanis solito alveo demotus in amnem Arnum transferretur idque ipsis perniciem adferret. Congruentia his Interamnates disseruere: pessum ituros fecundissimos Italiae campos, si amnis Nar (id enim parabatur) in rivos diductus superstagnavisset. Nec Reatini silebant, Velinum lacum, qua in Narem effunditur, obstrui recusantes, quippe in adiacentia erupturum; optume rebus mortalium consuluisse naturam, quae sua ora fluminibus, suos cursus utque originem, ita fines dederit; spectandas etiam religiones sociorum, qui sacra et lucos et aras patriis amnibus dicaverint: quin ipsum Tiberim nolle prorsus accolis fluviis orbatum minore gloria fluere.* This meeting of the senate probably took place in the summer of 15 AD. See Le Gall 2005: 141.

[37] For modern attempts to reconstruct and evaluate the project of 15 AD, see Le Gall 2005: 137–41, followed by Clementoni 1989: 173–4; Aldrete 2007: 185–8; Leveau 2008: 139–42; Cappelletti 2009: 242–4.

[38] For a comment on this Tacitean passage, Fedeli 1990: 64–5; Cappelletti 2009: 245–6.

apparently shared by senators.[39] However it was, it is certain that Tiberius did not renounce the care of the Tiber and, remedying the rejection of the project, decided to create at least a permanent magistracy of five senators. Their duty would be the general supervision of Tiber's bed and its banks, the *cura riparum et alvei Tiberis*. An allusion to this can be found in Cassius Dio's statement that the emperor, well aware of the direct cause of the last flood (νομίσας ἐκ πολυπληθίας ναμάτων αὐτὸ γεγονέναι), wanted to take action to ensure a regular hydrographical flow of the river and therefore instituted a board of five men, chosen by lot among the senators (πέντε ... βουλευτὰς κληρωτούς).[40] As a matter of fact, many inscribed boundary stones (*cippi*), used to delimit the banks of the Tiber and dated from the Tiberian age onwards, attest to the succession of boards constituted by five *curatores*, who had among their duties the specific task of the *terminatio riparum*.[41]

In order to better understand the origin and development of this magistracy, it is important to note that the oldest *cippi* belong to the previous century and were restored precisely in 15 AD, thanks to the explicit intervention of the *curatores riparum qui primi terminaverunt*, sometimes styled as *curatores riparum qui primi fuerunt*.[42] Contrary to the general opinion of scholars, this means that the innovation introduced by Tiberius was not the institution of the *cura Tiberis* as a whole, but the institution of the *cura riparum et alvei Tiberis*.[43] The new office encompassed in a single magistracy the existing *cura alvei Tiberis* and the duties connected to the *cura riparum*, including the *terminatio riparum* that had not yet been assigned to an *ad hoc* magistracy. As already noted, it was in fact Augustus who set up the first *curatores alvei Tiberis*, whose duty was simply to keep the bed of the Tiber clean, whereas the *terminatio riparum*, when necessary, was committed to other magistrates such as the consuls. This was the case of 8 BC, when the consuls in office operated a *terminatio riparum*, completed the following year by Augustus himself.[44] It is thus inevitable that the *cura Tiberis* was an invention of Augustus, even if it reached the height of its functions only

---

[39] Tac., *ann.* 1.79.4: *Seu preces coloniarum seu difficultas operum sive superstitio valuit, ut in sententiam Pisonis concederetur, qui nil mutandum censuerat.* According to FEDELI 1990: 65, the terms '*preces coloniarum*', '*difficultas operum*' and '*superstitio*' form a *climax*; Tacitus would have seen in *superstitio* the major obstacle to the realization of the project. On Cn. Calpurnius Piso, who opposed the project because he had many *clientes* in the territory of Interamna, Tac., *ann.* 3.9, and CLEMENTONI 1989: 174–5 n. 26.

[40] Cass. Dio 57.14.8: ἐκεῖνος δὲ δὴ [sc. ὁ Τιβέριος] νομίσας ἐκ πολυπληθίας ναμάτων αὐτὸ γεγονέναι πέντε ἀεὶ βουλευτὰς κληρωτοὺς ἐπιμελεῖσθαι τοῦ ποταμοῦ προσέταξεν, ἵνα μήτε τοῦ χειμῶνος πλεονάζῃ μήτε τοῦ θέρους ἐλλείπῃ, ἀλλ᾽ ἴσος ὅτι μάλιστα ἀεὶ ῥέῃ. See LE GALL 2005: 156 and 168; ALDRETE 2007: 199; CAPPELLETTI 2009: 238–40; MAGANZANI 2012: 95.

[41] *CIL* 6.31544 (C. Vibius Rufus, Sex. Sotidius Strabo Libuscidianus, C. Calpetanus Statius Rufus, M. Claudius Marcellus, L. Visellius Varro: 16–24 AD); *CIL* 6.31545 (Paullus Fabius Persicus, C. Eggius Marullus, L. Sergius Paullus, C. Obellius Rufus, L. Scribonius Libo: 41–4 AD). For a comment, see LE GALL 2005: 157–9 and 180–1; LONARDI 2013: 91–6 and 100–4.

[42] *CIL* 6.31540 *l*, 31541 *g, h, o, r, s, u*, 31542 *s*, 31557 *d* (and *CIL* 6.31541 *f., l, k, m, t*). Unfortunately the *curatores riparum qui primi terminaverunt* did not specify their names: LE GALL 2005: 179–80; LONARDI 2013: 164–6.

[43] See MOMMSEN 1896: 346; LE GALL 2005: 155–6; PALMA 1980: 232–3; GOODYEAR 1981: 171–2; GASCOU 1984: 643; ROBINSON 1992: 87–8; TAYLOR 2000: 82; LOUIS 2010: 261; LONARDI 2013: 14–5.

[44] See *supra*, n. 25. Also considering the years before the Augustan age, we know that a *terminatio riparum*, carried out in 54 BC, had been committed to the censors P. Servilius Isauricus and M. Valerius Messalla. See *CIL* 6.31540 and LE GALL 2005: 173–6.

under the reign of Tiberius, responding to the administrative needs matured in the meantime. Unfortunately, as for the features of the original *cura alvei Tiberis*, we know nothing apart from the fact that Augustus committed it to senators. Accordingly, we cannot exclude that it was a temporary magistracy, which was established only to better manage the crisis following the floods of 23 and 22 BC.[45] There is instead no doubt that the *cura riparum et alvei Tiberis* introduced by Tiberius was a permanent senatorial office, as is suggested by Cassius Dio's account.[46] The inscribed *cippi* demonstrate that the magistracy was regularly covered during the whole Julio-Claudian age and that its senatorial members, belonging to consular and praetorial rank, could be in office for variable terms, from just one to many years, since no specific duration seems ever to have been fixed.[47]

After 15 AD, it is difficult to say whether, thanks to such a permanent magistracy, the Romans performed more regular maintenance of the river.[48] It is only certain that, in the remaining years of Tiberius' principate, one flood occurred in 36 AD, when, according to Cassius Dio, many prodigies appeared in Rome and, among them, the Tiber flooded and made a large part of the city navigable for boats.[49] But the material consequences of this flood for the population of Rome were probably not very serious and, as far as we know, did not produce a political debate.[50] This situation is all the more significant since, after 36 AD, the overflows of the Tiber became in general a phenomenon which recurred less often. According to the ancient sources, in fact, the next episode of flooding occurred in mid-March 69 AD, when the Julio-Claudian dynasty was already over and Otho was going to combat the forces of his rival Vitellius. This means that for more than thirty years Rome was not afflicted by floods of the Tiber.[51]

In the meanwhile, at least one tangible intervention was realized on the Tiber, precisely in the lower course of the river, and no opposition apparently emerged against it. The point is that the emperor Claudius did not stop at beginning the construction of a new harbour at

---

[45] Suet., *Aug.* 37 (*supra*, n. 20): … [sc. *Augustus*] *excogitavit … curam … alvei Tiberis … q u o t i e n s q u e o p u s   e s s e t.*

[46] Cass. Dio 57.14.8 (*supra*, n. 40): … πέντε ἀεὶ βουλευτὰς κληρωτοὺς ἐπιμελεῖσθαι τοῦ ποταμοῦ [sc. ὁ Τιβέριος] προσέταξεν.

[47] Le Gall 2005: 157–9, with epigraphical references; Palma 1980: 237; Robinson 1992: 88.

[48] It is not clear whether the jurisdiction of the *curatores* was limited to the Roman section of the Tiber or extended throughout its course. See Le Gall 2005: 200, according to whom their jurisdiction extended with certainty from Rome to Ostia, but probably also to all the upper course of the Tiber. For different points of view: Robinson 1992: 91–3; Aldrete 2007: 199–200; Montero 2012: 305–6.

[49] Cass. Dio 58.26.5: Σέξτου δὲ δὴ Παπινίου μετὰ Κυΐντου Πλαυτίου ὑπατεύσαντος ὅ τε Τίβερις πολλὰ τῆς πόλεως ἐπέκλυσεν ὥστε πλευσθῆναι, καὶ πυρὶ πολὺ πλείω περί τε τὸν ἱππόδρομον καὶ περὶ τὸν Ἀουεντῖνον ἐφθάρη, ὥστε τὸν Τιβέριον δισχιλίας καὶ πεντακοσίας μυριάδας τοῖς ζημιωθεῖσί τι ἀπ' αὐτοῦ δοῦναι. Evidently the major damage was caused by the fire that destroyed the area around the *Circus Maximus* and the Aventine: confirmation comes from Tacitus (*ann.*, 6.45.1–2), who does not speak of the flood, but mentions the fire and the consequent donation of Tiberius. See Clementoni 1989: 170.

[50] Unconvincingly Cappelletti 2009: 248, speaks of 'danni nuovamente di proporzioni ingenti'.

[51] Tac., *hist.* 1.86.2; Suet., *Otho* 8.3; Plut., *Otho* 4.5: in 69 AD the *pons Sublicius* was destroyed once again. On this episode Aldrete 2007: 67; Montero 2012: 281–2.

4 kilometres north of Ostia (later inaugurated by Nero as *portus Augusti*),[52] but also wanted to dig a system of artificial canals, the so called *fossae Claudianae*, which would have provided a complex hydrographical reorganization next to the mouth of the Tiber.[53] An inscription, probably coming from the travertine *porticus* of the new harbour and datable to 46 AD, mentions the inauguration of the *fossae* and, above all, celebrates their usefulness: *Ti(berius) Claudius Drusi f(ilius) Caesar / Aug(ustus) Germanicus, pontif(ex) max(imus), / trib(unicia) potest(ate) VI, co(n)s(ul) design(atus) IIII, imp(erator) XII, p(ater) p(atriae), / fossis ductis a Tiberi operis portu[s] / caussa emissisque in mare Urbem / inundationis periculo liberavit.*[54] In this respect, it is worth noting that Claudius boasted of having liberated the city from floods, since the *fossae*, built in conjunction with the new harbour, would have allowed a better discharge of water into the Tyrrhenian Sea.

Of course, in the light of the inundation of 69 AD, Rome was not completely released from the danger of flooding. Furthermore, at a distance of nearly fifty years from the inauguration of the *fossae Claudianae*, the emperor Trajan intervened again in the lower course of the Tiber, after an episode of flooding under Nerva's principate, and decided not only to restructure the existing Claudian harbour, but also to build a new artificial canal for the drainage of the Tiber, the *fossa Traiana*, corresponding to the modern-day Fiumicino.[55] In this case another inscription, found in excavations of the *portus Traiani* and not datable with certainty, attests that also Trajan boasted of having liberated the city from the danger of floods, implicitly disproving Claudius.[56] Thus Joël Le Gall is probably right when arguing that Trajan did more and better than Claudius had done. Nevertheless, despite their long-term effects, we cannot disregard the importance of the works initiated by Claudius.[57] He was, in fact, the first emperor of Rome who was able to divert the course of the Tiber, after Caesar had designed but not realized such a project in 45 BC.[58]

---

[52] The need for a new *portus* derived from the fact that the existing harbour at Ostia was not a safe anchorage and ships were often unloaded in the superior port facilities of Puteoli. See Le Gall 2005: 148–50 and 153–4; Meiggs 1973: 51–8; Aldrete 2007: 184; Montero 2012: 308. The *portus Augusti* was probably inaugurated in 64–5: *RIC²* Nero 178–83.

[53] See Le Gall 2005: 150–1; Meiggs 1973: 153–61; Levick 1990: 110; Aldrete 2007: 184; Leveau 2008: 142; Montero 2012: 308–9.

[54] *CIL* 14.85 = *ILS* 207. On this inscription see also Meiggs 1973: 159; *EDR*094023 (F. Feraudi).

[55] On the *portus Traiani* and the *fossa Traiana*, Le Gall 2005: 150–1; Meiggs 1973: 58–62 and 162–7; Aldrete 2007: 184–5; Montero 2012: 310; *RIC* Trajan 471. The construction of the *fossa Traiana* is confirmed by Plin. Iun., *epist.* 8.17 and *CIL* 6.964 (see *infra*); the flood under Nerva is mentioned by Aur. Vict., *Caes.* 13.12 (the date cannot be fixed).

[56] *CIL* 6.964 (= 14, 88) = *ILS* 5797a: *[Imp(erator) Caes(ar) Divi] / Ne[rvae f(ilius) Nerva] / Tra[ianus Aug(ustus) Germ(anicus)] / Dac[icus, trib(unicia) pot(estate) ---] / im[p(erator) --- co(n)s(ul) --- p(ater) p(atriae)] / fossam [fecit, / q]ua inun[dationes Tiberis / a]dsidue u[rbem vexantes / rivo p]eren[ni arcerentur].* On the problems concerning the restoration of this text, see also Meiggs 1973: 488; *EDR*094033 (C. Cenati).

[57] Note that also the boasting of Trajan was proved wrong, even by a flood occurring under his principate (the date is unknown): Plin. Iun., *epist.* 8.17 and Aur. Vict., *Caes.* 13.12; Le Gall 2005: 153; Aldrete 2007: 28–9.

[58] The difference is that Caesar had proposed to divert the Tiber just north of Rome and to redirect it through an artificial channel around the Vatican hills: Cic., *ad Att.* 13.33a; Le Gall 2005: 130–2; Aldrete 2007: 182–3; Maganzani 2012: 94–5. On the other hand, Claudius realized another ma-

Conclusion

After having gained a sense of the attitude shown by (almost) every single Julio-Claudian emperor towards the Tiber, we can now provide an overall evaluation of their policy about the problem of floods.[59] First of all, it is evident that, from Augustus onwards, the populace of Rome suffered several episodes of flooding, whose frequency and magnitude made it necessary for the Roman rulers to intervene decisively. This is the reason why a board of senators appointed to the supervision of the river was first instituted by Augustus, probably limited in duties and duration, and then, after the flood of 15 AD, was made permanent by Tiberius. But the Julio-Claudian emperors and their advisors, far from limiting themselves to taking care of the river bed and its banks, were also able to define complex engineering projects aimed at diverting the course of the Tiber and that of its tributaries, so that the flow of water could be more regular in Rome. In other words, they proposed 'scientific' solutions to the problem of floods, in opposition to a more common tendency looking at floods as negative prodigies that needed to be expiated in a religious way.

Hence, can we say that the first emperors of Rome were inspired by a really 'rationalist' attitude? To answer this question, it is necessary to remember that Augustus had not completely disregarded the religious interpretation of floods, but had returned to it when it was politically useful to him. Indeed, in the case of 27 BC, Augustus did not hesitate in accepting the favourable response of the μάντεις in order to strengthen his *auctoritas*. Nevertheless, Augustus knew that a similar interpretation had to be unique. Furthermore, the succession of five other floods under his principate, as well as the memory of the unhappy end of the legate A. Gabinius after the inundation of 54 BC, made him find a strategy to prevent the flood emergencies from being used by his opponents in the senate for their political purposes. This fact explains why not only Augustus but also Tiberius carefully avoided the potentially dangerous consultation of the Sibylline Books. As a consequence, the adoption of 'scientific' measures for flood control was their active reaction to those senators who tried to destabilize the principate by means of religious procedures.

Bibliography

ALDRETE, G. S. 2007. *Floods of the Tiber in Ancient Rome*. Baltimore.

BECHER, I. 1985. 'Tiberüberschwemmungen. Die Interpretation von Prodigien in Augusteischer Zeit', *Klio* 67, 471–9.

CAPPELLETTI, S. 2009. 'Il progetto sull'esondazione del Tevere. Tacito, *Annales* 1.76 e 1.79', *Acme* 62, 235–53.

CLEMENTONI, G. 1989. 'Tiberio e il problema della protezione civile', in M. SORDI (ed.), *Fenomeni naturali e avvenimenti storici nell'antichità* (Contributi dell'Istituto di Storia Antica, 15). Milano, 167–83.

DAGUET-GAGEY, A. 2011. 'Auguste et la naissance des services publics à Rome. À propos de Suétone, Vie d'Auguste, 37, 1', in S. BENOIST, A. DAGUET-GAGEY and C. HOËT-VAN CAUWENBERGHE (eds.), *Fi-*

---

jor project already contemplated by Caesar, the draining of the Fucine Lake into the Liris. See LEVICK 1990: 110–1.

[59] Caligula apparently did not concern himself with the care of the Tiber, whereas Nero limited himself to completing and inaugurating the *portus Augusti* in 64–5 (see *supra*).

gures d'empire, fragments de mémoire: pouvoirs et identités dans le monde romain impérial (Ier s. av. n. è.-Ve s. de n. è.). Lille, 341–60.

DESNIER, J.-L. 1998. 'Les débordements du Fleuve', *Latomus* 57, 513–22.

FEDELI, P. 1990. *La natura violata. Ecologia e mondo romano.* Palermo.

FREYBURGER, M.-L. 1999. 'L'Etrusca disciplina chez Dion Cassius', in *Les écrivains du Troisième Siècle et l'Etrusca disciplina,* actes de la Table-Ronde de Paris, 24 et 25 octobre 1997 (*Caesarodunum*, supplément 66). Paris, 17–32.

GASCOU, J. 1984. *Suéton historien* (BEFAR 255). Rome.

GIARDINA, A. 1997. 'Allevamento ed economia della selva in Italia meridionale', in A. GIARDINA, *L'Italia romana. Storie di un'identità incompiuta.* Roma-Bari, 139–92.

GOODYEAR, F. R. D. 1981. *The Annals of Tacitus*, vol. 2. Cambridge.

GRIFFITH, A. B. 2009. 'The pons Sublicius in context: revisiting Rome's first public work', *Phoenix* 63, 296–321.

LE GALL, J. 1953. *Le Tibre, fleuve de Rome dans l'antiquité.* Paris.

LE GALL, J. 2005. *Il Tevere, fiume di Roma nell'antichità,* revised edition by C. MOCCHEGIANI CARPANO and G. PISANI SARTORIO. Roma.

LEVEAU, P. 2008. 'Les inondations du Tibre à Rome: politique publique et variations climatiques à l'époque romaine', in E. HERMON (ed.), *Vers une gestion intégrée de l'eau dans l'empire romain*, Actes du Colloque international, Université Laval, octobre 2006. Roma, 137–46.

LEVICK, B. 1976. *Tiberius the Politician.* London.

LEVICK, B. 1990. *Claudius.* London.

LONARDI, A. 2013. *La cura riparum et alvei Tiberis. Storiografia, prosopografia e fonti epigrafiche* (BAR IntSer, 2464). Oxford.

LOUIS, N. 2010. *Commentaire historique et traduction du Divus Augustus de Suétone* (Coll. Lat. 324). Bruxelles.

MAGANZANI, L. 2012. 'Le inondazioni fluviali e le loro ricadute sulle città romane: considerazioni storico-giuridiche', in G. PURPURA (ed.), *Revisione ed integrazione dei Fontes Iuris Romani Anteiustiniani (FIRA). Studi preparatori I Leges.* Torino, 93–102.

MASI DORIA, C. and C. CASCIONE, 2010. '*Cura riparum*', in E. HERMON (ed.), *Riparia dans l'Empire romain, pour la définition du concept* (BAR IntSer, 2066). Oxford, 283–94.

MAZZARINO, S. 1966. 'Le alluvioni 54 a. C./23 a. C., il cognome *Augustus*, e la data di Hor. *Carm.* I 2', *Helikon* 6, 621–4.

MEIGGS, R. 1973. *Roman Ostia*, second edition. New York.

MOMMSEN, TH. 1896. *Le droit publique romain*, vol. 5. Paris.

MONTERO, S. 2012. *El emperador y los ríos. Religión, ingeniería y política en el Imperio Romano.* Madrid.

ORLANDI, S. 2008. 'Le iscrizioni del ponte Fabricio viste da vicino', in M. L. CALDELLI, G. L. GREGORI and S. ORLANDI (eds.), *Epigrafia 2006. Atti della XIVe Rencontre sur l'Épigraphie in onore di S. Panciera, con altri contributi di colleghi, allievi e collaboratori.* Roma, 177–86.

PALMA, A. 1980. *Le curae pubbliche. Studi sulle strutture amministrative romane.* Napoli.

RICH, J. W. 1990. *Cassius Dio, The Augustan Settlement (Roman History 53–59.9).* Warminster.

ROBINSON, O. F. 1992. *Ancient Rome: City Planning and Administration.* London.

ROGERS, R. H. 1982. 'Curatores aquarum', *Harvard Studies in Classical Philology* 86, 171–80.

RÜPKE, J. and GLOCK, A. 2005. *Fasti sacerdotum. Die Mitglieder der Priesterschaften und das sakrale Funktionspersonal römischer, griechischer, orientalischer und jüdisch-christlicher Kulte in der Stadt Rom von 300 v. Chr. bis 499 n. Chr.* Wiesbaden.

SCHMITTHENNER, W. 1962. 'Augustus' spanischer Feldzug und der Kampf um den Prinzipat', *Historia* 4, 29–85.

TAYLOR, R. 2000. *Public Needs and Private Pleasures. Water Distribution, the Tiber River, and the Urban Development of Ancient Rome.* Rome.

VIGANÒ, R. 1972. 'Appunti sulla "cura riparum et alvei Tiberis": gestione diretta o indiretta?', in *Studi in onore di G. Scherillo*, vol. 2. Milano, 803–8.

Orietta Dora Cordovana

Pliny the Elder and Ancient Pollution

Premise

Despite the modern rift between science, religion, and philosophy, there is no doubt that moral, religious, and philosophical elements characterised the approach of the ancient intellectuals to the natural environment. Modern scholars are aware that philosophy, natural history, and scientific knowledge were deeply intermingled in ancient thought and there was no clear-cut distinction between philosophy and natural sciences.[1] Pliny the Elder is no exception to this general rule. In current scholarly debate, the *Naturalis Historia* has been classified and analysed primarily as a product of the encyclopaedic tradition and, more precisely, has been considered among one of the first examples of the technical literary genre known as encyclopaedism.[2] Furthermore, several aspects of this *Encyclopaedia* have been examined under methods related to history of science, political history, as well as administrative and socio-economic history. The general structure and content of the work clearly imply the possibility of different perspectives and various levels of reading. By such reasoning, Pliny's scientific method, empirical approach, and notions of Nature and Landscape have received accurate scrutiny. Some scholars, in particular, have criticised the compendium and highlighted a deficiency of scientific rigour; others also have detected a certain chauvinistic ideology aimed at supporting Roman imperialism amongst the main precepts of the whole series of books.[3] Nevertheless, there is evidence in the *Naturalis Historia* of much more and much less than this in terms of problems and complexity of structure. We can distinguish, for example, more about the social and cultural elements and less on the politics of Pliny's time; more on empirical observation and the technical knowledge of that society, but less about valid scientific method.[4] Many different aspects characterize Pliny's work, and these cannot be reduced into unity.

In this paper I shall not go back to analyse the views of landscapes and the idea of Nature which permeated the *Naturalis Historia*. These topics have been already fully investigated by scholars. Neither is it my intention to engage in the debate that labels this series of books as the product of Roman imperialism, allegedly permeated by the view of the rightful duty of Rome to dominate the Mediterranean. Rather, I prefer to focus on specific elements of

---

[1] See especially HARRISON 2011: 1–7.

[2] An overview of the *status quaestionis* in the current debate is in: SIGRÍDUR ARNAR 1990; MAG-GIULLI, 2000: 1–42; MURPHY 2004: 11–6; KÖNIG and WOOLF 2013a: 1–20; BEAGON 2013: 84–107.

[3] See CAREY 2003: 33–6; DOODY 2010; MURPHY 2004: 50–74; NAAS 2011: 57–70; KÖNIG and WOOLF 2013b: 23–63.

[4] See CITRONI MARCHETTI 2011: 13–30.

the *Naturalis Historia*, elements that may be principally useful in distinguishing the ancient roots of modern environmental awareness. This paper aims at offering evidence that not only a few passages in the books of the *Naturalis Historia*, but specifically the work as a whole encompasses certain environmental consciousness.

By focusing on Pliny's work and via comparison with other literary sources, the main purpose of this paper will focus on some crucial points related to the attitude of ancient intellectuals towards pollution and exploitation of natural resources. In particular, in the *Naturalis Historia* it is possible to detect definite answers for basic enquiries. One of the key issues is whether we can distinguish a certain awareness related to environmental problems in terms of ecological damage and environmental impact, which may characterize the compendium as a whole. It follows that, it also has to be defined in which sense Pliny refers to 'pollution', 'exploitation', and 'depletion' of natural resources; whether he gives evidence of human action for the protection of the environment and in which specific contexts. Though, before starting this kind of analysis, it may be useful to recapitulate some preliminary stated points concerning the compendium itself.

1. The dedication to Titus and the ἐγκύκλιος παιδεία

In the introductive *Epistula* to the emperor Titus, Pliny outlines the contents and aims of the thirty-seven books, which compose the *Naturalis Historia*, and illustrates a series of topics that, on his own admission, are part of the Greek *encyclopaedic* cultural tradition. The main theme is the description of Nature, which is synonymous with Life, and even the humblest issues of daily existence, the definitions of which are often given by 'foreign or barbarous terms'. He refers to those subjects that amongst the Greeks are parts of the ἐγκύκλιος παι-δεία, without apparently specifying the exact meaning and content of that which modern commentators usually ascribe to a precise literary genre.

> My subject is a barren one – the world of nature, or in other words life; and that subject in its least elevated department, and employing either rustic terms or foreign, nay barbarian, words that actually have to be introduced with an apology. (…) Deserving of treatment before all things are the subjects included by the Greeks under the name of 'Encyclic Culture' (ἐγκύκλιος παιδεία); and nevertheless they are unknown, or have been obscured by subtleties, whereas other subjects have been published so widely that they have become stale. It is a difficult task to give novelty to what is old, authority to what is new, brilliance to the common-place, light to the obscure, attraction to the stale, credibility to the doubtful, but nature to all things and all her properties to nature. Accordingly, even if we have not succeeded, it is honourable and glorious in the fullest measure to have resolved on the attempt. For my own part I am of opinion that a special place in learning belongs to those who have preferred the useful service of overcoming difficulties to the popularity of giving pleasure.[5]

---

[5] Plin., *nat. hist.*, *praef.* 13–6: *rerum natura, hoc est vita, narratur, et haec sordidissima sui parte ac plurimarum rerum aut rusticis vocabulis aut externis, immo barbaris etiam, cum honoris praefatione ponendis. (…) ante omnia attingenda quae graeci τῆς ἐγκυκλίου παιδείας vocant, et tamen ignota aut incerta ingeniis facta; alia vero ita multis prodita, ut in fastidium sint adducta. res ardua vetustis novitatem dare, novis auctoritatem, obsoletis nitorem, obscuris lucem, fastiditis gratiam, dubiis fidem, omnibus vero naturam et naturae sua omnia. itaque etiam non assecutis voluisse abunde pulchrum*

On the basis of these statements, the classification of the *Naturalis Historia* under the label of 'encyclopedic work' would be reductive and anachronistic.[6] By contrast, if we focus on the general context of these introductive paragraphs to his *encyclopaedia*, it is possible to better understand what Pliny meant by his definition. Firstly, he specifies that the work covers both well-known subjects from the past and new studies, which need to be accredited. By and large, in Pliny's words the books represent a summary of the Romans' common knowledge in various fields. The formal exposition is not an easy task, since different topics, old and new, have to be made engaging to a wide audience of a lower-middle cultural basis. The books of the *encyclopaedia* are not supposed to be addressed to scientists, philosophers, and professionals. What is important is to give 'nature to all things and all her properties to nature'. This is a specific choice either in terms of method or audience addressed. Indeed, in Greco-Roman society encyclopaedic education was the counter-part of the highest level of cultural education aimed at the finest specialism in subjects such as philosophy, rhetoric, mathematics, geometry, and physics.[7] Conversely, topics of the ἐγκύκλιος παιδεία cover ordinary subjects and encompass practical aspects of daily life. The treatise, then, 'was written for the common herd, the mob of farmers and of artisans, and after them for students who have nothing else to occupy their time', as Pliny openly states.[8] This assertion is perfectly consonant with the original meaning of ἐγκύκλιος παιδεία, which refers to ordinary and general education and does not investigate deeply any field of specialism.[9] Nevertheless, the apparent modesty on the value of his work conceals a very ambitious target, which was not simply aimed at a catalogue or a comprehensive data collection of human knowledge of that time. By reading the *Naturalis Historia*, it becomes evident that one of the main purposes was essentially pedagogical, with the aim of educating Roman citizens and in giving them useful notions for civic life.

This is a general premise on Pliny's work. Nonetheless, topics of general education for free born people of the Roman Empire are permeated in the *Naturalis Historia* by another fundamental component, which influenced and oriented its internal structure in a specific way. Stoic philosophical principles founded the clear and subtle Leitmotiv that pervades all the books. Indeed, these ethics largely inspired the treatise. Pliny's general aim is to show the natural order of the world and its delicate balance that is constantly under threat. On the one hand, following in the footsteps of Posidonius, physics becomes the main tool, a specific method to promote a moral practice in the audience of the *Naturalis Historia*.[10] On the other, Pliny is also deeply concerned with the separation of human behaviour from nature. The harmony between the humankind and the *cosmos* is a crucial focus. It follows that in

---

atque magnificum est. equidem ita sentio, peculiarem in studiis causam eorum esse, qui difficultatibus victis utilitatem iuvandi praetulerint gratiae placendi. (*Caius Plinius Secundus, Naturalis Historia libri XXXVII*, K. F. T. Mayhoff, L. Ian edd., Leipzig 1892–1909. Engl. transl. by H. Rackam, W. H. S. Jones, D. E. Eicholz, London-Cambridge MA 1938–1963).

[6] See especially Carey 2003: 17–20.

[7] On these aspects see Doody 2009: 1–21; Morello 2011: 147–65; the rhetoric of Pliny's modesty is highlighted by Schultze 2011: 167–86; Morgan 2013: 108–28.

[8] Plin., *nat. hist., praef.* 6: *Humili vulgo scripta sunt, agricolarum, opificum turbae, denique studiorum otiosis.*

[9] See also Dion. Halic., *comp.* 25; Plut., *Alex.* 7; Vitr., *de arch.* 1.12; Quint., *inst.* 1.10.1.

[10] These points are developed in particular by Paparazzo 2005: 363–76; Paparazzo 2011: 95–8, 102, 105.

the *Encyclopaedia* we should not search for any kind of scientific rigour and method in the modern sense of the meaning. Further, Pliny's statements on Roman power should not be extrapolated from their context, since one of the main targets of the author is to criticize the senatorial ruling class and highlight their excesses against Nature.

In Pliny's eyes Roman imperialism should consist in the preservation of Nature. The leading role of Rome in the Mediterranean is not a matter of mere exaltation of power from the conquerors' dominant point of view.[11] Rather, to Pliny this power should imply duties and responsibility in keeping a safe balance between man and natural environment. Even the common literary topic of *laus Italiae* is not an ordinary element of Roman imperialism, but should be interpreted in the specific perspective that is part of the long literary and scientific tradition of environmental determinism.[12] Pliny often complains of the Romans' behaviour and attempts against nature; respect and sense of environmental responsibility are widely missing in Roman ruling class. Many invectives and sarcastic tirades are often addressed in condemning the greed and luxury of the conquerors.

Therefore a historical evaluation of the *Naturalis Historia* should take into account this specific cultural background, as well as definite philosophical values that clearly inspired the work and were at the core of its intellectual structure.[13] Without considering its fundamental philosophical backbone, the manifold and multifaceted subjects of the *Naturalis Historia* can be misinterpreted and deceptive in the context of the corpus and in terms of its quality and aims.

These are major preconditions before engaging in any consideration about Pliny's attitude toward nature and the environment. Although it is a matter of fact that he was not a Stoic philosopher 'interested in the finer points of any doctrine' (BEAGON), it can be inferred that, among many philosophical influences, Stoic ethical values are a life-style and source of inspiration in his way of seeing life and the surrounding natural world.[14] On the one hand, the *Naturalis Historia* is a portrait of the Nature that Pliny sees as a constant miracle, a divine and immanent being. On the other, the whole series of books develop gradually and variously the story of man and his relationship with Nature.[15] Analyses of Pliny's encyclopaedic studies, therefore, should take into due account the double level of enquiry and awareness regarding natural history and human history being interlinked with natural environment. To Pliny this investigation was the product of a long life experience, as he travelled all over the Mediterranean empire of Rome to perform the various duties of his military and administrative career.[16]

---

[11] This perspective is prevailing in the works of MURPHY 2004; DOODY 2010; JONES-LEWIS 2012: 51–74; LAEHN, 2013: 60–2.

[12] The literary topic of *laudes Italiae* is also rooted in the general idea that climate and environment greatly influence physical and moral characters among populations. A pertinent interpretation of the environmental determinism in ancient culture is in FEDELI 1990: 17–27. A useful overview of sources is in PANESSA 1991: 123–53; FEDELI 1997: 317–30.

[13] See esp. GRIMAL 1986: 239–49, who identifies various philosophical influences and not only stoic principles; SALLMANN 1986: 251–66.

[14] BEAGON 1992: 26–8, 55–9; BEAGON 1996: 284–389.

[15] See also FRENCH 1994: 206–8; MALAVOLTA 2013: 93–115.

[16] Biographical outlines in CITRONI MARCHETTI 2011: 100–24.

2. A Manifesto

A long list of contents composes the entirety of the first book of the *Naturalis Historia*. In the opening section we can roughly identify the main subjects of the corpus and distinguish the detailed topics that develop throughout the following thirty-six books. The compendium encompasses particular themes, but observation and analysis proceed from a general framework towards gradually restricted episodes by an increasing focus on the single elements of the world. In the general structure of the *Naturalis Historia* this is a matter of fact and undermines the idea of a catalogue uncritically assembled. Indeed, the general description of the Sky and the Universe (*Cosmos*) deliberately and programmatically shifts to the Earth and her physical conformation (books 2–6 about Cosmology and Geography). Subsequently, the lens focuses on the natural environment and reviews all the living beings and elements on the planet. Firstly, the attention is drawn to all living organisms and their characteristics (books 7–11 about anthropology, ethnography, and zoology). Nevertheless, specific hierarchy is evident between man, animals, and plants. Secondly, in the general development the account of the specific distinction between natural species (books 12–16 about botany) and farmed plants (books 17–19, trees, grains, vegetables, and herbs) is noteworthy; especially that these species need particular care and cultivation techniques (sowing, grafting, and transplanting) in order to harvest edible products.

The following part is the result of the common medical and pharmacological knowledge at the time of Pliny.[17] The subject implies accurate scrutiny of remedies from plants and natural poisons (books 20–28). Remedies from animals and their characteristics are specifically the focus of books 29–32. Metals, stones, art, and handicraft are the object of books 33–37.

In the general ratios of the work, book 18th is the longest one and, more precisely, represents a crucial cross-point. It is not by chance, we can infer, that this book at the middle of the whole *Naturalis Historia* encompasses vegetal species farmed by man. The aims are not simply related to the compilation of an exhaustive catalogue. Accurate description of various species of grains, quality of soils, and agricultural techniques involves different topics of botany. Nevertheless, beyond the main intention of analysing cereals, their farming, and seasonal agrarian works, the book opens with a very penetrating insight into the environment from an ecological point of view. Especially since this book extensively describes the crucial problem of the human impact on nature, primarily via agriculture related to men's own survival. The manipulation of natural products, which is at the core of the section, inevitably produces specific effects on the environment; this is a fundamental topic, which pervades book 18 and underpins the following books of the *encyclopaedia*. Pliny's statements undeniably appear to be very close to modern environmental perception.

The introductive passages stimulate specific examination.

> And in this section it is our pleasant duty first of all to champion Earth's cause and to support her as the parent of all things, although we have already pleaded her defence in the opening part of this treatise (in 2.154–7). Nevertheless, now that our subject itself brings us to consider her also as the producer of noxious objects, they are our own crimes with which we charge her and our own faults which we impute to her. She has engendered poisons – but who discovered them except man? Birds

---

[17] See STANNARD 1986: 95–106.

and beasts are content merely to avoid them and keep away from them. (…) yet which of them (*scil.* the animals), excepting man also dips its weapons in poison? As for us, we even poison our arrows, and add to the destructive properties of iron itself; we dye even the rivers and the elemental substances of Nature, (*flumina inficimus et rerum naturae elementa*) and turn the very means of life into a bane. (…) Nor does any creature save man fight with poison borrowed from another (*veneno pugnatur alieno*). Let us therefore confess our guilt, we who are not content even with natural products, inasmuch as how far more numerous are the varieties of them made by the human hand![18]

The first part of the treatise (books 2–17) illustrates the world, its definite components, and natural characters. By contrast, this opening of the second section (books 18–37) introduces fundamental elements and concerns the world and nature manipulated by human action. Book 17 also concerns farmed trees, but it is consistently related to the previous topic on wild trees, as it is developed in the book 16. We can distinguish a proper manifesto precisely in this preface of book 18. A new aspect of the general topic is introduced after the long excursus that describes the natural environment of the world. Pliny urges us to act in defence of the Earth and as her advocate. The idea that Earth is 'the common parent of all' is not a new one in the Greek and Roman cultural tradition. In Pliny though, this literary topic becomes essential to highlight that the Earth herself produces essences in a variety of species and, consequently, several poisonous and even lethal substances are products of nature. Clearly, the Earth sounds in this case as being synonymous with Mother Nature. It is a sort of paradox, then, that men usually charge her with their crimes, and ascribe to her a guilt for which they are uniquely responsible. Man is the only living creature to be capable in using natural poisons for evil purposes. Even if we try to discharge our responsibility on Nature herself, in Pliny's view humankind is guilty of authentic crimes against Nature, environmental offences in current terminology. Amongst all the animals, man is the only living being to dip his weapons and arrows into natural poisons. Moreover, he also pollutes waters by using lethal substances; he adds destructive power to iron itself and infects various natural elements. Even the air, the main support of life, by human action becomes a tool of destruction (*in perniciem vertimus*). It is true that there are several natural poisons, but human hand prepares artificially by far the greater amount of lethal venoms.[19]

This picture looks like a post-industrial revolution one, and there is no need to underline how close it is to modern concerns for the environment. Nonetheless, beyond a definite moral attitude and evident didactical intent, this introduction appears like a real declaration of principle. Some specific elements have to be emphasized in it, and it is even possible to pinpoint them in a concise way.

---

[18] Plin., *nat. hist.*, 18.1: *qua in parte primum omnium patrocinari terrae et adesse cunctorum parenti iuvat, quamquam inter initia operis defensae. quoniam tamen ipsa materia accedimus ad reputationem eiusdem parientis et noxia: nostris eam criminibus urguemus nostramque culpam illi inputamus. genuit venena. set quis invenit illa praeter hominem? cavere ac refugere alitibus ferisque satis est.* (…) *quod tamen eorum excepto homine et tela sua venenis tinguit nos et sagittas tinguimus ac ferro ipsi nocentius aliquid damus, nos et flumina inficimus et rerum naturae elementa, ipsumque quo vivitur in perniciem vertimus.* (…) *nec ab ullo praeter hominem veneno pugnatur alieno. fateamur ergo culpam ne iis quidem, quae nascuntur, contenti; etenim quanto plura eorum genera humana manu fiunt!*

[19] See also on this introduction MAZZARINO 1970: 643–5; FEDELI 1990: 69–70.

(a) Man is responsible for manipulation and perverse usage of natural poisons.

(b) He is also responsible for the pollution of natural elements such as waters, air, soils, and the Earth herself.

(c) He manipulates Nature and the surrounding world; this also implies depletion of productivity and impoverishment of natural resources.

These concepts are not simple assertions in the preface of book 18. Indeed, they are fully developed in all books of the *Naturalis Historia* and supported by examples, which the author derived from practical observation. Pliny states his conception of the complex relationship between man and Nature. First of all, he introduces the fundamental difference between natural products (i. e. elements, substances, and materials existing in nature) and artificial products created by human hand. This simple distinction is based on empirical evidence but, inevitably, reveals implicit acknowledgement of the human impact on the environment, firstly via agriculture. Especially in agriculture, among human activities, seeds, the Earth herself, and 'all things have a limited period of fertility'; natural resources, then, are subjected to depletion.[20]

Likewise, man's ability to exploit even poisoning natural substances is part of a very delicate and precarious balance constantly at risk. In Pliny's words a specific consciousness about the concatenation of cause and effect in nature is evident; a chain of different phenomena is produced by human activity and by man's interaction with the natural environment. Apparently the tone of these introductive allegations to the whole book sounds general. Nevertheless, throughout the first and second section of the *Naturalis Historia* many cross-references concern practical examples, which strongly support his statements. Indeed, Pliny's style always proceeds through general assertions immediately followed by, and rooted in, tangible cases, which are observed and studied on a daily basis. It is important, then, to detect precisely some detailed examples and topics related to the main points of this manifesto together with assessments of other sources. Specific subjects may be privileged in the following order and scheme: (a) poisoning substances and their usage; (b) pollution of waters and natural elements; (c) exploitation and depletion of natural resources.

3. *Veneno pugnatur alieno*: poisonous substances and their usage

A whole section of the *Digesta* concerns poisoning. It regulates specific obnoxious charges caused by malpractice, ignorance, and premeditated intent of murder.[21] We can easily infer which specific social behaviours were in the background of the legal doctrine during the imperial period, and which were related to criminal offences by poisoning. Poisoning was a quite frequent phenomenon, either in common treatments of various diseases or in deliberate attempts against the life of obnoxious opponents. Lethal essences, after all, usually were the appropriate and inevitable items in the medical supply of kings and royal courts. Cato the Younger, for example, is said to be involved in a scandalous affaire, in trying to sell the cantharides supply of king Ptolemy of Cyprus at the price of 60.000 sesterces, in an auc-

---

[20] Plin., *nat. hist.* 18.195: *etenim omnium definita generatio est.*

[21] See Marc., *l.* 14 *inst.*: D. 48.8.1–3.

tion of which he was supervising.[22] In addition, king Mithridates VI Eupator of Pontos and queen Cleopatra VII of Egypt were also infamous for their expertise and deep knowledge in poisons. Mithridates also developed amazing resilience to venoms and created the *mithridatium*, the celebrated antidote with his name.[23] Apart from the famous king of Pontos, Pliny also refers of some populations of North Africa (*Psylli*) and central Italy (*Marsi*) who were renowned because they were immune to frogs and snakes' poisons.[24]

In this scenario, then, no wonder whether some norms of Marcianus clearly specify that the *lex Cornelia de sicariis et veneficiis* prosecuted anyone who prepared and sold poisons with the aim of killing other people. In addition to this law, the jurist also refers to two *senatus consulta*, which covered the same topic. Specific distinction is between 'baneful and not baneful poisons' (*mala venena* and *non mala venena*), as some of them have pharmacological properties and are essential to treat infections, but others can cause death. Obviously, in this concise picture Marcianus lists some of the most dangerous substances that dealers in unguents and paint-sellers usually manufactured. These venoms were among the most notorious of the time: hemlock, salamander, aconite, pine processionary moth, buprestis, mandrake, and cantharides.[25]

Depending on dosage and mixture, these substances have ambivalent usage, which may be both poisonous and therapeutic and it is accurately scrutinised in the whole corpus. The review of natural poisons and their lethal effects, as well as the curative and pharmacological strengths of some among these venoms, encompass several books of the *Naturalis Historia*. Specific examples are very insightful for our topic aimed at investigating the ancient roots of environmental consciousness. Pliny mentions these toxic plants and insects in various contexts. Precisely, by reason that the evil ability of killing via the usage of poisons is part of human nature, as well as the ability of creating new even worse venoms, Pliny's main intent is to give specific warnings and practical suggestions for the correct use of toxic substances in different daily life situations. For instance, poisoning caused by ingestion of extracts from salamander, pine processionary moth, buprestis, and cantharides can be treated either with goats' milk or white must, natural detoxifying agents which induce vomit.[26] The efficacy of the suggested remedies is questionable in many examples, but it will not be under enquiry here. Yet, man's responsibility for the manipulation of these lethal substances is an underlying element, as well as the empirical usage of poisons and their side effects, which are common threads that often interlink all the items of this allegedly botanical catalogue. Hellebore and *limeum* are toxic herbs, for example, but Pliny largely mentions their curative proper-

---

[22] It happened in 58 BC: Plin., *nat. hist.* 29.96; Sen. the Elder, *contr.* 6.4.3.

[23] Mithridates: Plin., *nat. hist.* 23.149; 25.5–7; 29.24–5; Cleopatra: Plin., *nat. hist.* 9.119–21; 21.12. On this specific topic see the recent works of MAYOR 2009; ROLLER DUANE 2010; BALLESTEROS PASTOR 2013.

[24] Plin., *nat. hist.* 7.14–5; 21.78; 25.123.

[25] Marc., *l.* 14 *inst.*: D. 48.8.3.3: *Alio senatus consulto effectum est, ut pigmentarii, si cui temere cicutam salamandram aconitum pituocampas aut bubrostim mandragoram et id, quod lustramenti causa dederit cantharidas, poena teneantur huius legis.* It is laid down by another *senatus consultum* that dealers in cosmetics are liable to the penalty of this law if they recklessly hand over to anyone hemlock, salamander, monkshood, pinegrubs, or a venomous beetle, mandragora, or, except for the purpose of purification, Spanish fly. (A. WATSON, Philadelphia 1985). On this passage see RIVES 2006: 47–67.

[26] Plin., *nat. hist.* 23.62; 28.128; also Celsus 5.27.12a; Scrib. Lon. 189. CAPITANI 1972: 120–40.

ties. Equally, he specifies the traditional and common practice amongst Gauls of dipping their weapons into these lethal poisons, especially to cause their enemies' and prey's imme-diate death in war and hunting.[27] The Scythians' habit of using a mixture of snake poison and human blood to taint their arrows is even worse and by far much more lethal.[28] There is no remedy for this venom, Pliny states, and death is instantaneous – even for those who are barely touched by it. These are patently not the only examples of causing human death by poisoning. Pliny extensively also describes the qualities and effects of hemlock, which notoriously was also the hideous drink for capital punishment in classical Athens.[29] Usage of natural poisons, then, and production of new lethal mixtures is a peculiar characteristic of man in the course of fighting his own species.

In his analysis of various poisons, though, a fundamental element is detectable espe-cially in the description of certain toxic honeys, which are farmed in North Africa and in some areas of the Black Sea. Biodiversity is a key element and Pliny meaningfully highlights the close relation between the natural environment and the specific flowers' nectar that the bees extract. These are crucial components related to the toxicity of those honeys and are precisely identified in a peculiar causal connection. The nectar of *aegolethron*, a plant which is to be identified perhaps with the Pontic rhododendron, as well as other unidentified es-sences (perhaps rose laurel, oleander, and azalea Pontica) are the main toxic ingredients of the poisoning honeys produced in Pontus, Persia, and in Mauretania Caesariensis. In the mentioned Black Sea woodland habitat (*situs*), specifically in the country of the *Sanni*, 'maddening' honey is also produced from rhododendron, which grows abundantly in those woods (*silvae*) and causes madness.[30] An incisive consideration closes these passages on exotic poisonous honeys. Why does Nature lay such kinds of traps and give us poisonous honey, which is produced by the bees themselves? This question is even more relevant, since Pliny strongly emphasises the role of the bees in the natural environment and praises the amazingly organised structure of their community in book 11 (chapters 11–29). The mo-tivation about the natural production of the poisoning honey can only be one: to render human beings more cautious and less greedy on the exploitation and depletion of natural resources (21.78). This answer only, is marked by apparent moralism. It reveals, by contrast, faith in the existence of a superior, general and natural order, which must be kept in balance and preserved. In other words, it is rooted in the principle that man should constantly be environmentally aware that *modum iudicem rerum omnium utilissimum*, moderation is the most valuable criterion of all things.[31] In the broad-spectrum ratio of the *encyclopedia* this thought also recalls the numerous examples related to man's falsifications of precious, ex-otic, and expensive essences which are part of the luxury market. Frauds of faked substances and *furta* for selling them at dishonest prices are topics of the twelfth book, and they are negative evidence of the dangerous and inappropriate exploitation of supplies, which, by

---

[27] Hellebore: Plin, *nat. hist.* 25.61; Pliny is the only author to mention *limeum* that has not been iden-tified by scholars: *nat. hist.* 27.101.

[28] Plin., *nat. hist.* 11.279.

[29] Plin., *nat. hist.* 25.151–4.

[30] Plin., *nat. hist.* 21.74–8.

[31] Plin., *nat. hist.* 18. 38; the concept is also very close to Hor., *sat.*, 1.1.106–7: *est modus in rebus*, there is a proper measure in all things.

contrast, should be preserved.[32] In this continuous line of thought, then, poisonous honey becomes a reminder and stimulus of environmental responsibility in the usage of natural resources and part of the general agenda, which is developed across the first and second section of the *Naturalis Historia*.

4. *Flumina inficimus et rerum naturae elementa*: pollution of waters and natural elements

Manipulation and production of poisons, together with waste disposal, affect inevitably water streams and springs. Besides the building of channel-networks, aqueducts, and sewers in the most important cities, Roman authorities were constantly concerned with public *salubritas* largely dependent on the purity of rivers and springs. A conspicuous amount of imperial edicts and local regulations ruled on the management of rivers and rainwaters, reclamation of swamps and marshes, and prevented streams' contamination. Public health and prevention of epidemics and plagues were also good tools for keeping public order. The evidence on this subject is sizeable and embodies inscriptions, legal provisions by emperors and local authorities, as well as specific discussions in literary sources.[33] It is worth noting, for example, the massive construction of a channel in the suburban area of Antiochia in 73–74 AD. The main aim, apparently, was to move the *fullonicae* and the polluting activity of the textile industry to the outskirts. Clothes and textile manufacture were the fundamental economic production of the city, in support of which a huge amount of water, mainly contaminated by ammonia, sulphur, earths, and various minerals was usually drained into the stream of the Orontes.[34] Several groups of workers – members of the *plintheia* – which probably were the local *collegia fabrorum* resident in the urban quarters, were in charge of building the new drainage system, the memory of which was recorded in many similar inscriptions.[35]

Further aspects concerning water and its management also are in Vitruvius, Frontinus, and Pliny the Younger, who testify how important the danger of water pollution was and how the central government tackled the problem. An interesting exchange of letters, for instance, between the emperor Trajan and the governor of Bithynia Pliny the Younger refers about some environmental problems in the Asian city of Amastris. The case of Amastris is highlighting about the concern of public administration to prevent the side effects which might have been produced by an open sewer. In that beautiful city of Asia Minor, the river that flowed close to the central street was transformed into a sewer by the waste. The gov-

---

[32] Especially in *nat. hist.* book 12: 28; 31; 36; 43; 46; 49; 65; 70–72; 76; 98; 107; 119; 121; 125. On this topic see CHEVALLIER 1986: 148–72.

[33] Legal evidence: Ulp., *l.* 17 *ad edic.*: *D.* 8.5.8.5 (discharge of smoke); Ulp., *l.* 81 *ad edic.*: *D.* 39.3.30; Ulp., *l.* 43 *ad edic.*: *D.* 39.3.1.1–5; 43.8.2.26–8 (management and liability for damages of rain- and drainage-waters); Ulp., *l.* 68 *ad edic.*: *D.* 43.8.2.8; 43.12.1.17–22 (illegal buildings on rivers and seashores). Literary and material evidence: ANDRÈ 1986: 45–52; FEDELI 1990: 66–72; FEDELI 1997: 317–30; PANCIERA 2000: 95–105; DE FEO, DE GISI, and HUNTER, 2014: 251–68.

[34] The technical process for colouration of textiles is described by Plin, *nat. hist.* 35.196–8. He also mentions a *lex Metilia de fullonibus* (third cent. BC), which regulated this activity. See WALLINGA 1996: 183–90.

[35] *SEG* 35, 1483; *AE* 1986, 694. FEISSEL 1985: 77–103; VAN NIJF 1997: 80–91.

ernor Pliny the Younger reported to Trajan the conditions of the city centre and asked for prudent arrangements to preserve public health. What followed was Trajan's prompt reply, who endorsed fast action.

> Gaius Plinius to the Emperor Trajan. Among the chief features of Amastris, Sir, (a city which is well built and laid out) is a long street of great beauty. Throughout the length of this, however, there runs what is called a stream, but is in fact a filthy sewer, a disgusting eyesore which gives off a noxious stench. The health and appearance alike of the city will benefit if it is covered in, and with your permission this shall be done. I will see that money is not lacking for a large-scale work of such importance.
>
> Trajan to Pliny. There is every reason, my dear Pliny, to cover the water which you say flows through the city of Amastris, if it is a danger to health while it remains uncovered. I am sure you will be active as always to ensure that there is no lack of money for this work.[36]

This evidence can also be combined with specific legal regulations referred by Ulpian. Distinction is made in Roman law between public and private sewers, the connection of which was not mandatory, rather more usually missing. Reparation, cleaning, and management of these private and public drainage-systems devolved respectively upon public administrators and private individuals.[37] During the third century, at Ulpian's time, the administrative practice was strongly consolidated by legal tradition and the lawyer clearly specified these juridical principles:

> The praetor has taken care by means of these interdicts for the cleaning and the repair of drains. Both pertain to the health of *civitates* and to safety. For drains choked with filth threaten pestilence of the atmosphere and ruin, if they are not repaired.[38]

Apart from this practical approach and habits of good administrative organization, contamination of waters is a relevant topic in literary sources and technical literature too.

Although it was not always a straightforward link between sanitation and public health of communities taken as a whole, in medical literature especially there was notion that environmental conditions could develop human diseases in single individuals. Latin language distinguishes between collective *salus* and individual *valetudo*; yet measures to preserve public environmental *salubritas* (according to diverse forms of sanitation) were very poor and neither were they the exclusive duty of a fully-aware public authority. *Stercorarii* and

---

[36] Plin. Iun., *ep.* 10.98–99: *C. Plinius Traiano Imperatori. Amastrianorum civitas, domine, et elegans et ornata habet inter praecipua opera pulcherrimam eandemque longissimam plateam; cuius a latere per spatium omne porrigitur nomine quidem flumen, re vera cloaca foedissima, ac sicut turpis immundissimo aspectu, ita pestilens odore taeterrimo. Quibus ex causis non minus salubritatis quam decoris interest eam contegi; quod fiet si permiseris curantibus nobis, ne desit quoque pecunia operi tam magno quam necessario.*
Traianus Plinio. Rationis est, mi Secunde carissime, contegi aquam istam, quae per civitatem Amastrianorum fluit, si intecta salubritati obest. Pecunia ne huic operi desit, curaturum te secundum diligentiam tuam certum habeo. (Engl. transl. by B. Radice, Cambridge MA 1969).

[37] Ulp., *ad edic. l.* 71: *D.* 43.23.1.3.

[38] Ulp., *ad edic. l.* 71: *D.* 43.23.1.2: *Curavit autem praetor per haec interdicta, ut cloacae et purgentur et reficiantur, quorum utrumque et ad salubritatem civitatium et ad tutelam pertinet: nam et caelum pestilens et ruinas minantur immunditiae cloacarum, si non reficiantur.*

foricarii were private workers who usually were in charge of empting latrines, cesspits, sewers, and urinals for a fee. At the end of the first century AD, Rome with her extremely dense population of about one million inhabitants was notoriously famous for being extremely dirty and smelly from the various miasmas in the air and soil. We can infer from literary evidence that it was a pretty normal practice, for example, to leave unburied corpses and animals' carcasses on the streets, to clean chamber pots on public roads, even if sewer networks and cesspits were supposed to remove domestic wastewater, rainwater, and surplus water from fountains and public bathrooms.[39] Smoke coming from the combustion of wood and charcoal (the only available material for heating, cooking, and any artisanal activity at that time) seriously degraded the quality of air.[40] Moreover, a specific concern was the pollution of waters by animal blood after the frequent ritual practices of sacrifices in the temples and sanctuaries of urban areas. Likewise, even human blood from battles was a well-known element of water contamination and pollution was clearly perceived as the effect of man's violence against Nature. There is very vivid evidence of this in many ancient authors, and Pliny certainly remembers the passage of Cicero in *Tusculan Disputations* when he refers to Darius' declaration that he drank the contaminated water of a river full of blood and corpses and greatly enjoyed it.[41] In a very detailed chapter, Pliny illustrates the 'powers of iron' which might effectively be defined as part of the specific 'iron economy' in Roman imperialistic society. The power of iron is stronger than gold, claims Pliny, since it feeds wars and massacres. It is a precious natural gift that man employs for domestic and working tools. Nevertheless, iron also has the power to pollute rivers and springs by the blood of thousands of soldiers during wars.[42]

In these sort of environmental conditions, no wonder that members of the aristocracy very often chose to live in luxurious villas in the countryside, since they commonly shared the idea, also rooted in medical literature, that miasmas from sewers, putrefying corpses, and carcasses, as well as stagnant waters, were deleterious for public health.[43] Frontinus refers that the emperor Nerva was trying to improve the quality of water and air in the *Urbs* by a series of different public works, which were aimed mainly at sanitation and environmental cleansing. The urban pollution seemed greatly reduced and there were good possibilities to

---

[39] Peculiar descriptions are in Suet., *Nero* 48; *Vesp.*, 5.4; Mart., *ep.* 10.5.11; Cass. Dio. 65.1; Petr. 134.1. See Panciera 2000: 95–105; Davies 2012: 67–80; De Feo, De Gisi, and Hunter 2014: 251–68; Vuorinen 2014: 429–36.

[40] Meaningful evidence in Ulp., *ad edic. l.* 17: *D.* 8.5.8.5–6.

[41] Pollution from sacrificial blood: Hor., *c.* 3.13, ll. 6–8; Plin., *nat. hist.* 31.55; Strabo, 6.2.4. The topographical and religious function of the *Cloaca Maxima* at the junctions with temples in Rome is highlighted by Hopkins 2012: 81–102. Pollution from human blood after battles in war: Cic., *tusc. disp.* 5.97; Sil. It., *pun.* 1.42–54.

[42] Plin., *nat. hist.* 34.138; Veg., *mil.* 3.2. See also Santini 1983: 88–90.

[43] From the second century AD onwards Galen's thesis had a very large echo. See Galen, *de san. tuen.* 1.11.15–18, *CMG* 5.4.2: whatever air is defiled from any sewer of those draining any large city or populous camp is harmful; and harmful too is that which is contaminated from any putrefaction of animals and vegetables or oils or manure; and it is not good which is cloudy from a neighbouring river or swamp; and likewise that which, in a hollow place, surrounded on all sides by lofty mountains, receives no breeze, for such air is stifling and foul, like that in certain closed houses in which it collects from putrefaction and lack of ventilation. Such air is injuriouus at all ages. (Engl. transl. Montraville Green 1951: 35–36).

eradicate it, but Frontinus, beyond any allegation of biased propaganda, didn't want to give many details before the effective achievements of all the works.

> Thus from day to day Rome, queen and mistress of the world, perceives the watchful care of her Emperor Nerva, and the wholesome environment of this same eternal City will be perceived all the more by an increased number of delivery-tanks, public works, *munera*, and basins. No smaller is the benefit for private parties which results from an increase in the number of grants made by the same emperor's favor. Those who once ran risks by drawing water illicitly are now untroubled because they enjoy it by right of official grants. Not even overflow waters are without useful purpose. There is now an entirely fresh guise of cleanliness and a cleaner air; gone are causes of unhealthy climate which often gave our City a poor reputation in former times. I am aware that this booklet should incorporate data which reflect the new pattern of delivery, but these we shall append along with those for the additional supply: one can understand, of course, that new figures are not to be set down until the various projects are completed.[44]

We do not know if Nerva also devised a proper centralized waste collection together with these public works and cleaning measures; apparently not. This evidence of Frontinus clearly is the arrival point of a longer tradition in terms of awareness and knowledge of polluting factors in the environment, which may go back at the least to the Hippocratic treaty *On Airs, Waters, and Places*. It is not my intention here to engage in the debate and analysis of environmental determinism, which largely characterized ancient sources and strongly influenced literary, technical, and medical writing. Yet it is noteworthy that even in the cultural and philosophical background of environmental determinism, objective elements of water pollution are clearly detectable in Roman technical literature. It may be unexpected by modern sensibilities that ancient authors also had some notion of natural bio-monitors, which usually mark polluted habitats, especially waters. On the basis of empirical observation, Vitruvius and Pliny in particular, realized that some species of plants and animals perform the function of bio-indicators. The presence of certain organisms in a certain habitat is a sign of their crucial role in revealing the health of specific ecosystems, specifically the purity of waters in springs, rivers, and ponds. Vitruvius states that some plants, mainly moss and reeds, if they grow near water streams and ponds, give incontrovertible evidence of pollution.[45] Pliny is even more precise in describing specific signs of water pollution and in fixing criteria which

---

[44] Front., *de aquaeduc.* 88: *Sentit hanc curam imperatoris piissimi Nervae principis sui regina et domina orbis in dies et magis sentiet salubritas eiusdem aucto castellorum, operum, munerum et lacuum numero. Nec minus ad privatos commodum ex incremento beneficiorum eius diffunditur; illi quoque qui timidi inlicitam aquam ducebant, securi nunc ex beneficiis fruuntur. Ne pereuntes quidem aquae otiosae sunt: alia munditiarum facies, purior spiritus, et causae gravioris caeli quibus apud veteres urbis infamis aer fuit, sunt remotae. Non praeterit me, deberi operi novae erogationis ordinationem; sed haec cum incremento adiunxerimus; intellegi oportet, non esse ea ponenda nisi consummata fuerint.* (Engl. transl. R. H. RODGERS, Cambridge 2003).

[45] Vitr., *de arch.* 8.4.2: *non minus etiam ipsa aqua quae erit in fonte si fuerit limpida et perlucida, quoque pervenerit aut profluxerit muscus non nascetur neque iuncus, neque inquinatus ab aliquo inquinamento is locus fuerit sed puram habuerit speciem, innuitur his signis esse tenuis et in summa salubritate.* Likewise, if the water itself in the spring is limpid and transparent and if wherever it comes or passes, neither moss nor reeds grow nor is the place defiled by any filth, but maintains a clear appearance, the water is indicated by these signs to be light and most wholesome. (Engl. transl. by F. GRANGER, Cambridge MA 1934). See also CAMPBELL 2012: 338–40.

are useful to distinguish the purity of water. Book 31 is entirely dedicated to watery habitats;
it encompasses an excursus of several detailed chapters on different kinds of waters and their
organoleptic characteristics. Mud especially reveals contamination. By contrast, if a river is
full of eels and pinworms, these are symptomatic of healthiness and freshness.[46] Another
good general rule to distinguish pure drinkable water is that it should neither taste nor smell;
it should be like air.[47] These suggestions conform to the general, practical, and pedagogical
aim of the compendium. Nevertheless, in the same section of the book, Pliny also engages
with the medical debate concerning qualities and the purity of different waters. In particular
he discusses the healthiness of rainwater and, surprisingly, directs his conclusions against
the *communis opinio* of physicians. Despite the view of those professionals, who are often
criticised by Pliny as being dishonest charlatans, his opinion reveals the veracity of scientific
principles.[48] Pliny is very critical about the assertion of disreputable practitioners who recommend drinking water from cisterns, since coming from rain it is supposed to be the lightest
and most pure water. His arguments are well documented and clearly articulated.

> It is a question debated by the physicians what kinds of water are most beneficial. They rightly
> condemn stagnant and sluggish waters, holding that running water is more beneficial, as it is made
> finer and more healthy by the mere agitation of the current. For this reason I am surprised that
> some physicians recommend highly water from cisterns. But these physicians put forward a reason;
> the lightest water, they say, is rain-water, seeing that it has been able to rise and to be suspended in
> the atmosphere. To refute this view is a matter of fact that is important to all men. For first of all,
> this lightness of water can be discovered with difficulty except by sensation, as the kinds of water
> differ practically nothing in weight. Nor is it proof of the lightness of rain water that it rose to the
> sky, since even stones are seen to do the same, and as it falls it is infected with exhalations from the
> earth. Hence it comes about that rain water is found to be full of dirt, for which reason this water
> becomes hot very quickly. (…) Rainwater, it is agreed becomes putrid very quickly, and it is the
> worst water to stand a voyage. (…) Yet it must be admitted, they hold, that river water is not ipso
> facto the most wholesome, nor yet that of any torrent whatsoever, while there are very many lakes
> that are wholesome. What water then, and of what kind, is best? It varies with the locality.[49]

---

46 Plin., *nat. hist.* 31.36: *limus aquarum vitium est. Si tamen idem amnis anguillis scateat, sallubritatis
   indicium habetur, sicuti frigoris taenias in fonte gigni.* Slime in water is bad. If however the same
   river is full of eels, it is held to be a sign of wholesomeness, as it is of coldness for worms to breed
   in a spring. See also ANDRÉ 1986: 45–52.

47 Plin., *nat. hist.* 31.37: *est etiamnum vitium non fetidae modo verum omnino quicquam resipientis
   (…). aquam salubrem aeris quam simillimam esse oportet. (…) de cetero aquarum salubrium sapor
   odorve nullus esse debet.* Not only too is fetid water bad, but also that which tastes of anything at all
   (…). Wholesome water ought to be very like air. (…) Apart from this, wholesome water should
   have no sort of taste or smell.

48 Polemic passages against physicians also in: *nat. hist.* 29.25–7; 33.124.

49 Plin., *nat. hist.* 31.31: *Quaeritur inter medicos cuius generis aquae sint utilissimae. stagnantes pigra-
   sque merito damnant, utiliores quae profluunt existimantes, cursu enim percussuque ipso extenuari
   atque proficere, eoque miror cisternarum ab aliquis maxime probari. sed hi rationem adferunt, quoniam levissima sit imbrium, ut quae subire potuerit ac pendere in aere. (…) horum sententiam refelli
   interest vitae. in primis enim levitas illa deprehendi aliter quam sensu vix potest, nullo paene momento
   ponderis aquis inter se distantibus. nec levitatis in pluvia aqua argumentum est subisse eam in caelum,
   cum etiam lapides subire appareat cadensque inficiatur halitu terrae, quo fit ut pluviae aquae sordium
   plurimum inesse sentiatur citissimeque ideo calefiat aqua pluvia. (…) pluvias quidem aquas celerrime
   putrescere convenit minimeque durare in navigatione. (…) Item confitendum habent nec statim am-*

This passage is rooted in the authority of Greek medicine, following especially in the footsteps of Theophrastus and Hippocrates.[50] Pliny wonders, then, which kind of water among so many is the best and *saluber*. The answer is clear and, quite surprisingly, conscious of the natural process involved in the regeneration of water. Pliny has no doubt that it depends on locality, but usually the best water is well-water. This conclusion, though, is only in the case of wells in which water is kept in constant agitation by repeated drawing, and when the well-waters are 'those where due thinness is obtained by filtering through the earth', *illa tenuitas colante terra* (31.23). More importantly, he is aware of the various elements that can affect purity of water, and recognizes that the earth and deep rocks, in particular, act as natural filters. Modern chemistry and geological sciences, indeed, taught us that purest water flows through rocks and is from deep groundwater aquifer. Pliny, therefore, also distinguishes the reasons by which rain-water is rightly among the most polluted, since it is not filtered by rocks and sand and it lacks that fundamental chemical and physical process of filtration which develops purity.[51] Obviously, at his time there was no notion of water molecules, which react with polluting agents in the atmosphere and can become the notorious phenomenon of acid rain. Nevertheless, such understanding of the evaporation cycle and rainfall is noteworthy. It is remarkable intuition to conclude that exhalations and polluting agents from the earth affect purity of water, especially rain water during the natural cycle of evaporation from the ground, in the air and, again, when it turns into new rainfalls.

5. *Didicit homo naturam provocare:* building materials, metals, exploitation and depletion of natural resources

Contamination of air and waters, production of garbage and any kind of waste are only some of the noxious human activities in the environment. Violence against Nature and her internal order is also pursued by irrational exploitation and depletion of resources. It is fascinating to detect in which terms ancient intellectuals stressed the environmental dangers related to irresponsible human behaviour. Several references and grievances, for example, are directed against the upper-class' fashion of building sumptuous villas on the sea, facilities in which (baths, porticoes, and even *piscinae* for fish farming) often intruded into seawaters and altered the coastline. The topic receives particularly vibrant descriptions in Sallust and Varro's prose, in the writings of Valerius Maximus, as well as in Seneca.[52] This literary tradition in the spirit of moralistic invectives against plutocrats is clearly in the cultural background of Pliny's compendium. Moreover, a straight intellectual connection can also be detected between Horace's elegant poetics and Pliny's technical writings, especially in the last books of the *Naturalis Historia*. Apart from the shared moralistic commitment against any excess of luxury in imperial Rome, both of them reveal understanding of the environmental damages that may be involved in depletion of natural resources. For Horace and Pliny Na-

---

nium *utilissimas esse, sicuti nec torrentium ullius, lacusque plurimos salubres. Quaenam igitur et cuius generis aptissimae? Aliae alibi.*

[50] See Athen., *deipn.* 2.41 f.; 2.46d.

[51] *Contra:* Vitr., *de arch.* 8.2.1, who claims that the most *saluber* and finest water is rain-water.

[52] Varro, *de re rus.* 3.17.9; Sall., *Cat.* 13.1–2; 20.11; Val. Max., *memor.* 9.1; Sen., *ep.* 89.21. See also FEDELI 1990: 45, 55; FEDELI 1997: 327–8.

ture and natural order, often impiously subverted, are main key elements in the texture of ethical discourses against lavishness. There is no doubt that Horace is part of Pliny's cultural substratum, which enacts and intensifies claims in favour of Nature. A certain intellectual disdain and sarcastic detachment characterize Horace's point of view, when he refers to environmental problems in some of the *Carmina* and in the *Satyrae*. Environmental themes are various and imply certain ecological considerations regarding uncontrolled building, depletion of natural resources by the exploitation of mining and marble extraction, as well as the drastic reduction of arable land. In the beautiful poem to the *fons Bandusiae*, Horace blames the common fashion within the ruling classes of building lavish villas on the shore line. Their basements consist of huge rocky and concrete platforms (*caementa*) that human rapacity outrageously throws amongst the sea waves. Fishes have perception of the narrowing of the sea streams, when a wealthy *dominus*, disdainful (*fastidiosus*) for dry land, supervises the work of his servants and fills the sea depths.[53] This realistic and pretty topical image may also go in hand with the theme of another poem on the vacuity of human life, when it focuses excessively on accumulation of natural precious gifts (ivory, gold, Hymettian and African marbles). The cut of marbles should be a memento of graves, but their usage is mainly for building houses 'to move back the coastline where the sea roars in protest at Baiae'. This selfish attitude and careless behaviour also is a patent mark of deep social inequality and distress in the management and distribution of resources. Everyone should keep in mind that death is the destiny alike for all, poor and children of kings.[54]

We find the same themes in Pliny. These are even more developed and plainly unequivocal. Exploitation and depletion of natural resources, as well as deep alteration of natural landscapes and habitats are the fundamental and linking elements in the detailed descriptions of the last section of the compendium. The passionate introductions to the books on 'Arts' (mainly in books 33 and 36) reinvigorate the 'Manifesto' of book 18. Especially on those subjects concerning metals and mining activity, marbles and quarries, mineral colours and paintings, precious stones and gems, it is even more evident the sacrilegious efforts man performs against Nature. The search for treasures in Earth's depths is a desecration, and yet we are astonished when 'occasionally she gapes open or begins to tremble'.[55] We penetrate

---

[53] Hor., *car.* 3.1.33–7: *contracta pisces aequora sentiunt iactis in altum molibus: huc frequens caementa demittit redemptor cum famulis dominusque terrae fastidiosus*. The fish feel their waters shrinking as pier after pier is pushed into the sea. The contractor with his workmen repeatedly tips in rubble; at his side is the owner who is bored with living on land. (Engl. transl. by N. Rudd, Cambridge MA 2004). See also Fedeli 1997: 327–9. Archaeological evidence can be also compared: see Gazda and Clark 2016.

[54] Hor., *car.* 2.18.17–22: *tu secanda marmora locas sub ipsum funus et sepulcri inmemor struis domos marisque Bais obstrepentis urges submovere litora, parum locuples continente ripa quid quod usque proximos revellis agri terminos et ultra limites clientium salis avarus? (…) aequa tellus pauperi recluditur regumque pueris*. But you, though in the very shadow of death, place contracts for cutting marble slabs, and build houses without giving a thought to your tomb. You press on to move back the coastline where the sea roars in protest at Baiae, for you have insufficient property as long as the shore hems you in. What of the fact that you repeatedly tear up the stones that mark your neighbour's farm, and in your greed leap over your tenants' boundaries? (…) The earth opens impartially for the poor and for the sons of princes. (Engl. transl. by N. Rudd, Cambridge MA 2004).

[55] Plin., *nat. hist.* 33.1: *persequimur omnes eius fibras vivimusque super excavatam, mirantes dehiscere aliquando aut intremescere illam, ceu vero non hoc indignatione sacrae parentis exprimi possit*. We

into her bowels not with the intent of finding new medical remedies, but only in search of treasures, the formation of which requires much processing.[56] Objects, like vases of crystal and myrrhine for instance, are enormously precious, precisely by virtue of their extreme fragility. Since the Earth does not generate these gifts in one moment, we should be aware and more thoughtful that there might be an end to all these valuable supplies at some stage in the future. The impossibility of trusting in unlimited natural resources, the necessity of distinguishing between eco-sustainable behaviours and the depletion of reserves, are here not part of an ordinary moralistic invective. These claims induce a certain style of living. The following two passages, which are so closely connected at a conceptual level, can be read as a useful comparison. Both the authors Pliny and Horace, imply ominous future visions; both suggest possible alternative choices.

The things that she has concealed and hidden underground, those that do not quickly come to birth, are the things that destroy us and drive us to the depths below; so that suddenly the mind soars aloft into the void and ponders what finally will be the end of draining her dry in all the ages, what will be the point to which avarice will penetrate?[57]	But how much better – how utterly at variance with this – is the course that nature, rich in her own resources, prompts, if you would only manage wisely, and not confound what is to be avoided with what is to be desired![58]

Behind the rhetoric, Pliny and Horace presage what we *a posteriori* know in the light of nowadays technology, statistical data, and deeper scientific knowledge. Man penetrates into the bowels of the Earth and excavates her treasures for his own ruin. This spasmodic search and exploitation of resources will exhaust the Earth. A more balanced ecological approach would be desirable, and man should be able to keep a distinction between what is avoidable and what can be pursued in Nature (*non fugienda petendis inmiscere*). Yet, greed is by far the prevailing force in man's environmental impact. Indeed, the concept of rapacity/*avaritia* summarizes and encompasses all forms of economic benefits that man pursues over the centuries. In the compendium Pliny offers great amount of detail concerning prices and quantifies precise exchange values for both authentic products and forgeries. Ancient societies, for example, had some notion of the damages for public health caused by the usage of lead in water pipes.[59] Pliny also knew about the high level of toxicity in fumes coming from extraction of *minium* (cinnabar) and quicksilver, as well as more generally in all mining

---

trace out all the fibres of the earth, and live above the hollows we have made in her, marveling that occasionally she gapes open or begins to tremble – as if forsooth it were not possible that this may be an expression of the indignation of our holy parent!

[56] The violation of the Earth is a deliberate innuendo in Pliny as well as in other ancient sources.

[57] Plin., *nat. hist.* 33.3: *illa nos peremunt, illa nos ad inferos agunt, quae occultavit atque demersit, illa, quae non nascuntur repente, ut mens ad inane evolans reputet, quae deinde futura sit finis omnibus saeculis exhauriendi eam, quo usque penetratura avaritia.* See also 2.158–9.

[58] Hor., *sat.* 1.2.74–6: *at quanto meliora monet pugnantiaque istis dives opis natura suae, tu si modo recte dispensare velis ac non fugienda petendis inmiscere.* (Engl. transl. by H. Rushton Fairclough, Cambridge MA 1926).

[59] Vitr., *de arch.* 8.6.10–1 (lead pipes and symptoms of saturnism); Plin., *nat. hist.* 34.167. Detailed description of data in Alain Bresson's contribution to this volume.

activities related to the industrial manipulation of metals.[60] Nevertheless we might infer that lead pipes were preferred to clay pipes and masonry conduits probably by reason of their lower costs for installation and maintenance. Likewise, in a society deeply rooted in slavery, economic incomes usually were more relevant than the lives of slaves and working class people, especially for wealthy owners of mines and quarries and even for the State. In the end: *didicit homo naturam provocare*, man has learned to challenge Nature in competition (33.4). The massive exploitation of building materials reduces even mountains into flat areas. The opening of book 36 is a vibrant description of the natural function of rocks and mountainous chains on the earth. Man constantly threatens heights and slopes of mountains that are depredated and transformed into quarries.

> Mountains, however, were made by Nature for herself to serve as a kind of framework for holding firmly together the inner parts of the earth, and at the same time to enable her to subdue the violence of rivers, to break the force of heavy seas and so to curb her most restless elements with the hardest material of which she is made. We quarry these mountains and haul them away for a mere whim; and yet there was a time when it seemed remarkable even to have succeeded in crossing them. (…) Now these selfsame Alps are quarried into marble of a thousand varieties.
> Headlands are laid open to the sea, and nature is flattened. We remove the barriers created to serve as the boundaries of nations, and ships are built specially for marble. And so, over the waves of the sea, Nature's wildest element, mountain ranges are transported to and fro.[61]

There is an internal balance in Nature. Mountains were created to shield the bowels of the Earth, as well as to restrain the violence of rivers and the fury of the sea-waves. They are the barrier and control upon natural, never resting, elements such as water and subterranean magma. Man deeply subverts this natural order by quarrying any kind of rocks, stones, and marbles. The hectic and restless transport of marbles 'to and fro' on ships (*naves lapidariae*) materializes a lively image and it sounds like a fairly sarcastic mockery of the human anxiety of possession.[62] This very intense invective condemns the indiscriminate usage of natural resources. At that time, there was no notion of the close interrelations, for example, between deforestation, quarry and building activities, landslides, and accelerated erosion of seashores. By reason of this, Pliny's statements and intuitions of possible natural disasters caused by reducing Nature to a single flat level sounds even more ominous. He admires the human genius that is *artifex* of wonderful masterpieces, but is deeply persuaded that that

---

[60] Mercure is a poisoning metal, but its usage was fundamental to purify gold: Vitr., *de arch.* 7.8.1; Plin., *nat. hist.* 33.99. In the extraction and manipulation of minium workers made use of some masks of loos bladder-skins to prevent inhalation in breathing: Plin., *nat. hist.* 33.122. On the exploitation of mines in the Roman world and the specific staff involved see esp. Hirt 2010: 48–82, 202–62.

[61] Plin., *nat. hist.* 36.1–2: *montes natura sibi fecerat ut quasdam compages telluris visceribus densandis, simul ad fluminum impetus domandos fluctusque frangendos ac minime quietas partes coercendas durissima sui materia. caedimus hos trahimusque nulla alia quam deliciarum causa, quos transcendisse quoque mirum fuit. (…) nunc ipsae caeduntur in mille genera marmorum. promunturia aperiuntur mari, et rerum natura agitur in planum; evehimus ea, quae separandis gentibus pro terminis constituta erant, navesque marmorum causa fiunt, ac per fluctus, saevissimam rerum naturae partem, huc illuc portantur iuga.*

[62] See also Strabo 5.2.5.222. More details and literature in Rohleder 2001: 53–135; Russel 2013: 129–31.

genius should be developed in accordance with preservation and it should be mainly characterized by measure, *modus et moderatio*, in dealing with the superb and constant miracles of Nature.[63]

***

Definitely, it is a matter of fact that the *Naturalis Historia* often provides general assessments about pollution and depletion of resources. This inclination and sensitivity cannot only be valued in terms of rhetorical and moralistic affectation. Furthermore, it is even difficult to distinguish in the corpus a prevailing ideologically oriented trend in supporting Roman imperialism. The complexity of the compendium is dense; its structure is problematic. On the basis of these fundamental characteristics, modern commentators should not reduce the *Naturalis Historia* into a unity. Pliny's statements are frequently supported by the real evidence of human action and its impact on nature. This peculiar environmental practice and observation of nature makes Pliny a *sui generis* forerunner, an environmentalist *ante litteram*. He did not have specific knowledge of the reasons which determine the accelerated sea erosion of coasts. Neither knew he the interrelation between the process of deforestation, quarry activity, and phenomena of avalanche and landslide. No notion at that time of the possible climatic consequences of the reduction of forests and melting of glaciers on the planet. These are our conquests of knowledge in the general economy of human history and human limits.

Bibliography

ANDRÈ, J.-M. 1986. 'L'épidémiologie de Pline', *Helmantica* 37, 45–52.

BALLESTEROS PASTOR, L. 2013. *Pompeyo Trogo, Justino y Mitridates: comentario al Epítome de las Historías Filípicas (37, 1,6–38, 8,1)* (Spudasmata 154). Hildesheim.

BEAGON, M. 1992. *Roman Nature: the Thought of Pliny the Elder*. Oxford.

BEAGON, M. 1996. 'Nature and views of her landscapes in Pliny the Elder', in G. SHIPLEY and J. SALMON (eds.), *Human Landscapes in Classical Antiquity*. London-New York, 284–389.

BEAGON, M. 2013. '*Labores pro bono publico*. The burdensome mission of Pliny's *Natural History*', in J. KÖNIG and G. WOOLF (eds.), *Encyclopaedism from Antiquity to the Renaissance*. Cambridge-New York, 84–107.

CAMPBELL, B. 2012. *Rivers and the Power of Ancient Rome*. Chapel Hill.

CAPITANI, U. 1972. 'Celso, Scribonio Largo, Plinio il Vecchio e il loro atteggiamento nei confronti della medicina popolare', *Maia* 24, 120–40.

CAREY, S. 2003. *Pliny's Catalogue of Culture. Art and Empire in the Natural History*. Oxford.

CHEVALLIER, R. 1986. 'Le bois, l'arbre et la forêt chez Pline', *Helmantica* 37, 148–72.

CITRONI MARCHETTI, S. 2011. *La scienza della natura per un intellettuale romano*. Pisa-Roma.

DAVIES, P. J. E. 2012. 'Pollution, propriety and urbanism in Republican Rome', in M. BRADLEY and K. STOW (eds.), *Rome, Pollution and Propriety. Dirt, Disease and Hygiene in the Eternal City from Antiquity to Modernity*. Cambridge, 67–80.

---

[63] His disgust and disdain is patent, for example, when he describes the famous episode of Cleopatra's dinner consisting in a 10 millions sesterces pearl, a perfect and unique *naturae opus*, which was destroyed for a simple bet: *nat. hist.* 9.119–21.

De Feo, G., S. De Gisi, and M. Hunter. 2014. 'Sanitation and wastewater technologies in ancient Roman cities', in A. N. Angelakis and J. B. Rose (eds.), *Evolution of Sanitation and Wastewater Technologies through the Centuries*. London, 251–68.

Doody, A. 2009. 'Pliny's Natural History: Enkuklos Paideia and the Ancient Encyclopedia', *Journal of the History of the Ideas* 70.1, 1–21.

Doody, A. 2010. *Pliny's Encyclopedia. The Reception of the Natural History*. Cambridge.

Fedeli, P. 1990. *La natura violata. Ecologia e mondo romano*. Palermo.

Fedeli, P. 1997. '*Nos et flumina inficimus* (Plin. *nat.*, 18,3) Uomo, acque, paesaggio nella letteratura di Roma antica' in S. Quilici Gigli (ed.), *Uomo Acqua e Paesaggio. Atti dell'incontro sul tema Irreggimentazione delle acque e trasformazione del paesaggio antico*, S. Maria Capua Vetere 22–23 Novembre 1996. Roma, 317–30.

Feissel, D. 1985. 'Deux listes de quartiers d'Antioche astreints au creusement d'un canal (73–74 après J.-C.)', *Syria* 62, 77–103.

French, R. 1994. *Ancient Natural History. Histories of nature*. London-New York.

Gazda E. K., and J. R. Clarke (ed.) 2016. *Leisure and Luxury in the Age of Nero: The Villas of Oplontis near Pompeii* (Kelsey Museum publication 14). Ann Arbor MI.

Grimal, P. 1986. 'Pline et les philosophes', *Helmantica* 37, 239–49.

Harrison, P. 2011. 'Introduction', in P. Harrison, R. L. Numbers and M. H. Shank (eds.), *Wrestling with Nature. From Omens to Science*. Chicago, 1–7.

Hirt, A. M. 2010. *Imperial Mines and Quarries in the Roman World. Organizational Aspects 27 BC–AD 235*. Oxford.

Hopkins, J. 2012. 'The 'sacred sewer': tradition and religion in the Cloaca Maxima', in M. Bradley and K. Stow (eds.), *Rome, Pollution and Propriety. Dirt, Disease and Hygiene in the Eternal City from Antiquity to Modernity*. Cambridge, 81–102.

König, J. and G. Woolf. 2013a. 'Introduction', in J. König and G. Woolf (eds.), *Encyclopaedism from Antiquity to the Renaissance*. Cambridge-New York, 1–20.

König, J. and G. Woolf. 2013b. 'Encyclopedism in the Roman Empire', in J. König and G. Woolf (eds.), *Encyclopaedism from Antiquity to the Renaissance*. Cambridge-New York, 23–63.

Jones-Lewis, M. A. 2012. 'Poison: Nature's Argument for the Roman Empire in Pliny the Elder's *Naturalis Historia*', *Classical World* 106.1, 51–74.

Laehn, T. R. 2013. *Pliny's Defense of Empire*. London-New York.

Maggiulli, G. 2000. 'Natura, violazione e difesa dell'ambiente: l'eredità di Plinio il Vecchio tra scienza e divulgazione', in G. Maggiulli et alii (eds.), *Tradizione enciclopedica e divulgazione in età imperiale* (Serta Antiqua et Mediaevalia 2). Roma, 1–42.

Malavolta, M. 2013. 'Il concetto di tutela dell'ambiente nell'antica Roma tra religione, filosofia e diritto', in M. Malavolta, *Fra Antichità e Storia*. Roma, 93–115.

Mayor A., 2009. *The Poison King: The Life and Legend of Mithradates, Rome's Deadliest Enemy*. Princeton.

Mazzarino, A. 1969–1970. 'Un testo antico sull'inquinamento', *Helikon* 9–10, 643–5.

Montraville Green, R. 1951. *A translation of Galen's Hygiene* (De Sanitate Tuenda), with an Introduction by H. E. Sigerist. Springfield.

Morello, R. 2011. 'Pliny and the Encyclopaedic Addressee', in R. K. Gibson and R. Morello (eds.), *Pliny the Elder. Themes and Contexts*. Leiden, 147–65.

Morgan, T. 2013. 'Encyclopaedias of virtue? Collections of sayings and stories about wise men in Greek', in J. König and G. Woolf (eds.), *Encyclopaedism from Antiquity to the Renaissance*. Cambridge-New York, 108–28.

Murphy, T. 2004. *Pliny the Elder's Natural History. The Empire in the Encyclopedia*. Oxford.

Naas, V. 2011. 'Imperialism, *mirabilia* and knowledge: some paradoxes in the *Naturalis Historia*, in R. K. Gibson and R. Morello (eds.), *Pliny the Elder. Themes and Contexts*. Leiden, 57–70.

Panciera, S. 2000. 'Nettezza urbana a Roma organizzazione e responsabili', in J. A. Remolà y X. Dupré Raventós (eds.), *Sordes urbis: la eliminación de residuos en la ciudad romana*, Actas de la Reunión de Roma, 15–16 de noviembre de 1996. Roma, 95–105.

PANESSA, G. 1991. *Fonti greche e latine per la storia dell'ambiente e del clima nel mondo Greco*. Pisa.

PAPARAZZO, E. 2005. 'The Elder Pliny, Posidonius and Surfaces', *The British Journal for the Philosophy of Science* 56.2, 363–76.

PAPARAZZO, E. 2011. 'Philosophy and Science in the Elder Pliny's *Naturalis Historia*', in R. K. GIBSON and R. MORELLO (eds.), *Pliny the Elder. Themes and Contexts*. Leiden, 89–112.

RIVES, J. B. 2006. 'Magic, Religion, and Law: The Case of the *Lex Cornelia de sicariis et veneficiis*', in C. ANDO and J. RÜPKE (eds.), *Religion and Law in Classical and Christian Rome*. Stuttgart, 47–67.

ROHLEDER, J. 2001. 'The cultural history of limestone', in F. W. TEGETHOFF, J. ROHLEDER and E. KROKER (eds.), *Calcium Carbonate from the Cretaceous Period into 21st Century*. Springer Basel, 53–135.

ROLLER DUANE, W. 2010. *Cleopatra: a biography*. Oxford.

RUSSELL, B. 2013. *The Economics of the Roman Stone Trade* (Oxford Studies in Roman Economy). Oxford.

SALLMANN, K. 1986. 'La responsabilité de l'homme face à la nature', *Helmantica* 37, 251–66.

SANTINI, C. 1983. *La cognizione del passato in Silio Italico*. Roma.

SCHULTZE, C. 'Encyclopaedic exemplarity in Pliny the Elder', in R. K. GIBSON and R. MORELLO (eds.), *Pliny the Elder. Themes and Contexts*. Leiden, 167–86.

SIGRÍDUR ARNAR, A. 1990. *Encyclopedism from Pliny to Borges*. Chicago.

STANNARD, J. 1986. 'Herbal medicine and herbal magic in Pliny's time', *Helmantica* 37, 95–106.

VAN NIJF, O. 1997. *The Civic World of Professional Associations in the Roman East*. Amsterdam.

VUORINEN, H. S. 2014. 'Ancient Greek and Roman authors on health and sanitation', in A. N. ANGELAKIS and J. B. ROSE (eds.), *Evolution of Sanitation and Wastewater Technologies through the Centuries*. London, 429–36.

WALLINGA, H. T. 1996. 'Official Roman Washing and Finishing Directions *Lex Metilia Fullonibus Dicta*', *The Legal History Review* 64.2, 183–90.

Luca Montecchio

La cultura dell'ambiente in ambito monastico tra V e VIII secolo

Con questo contributo si vuole dimostrare quanto sia grande lo iato tra la percezione che i Romani avevano dell'ambiente, del *kosmos*, e quella alto medievale. Quest'ultima risentì, infatti, in modo profondo dell'influsso del pensiero ecclesiastico. Le ragioni che portarono a tali differenze sono molteplici e, come avremo modo di approfondire, fondamentale fu il ruolo del mondo religioso per quanto concerne l'atteggiamento cristiano nei confronti del Creato. Almeno nell'antichità tarda e per tutto l'alto medioevo, gli alti prelati paiono maggiormente concentrati sulla salvezza delle anime piuttosto che sul legame, comunque inscindibile, tra uomo e natura. Si vedrà anzi come la natura e i temi inerenti a essa, e cioè tutto ciò che non riguarda esplicitamente e direttamente la salvezza del genere umano, vengano messi da parte e, a volte, non trattati.

Come afferma il DELORT, all'epoca del grande sviluppo demografico, ci fu un vero e proprio 'assalto generale alla foresta fin dai secoli settimo ed ottavo per la necessità di "terre vergini"'.[1] La distruzione indiscriminata degli alberi favorì l'erosione che divenne in taluni casi irreversibile. Sempre il DELORT osserva come 'il paesaggio occidentale come lo vediamo oggi è stato innegabilmente foggiato dall'opera ininterrotta del Medioevo; è stata questa a sconvolgere in modo definitivo la struttura e la dinamica degli ecosistemi naturali, da quell'epoca sostituiti con l'agrosistema occidentale... Ad animare in profondità la vita occidentale non è stata la certezza della vittoria, ma il senso della lotta quotidiana contro un ambiente ostile o ribelle, con risultati sempre rimessi in discussione'.[2]

Se il mondo romano, così come in precedenza quello greco, si pose il problema del rapporto tra l'uomo e la realtà che lo circondava, interrogandosi su quale fosse il modo più adeguato di convivere, il mondo alto medievale sembra essere immerso in altri pensieri. Certamente, caduta la *pars Occidentis* dell'impero romano, durante il periodo dei cosiddetti regni romano barbarici venne meno quella che era stata, per secoli, l'idea di Stato. Vero è che si era vissuto un secolo, il terzo, durante il quale l'incertezza aveva regnato sovrana, ma le genti che abitavano l'impero avevano comunque la percezione di una presenza dello Stato, di una compagine statale, di una struttura statale. Alla fine del secolo quinto e poi, nei secoli immediatamente successivi, quella idea che si era radicata nelle coscienze dei sudditi imperiali inevitabilmente venne meno. La situazione era cambiata anche se, almeno in un primo tempo, rimanendo in auge la burocrazia romana, i gallo romani, gli ispano romani, gli italici e gli altri popoli dell'ormai caduto impero non avvertirono il sostanziale cambiamento che pure c'era stato.

---

[1] DELORT 2014: 19–20.
[2] DELORT 2014: 21.

Un nuovo concetto dello stato tornò sul finire del millennio medievale con il Sacro Romano Impero, all'alba dell'evo moderno. Nel cosiddetto alto medioevo, uno stato significava un piccolo regno, insicuro, su cui nemmeno il signore poteva essere garante della vita dei sudditi, su cui gravava inesorabilmente una costante instabilità che impediva anche lo sviluppo dei commerci, degli stessi rapporti umani, propri dell'epoca romana. La popolazione europea dovette impegnarsi in una dura lotta per la sopravvivenza, perdendo, come dicemmo, il concetto di 'Stato', di *res publica*. Roma di certo non fu un'esperienza latrice di benessere per tutti, però aveva permesso ai popoli da essa governati di bere, di mangiare, di lavarsi, di commerciare. Insomma, caratteristica precipua di Roma sono gli acquedotti, le strade (sostanzialmente sicure), la possibilità concreta di far affluire i prodotti agricoli dalle campagne alle città.[3] Caduto l'impero occidentale, tutto questo finì. La frantumazione, poi, di alcuni regni romano barbarici – si pensi a quello dei Franchi – aumentò vieppiù le incertezze dei sudditi. La precarietà del momento vissuto, un aumento esponenziale dei pericoli, le difficoltà di trovare zone sicure ove coltivare campi contribuirono a un calo demografico che ridusse in modo significativo la popolazione della *pars Occidentis*. Naturali conseguenze furono il prevalere della sfiducia e la chiusura nei confronti di chi non apparteneva al proprio villaggio o alla propria città. Con ogni evidenza un uomo di quel periodo non avrebbe potuto focalizzare la propria attenzione sulla natura, anche se continuava a vivere immerso nella natura. Essa veniva vista come nemica, non più come una sorta di compagna di viaggio. Va infine considerato che il calo demografico influì senz'altro sullo sfruttamento dell'ecosistema europeo. Esso fu preservato rispetto al periodo del dominio romano quando, invece, le foreste, i corsi d'acqua e, in generale, le risorse naturali vennero ampiamente sfruttate.[4] Nella Tarda Antichità se, da una parte, le crescenti e costanti instabilità aumentarono il sentimento di incertezza, anche il pensiero della Chiesa non fu estraneo a tale convincimento. I Padri della Chiesa erano impegnati nella diffusione del Verbo e si prodigavano nella spiegazione dei precetti cristiani. Quando essi parlavano di Creato, di natura dunque, non intendevano, come succederà in tempi successivi, porre la loro attenzione sul rapporto natura-uomo, bensì intendevano esaltare l'azione creatrice di Dio. Di conseguenza lo sguardo dell'uomo era teso verso l'alto, non verso la Terra. Ma iniziamo a vedere come il mondo classico e, in particolar modo, quello romano hanno posto l'accento sulla natura e, di conseguenza, sul rapporto dell'umanità con la stessa.

1. La cultura dell'ambiente a Roma

Innanzi tutto si consideri come Roma, dagli albori della sua civiltà, iniziò una costante opera di costruzione, non solo di palazzi per il ceto abbiente della popolazione, ma anche di abitazioni per tutti. Grande e significativa differenza con la civiltà greca, ad esempio. Inoltre vennero costruite strade, acquedotti e terme. Costruire significò inevitabilmente un confronto continuo con la natura. Di qui l'attenzione posta dal mondo romano sulla stessa.

---

[3] Tra i recenti lavori sulla Tarda Antichità è da segnalare Wickam 2014; Montecchio 2012; Ward-Perkins 2008.

[4] Per quanto concerne il rapporto tra il mondo antico e l'ambiente riporto a Thommen 2014; si consideri poi Brückner 1986: 7–17.

Le testimonianze relative ai problemi dell'urbanizzazione sono particolarmente ampie per Roma: ai consueti problemi di affollamento, rumore, traffico, inquinamento da fumo e polvere, pericolo di crolli e incendi, deficienze del sistema fognario, va aggiunto il pericolo derivante, per l'acqua potabile, dall'uso di canalizzazioni di piombo (di cui era peraltro nota la tossicità, come attesta Vitruvio che proponeva di sostituire le tubazioni in piombo con quelle in terracotta).[5] La sensibilità verso questi problemi è acuita presso i Romani dal fatto che essi sentono in genere fortemente la preoccupazione per la salubrità dell'ambiente, per esempio nella scelta di luoghi per la costruzione di città, case, fattorie, edifici pubblici; nelle fonti troviamo spesso la denuncia della speculazione edilizia, che pone gli insediamenti umani in contrasto con l'attività agricola e con l'ordine naturale. Su questa stessa linea si inserisce la contrapposizione tra città e campagna, con la tipica idealizzazione del *locus amoenus*: non diversamente dalla poesia ellenistica, la natura viene vista nella prospettiva idealizzata tipica delle società urbane, che contrappone la vita serena della campagna a quella, insana e caotica, della città; ma accanto a questa idealizzazione è presente anche il rifiuto del mondo selvaggio, il *locus horridus* in cui la natura assume contorni drammatici e ostili. L'atteggiamento teorico e pratico degli antichi di fronte alla questione ambientale sembra muoversi dunque con significativa continuità tra mondo greco e mondo romano, tra il bisogno di controllare la natura e di piegarla alle esigenze umane e la percezione dei limiti di quegli stili di vita che apparivano non più in armonia con la natura stessa. Così Plinio il Vecchio, dopo aver stigmatizzato il comportamento degli uomini, che inquinano i fiumi e gli elementi naturali e avvelenano persino l'aria che è loro indispensabile per vivere, confida nella grandezza e nella magnanimità della natura per sostenere la necessità dell'uomo di continuare comunque a perseguire il progresso, impegnandosi per 'rendere migliore la vita'.[6] I Romani, come scrive il FEDELI, hanno 'nutrito una vigile cura nei confronti dell'ambiente' e si sono 'posti il problema del rapporto fra uomo e suo 'habitat' naturale'. Il FEDELI in proposito sottolinea lo scrupolo che essi avevano rispetto ai poderi e quanto riflettevano sull'importanza della loro collocazione 'prima di procedere ad acquisti incauti'.[7] Lo stesso autore ricorda come Catone nel *De agricultura* parli esplicitamente di come l'impatto ambientale dovesse essere la preoccupazione precipua dell'uomo.[8] Nell'esaminare un fondo, prima dell'acquisto, si doveva valutare la salubrità del clima e la fertilità della terra. Sempre il FEDELI osserva che 'se, come Catone, tutti gli scrittori di agricoltura insistono sulla necessità di scegliere luoghi salubri oltre che fertili, ciò dipende dal fatto che esistevano *loca pestilentia*, in cui regnava la

---

[5] Vitr., *de arch.* 8.6.10–11: *Habent autem tubulorum ductiones ea commoda. Primum in opere quod si quod vitium factum fuerit, quilibet id potest reficere. Etiamque multo salubrior est ex tubulis aqua quam per fistulas, quod per plumbum videtur esse ideo vitiosum, quod ex eo cerussa nascitur; haec autem dicitur esse nocens corporibus humanis. Ita quod ex eo procreatur, <si> id est vitiosum, non est dubium, quin ipsum quoque non sit salubre. Exemplar autem ab artificibus plumbariis possumus accipere, quod palloribus occupatos habent corporis colores. Namque cum fundendo plumbum flatur, vapor ex eo insidens corporis artus et inde exurens eripit ex membris eorum sanguinis virtutes. Itaque minime fistulis plumbeis aqua duci videtur, si volumus eam habere salubrem. Saporemque meliorem ex tubulis esse cotidianus potest indicare victus, quod omnes, et structas cum habeant vasorum argenteorum mensas, tamen propter saporis integritatem fictilibus utuntur.*

[6] Plin., *nat. hist.* 18.1.2–5.

[7] FEDELI 1990: 30.

[8] Cato, *de agr.* 1.1–3.

malaria, o zone difficilmente abitabili, come quelle della Campania ammorbate dalle esalazioni delle solfatare'.[9]

È Varrone a parlare in modo esplicito di come sia la natura a concedere all'uomo luoghi salubri ove abitare. L'uomo, nondimeno, può fare molto per migliorare la condizione dei luoghi e attenuare i pericoli per la propria salute grazie alla scienza. Proprio il letterato del secolo I avanti Cristo ci spiega dove e come, a suo giudizio, dovesse venire costruita una villa.

> Devi badare che la villa sia situata specialmente alle falde di un colle boscoso, dove i pascoli siano ricchi, e – in pari modo – che sia esposta ai venti che saluberrimi spireranno sulla campagna. Adattissima è quella che guarda all'est equinoziale, perché d'estate ha l'ombra, di inverno il sole. Nel caso che tu sia costretto a edificarla lungo un fiume, bisogna stare attenti a non costruirla di fronte ad esso, perché d'inverno sarebbe assai fredda e di estate assai malsana. Bisogna anche badare che non ci siano zone paludose, e per le stesse ragioni e perché non si formano dei microbi, che non si possono vedere ad occhio nudo, ma penetrano nell'organismo attraverso la bocca e il naso con la respirazione e causano gravi malattie.[10]

Con Vitruvio si ha piena consapevolezza 'della necessità di un complesso sistema d'interventi per rendere migliori, con l'ambiente, le condizioni di vita dell'uomo. L'architetto stesso, nell'ambito della sua formazione poliedrica, pur non potendo aspirare a divenir medico come Ippocrate, non dovrà in ogni caso restare a digiuno di condizioni igienico sanitarie'.[11] Vitruvio non si limita a questo. Anzi quando affronta il tema della costruzione di nuove città avverte come sia necessario 'scegliere un luogo molto salubre, che sia cioè elevato, esente da nebbie e da gelate notturne, lontano da paludi, che si affacci verso regioni temperate da eccessi di caldo e di freddo'. Non solo le persone soffrirebbero di una collocazione errata ma anche le derrate alimentari sarebbero sistemate in modo sbagliato e cioè non si conserverebbero bene. Se, infatti, i granai fossero esposti al sole, il grano si rovinerebbe. I viveri e la frutta, pertanto 'se non vengono riposti in un luogo non soleggiato, non si conservano a lungo'.[12]

Quando Roma e il suo impero lasciarono il posto ai cosiddetti regni romano barbarici era iniziato un evo nuovo. Erano ormai almeno un paio di secoli che il cristianesimo aveva preso piede all'interno dell'impero, ma tra i secoli quarto e quinto il cambiamento divenne irreversibile. Nel mondo rurale si era affermato anche in Occidente il monachesimo che però, è noto, aveva caratteristiche differenti rispetto a quello orientale. Pure si può considerare simile l'atteggiamento monastico rispetto alla natura. Il cristianesimo pensa alla natura come all'insieme dell'habitat biologico e cosmico che costituisce l'ambiente dell'uomo, nel quale egli è immerso e con cui entra in simbiosi con il suo corpo in una relazione viva e in costante dialogo. L'ambiente però viene visto come totalmente distinto da Dio e, pertanto, dissacrato. In realtà la natura è creatura, ma proprio la *Genesi* indica nell'uomo il dominatore dell'Universo e quindi anche della natura.[13] Discorsi sulla Natura, quindi, sulla sua importanza, sull'ambiente, vennero inevitabilmente visti come privi di senso per il fine ultimo

---

[9] Columella, rifacendosi a Varrone, sottolinea quanto fosse importante porre la giusta attenzione sul fondo che si voleva acquistare. FEDELI 1990: 31

[10] Varro, *de re rust.* 1.12.1–4.

[11] FEDELI 1990: 33.

[12] Vitr., *de arch.* 1.4.1–5.

[13] *Genesi*, 1.26–9.

dell'Umanità tutta, che è quello di tendere a Dio. Vivere immersi nella natura, fare parte di essa non viene considerato dai Padri e nemmeno dai cristiani di generazioni successive. Essi dovevano perseguire la Salvezza, si sentivano facenti parte del Corpo Mistico, più che della natura. Il cristiano sembra voler sorvolare sulla punizione che Dio ci ha inflitto e cioè di dover trarre il cibo con dolore e fatica proprio dalla terra. Vedremo come il monachesimo accetterà il lavoro manuale e, pertanto, il lavoro agricolo, perché necessario, ma non vedrà in esso una sorta di rapporto con la terra genitrice anche nostra ('polvere sei e in polvere tornerai').[14]

La maggior parte dei pagani considerava la natura in modo istintivo, parte del mondo divino. Essi dunque avevano nei suoi confronti 'un atteggiamento caratterizzato da una sorta di sottomissione a un ordine, a un destino inesorabilmente subìto e accettato con fatalismo'.[15] Contemporaneamente ci sono degli studiosi latini, Boezio per esempio, che portano avanti il pensiero aristotelico, mentre Agostino ragionerà sul pensiero platonico rivisitato ('l'anima ragionevole non deve adorare come suo dio le cose che per natura le sono inferiori né deve considerare superiori a sé come dèi le cose perché il vero Dio l'ha creata ad esse superiore').[16] Tra la Tarda Antichità e l'Alto Medioevo, sembrava che gli studiosi, gli ecclesiastici, non si ponessero specifiche domande sulle questioni inerenti la natura, pur elencando tutto ciò che veniva loro tramandato dall'antichità classica.[17] In epoca carolingia, al contrario, abbiamo esempi in libri anglosassoni di come si conoscesse in modo approfondito la flora.[18]

I Padri della Chiesa erano impregnati di cultura romana, dunque portavano con sé molti aspetti di quel modo di vedere il *kosmos*. Riuscivano, però, ad andare oltre e, in qualche modo, questo 'oltre' significò un allontanarsi dai problemi che proprio il rapporto tra l'uomo e il *kosmos* comportava. Come abbiamo appena detto, la natura viene distinta da Dio perché anch'essa 'creatura' mentre l'uomo deve proiettarsi verso il Creatore. I monaci non si discosteranno certo da queste premesse.

2. Il rapporto del monaco con la natura

Il monaco, anche quello occidentale, pur mostrando un'attenzione diremmo necessaria nei confronti della natura, si discostò non poco dal punto di vista del mondo latino. Si vuole dimostrare come in ambito monastico, durante la Tarda Antichità, ci fosse una qualche attenzione rispetto all'ambiente. Tale attenzione era senz'altro diversa da quella dimostrata dai Greci prima e dai Romani poi; tuttavia qualcosa vi era. Abati, monaci esperti, novizi, tutti mostravano un naturale slancio verso la natura. Ciò non era scontato in ambito ecclesiastico se è vero che, subito dopo, nel cosiddetto Alto Medioevo, gli aspetti inerenti la natura verranno sostanzialmente ignorati. La Chiesa dovrà attendere l'avvento di san Francesco perché qualcuno torni a parlare degli aspetti del Creato come qualcosa che andava al di là dell'uomo, pur essendo legati all'uomo stesso.

---

[14] *Genesi*, 3.17–9.
[15] DELORT e WALTER 2002: 71.
[16] August., *de civ. Dei* 8.5.
[17] GRANT 2007: 105.
[18] EHRSHAM VOIGT 1979: 250–68, partic. 266–8.

Con Benedetto da Norcia il monachesimo occidentale si trovò ad affrontare la questione del rapporto con la natura. Non fu frutto di un qualche ragionamento, ma semplicemente il giusto approccio – diremmo 'naturale' – alla realtà. La regola benedettina recita *ora et labora* e tale impegno, ovvero il *laborare*, che andava al di là della preghiera in senso stretto, era legato alla natura e alle suppliche a Dio. L'essere umano, infatti, si è sempre dovuto confrontare con gli aspetti naturali dell'ambiente ove viveva per cercare di vivere oltre che di sopravvivere. Dagli uomini di Chiesa la natura viene sempre vista in modo quasi superstizioso, almeno, come dicemmo, sino a san Francesco. La natura, parte del creato, viene utilizzata per vivere, ma con circospezione. Quando l'idea monastica viene portata in Occidente si era verso la fine dell'epoca romana. I monaci, pertanto, erano comunque romani e la loro cultura era latina.[19]

Paolino di Nola, di famiglia senatoria, in alcuni versi dei suoi *Carmina*, pur non esprimendo concetti pregnanti sull'ambiente, dimostrava, nei fatti, una cura particolare per il Creato.

> O sorgente della parola,
> Verbo Dio,
> ascolta e fammi ora canoro con dolce voce
> come il famoso uccello della primavera
> che, nascosto sotto il verde fogliame,
> suole con molteplici ritmi
> addolcire
> i campi deserti ed effondere
> con una sola lingua molte voci cambiando melodia,
> uccello di un solo colore nelle penne,
> ma variopinto nei canti ...[20]

La delicatezza del monaco nel descrivere il cinguettio di un usignolo indica quel genere di attenzione non dovuta soltanto ad un rapporto più intimo con la natura rispetto ai tempi moderni, bensì una sensibilità rara di un uomo verso il Creato. Paolino era però di cultura latina, aveva abbracciato il cristianesimo da adulto. Diversa, insomma, la sua predisposizione verso l'ambiente rispetto ai monaci di secoli successivi. Ma vediamo altri esempi del suo modo di osservare l'ambiente naturale. Nell'epistola 23 splendido il paragone delle opere dei fedeli con la rugiada:

> ... ora, invece, nella luce della Chiesa,
> come nel massimo splendore della luna piena,
> e negli uomini santi, come stelle purissime nel cielo sereno,
> stillano le opere dei fedeli
> simili alle rugiade nella notte di questo mondo.[21]

Il *topos* lettarario di Cristo quale 'sorgente' è, diremmo, abusato nei testi sacri. Nondimeno la poesia di Paolino lo esalta vieppiù rendendo ancor più viva tale metafora:

---

[19] HARRISON, NUMBERS e SHANK 2011.
[20] Paul. Nol., *Preghiere*, carmen 23.20–37 (GIULIANO 2008).
[21] Paul. Nol., *Preghiere*, ep. 23; *A Severo*, 33.37–48.

Tu, o Cristo sorgente, ti prego,
vieni a nascere nel mio cuore,
perché viva zampilli per me la vena della tua acqua…
Coloro che berranno di te, o Cristo,
ristorati dal dolce torrente,
non avranno più sete…[22]

Coevo di Paolino è Giovanni Cassiano il quale, nelle sue *Istituzioni cenobitiche*, trattato dove si descrivono le usanze dei monaci orientali, ammonisce circa l'importanza del lavoro, inteso come lavoro dei campi: 'una volta compiuti questi doveri spirituali in tempo ragionevole, non si trascurino i necessari obblighi del lavoro'.[23] A tal proposito interessante leggere le sue parole che esaltano il lavoro manuale, almeno rispetto ad altre *Regulae* del monachesimo occidentale che, come vedremo, accettano tale lavoro perché indispensabile per la sopravvivenza, ma accessorio ai fini spirituali.

durante queste veglie si dedicano al lavoro, per evitare che, restando in ozio, il sonno li possa sorprendere. Come infatti non concedono quasi alcun tempo all'ozio, così pure non pongono alcun limite alla meditazione spirituale, ed esercitando simultaneamente le virtù del corpo e dell'anima, fanno sì che al profitto dell'uomo interiore corrisponda quello dell'uomo esteriore: nei movimenti impuri del corpo e nell'instabile ondeggiamento del cuore gettano il peso del loro lavoro come un'ancora salda e immobile, e con tale vincolo riescono a trattenere all'interno dei muri della cella l'instabilità e le divagazioni del cuore, come in un porto sicurissimo. E così, rivolgendo la propria attenzione unicamente alla meditazione spirituale e alla custodia dei pensieri, non solo non impediscono che la mente, che rimane sempre vigilante, si lasci trascinare a dare il proprio consenso a una qualche suggestione malvagia, ma la preservano anche da ogni pensiero inutile e ozioso. Perciò, non è facile discernere quale delle due cose dipenda dall'altra: se cioè essi pratichino un lavoro manuale incessante grazie alla meditazione spirituale, oppure se sia proprio grazie alla continuità del lavoro che essi riescono a raggiungere un grado di progresso spirituale così avanzato e una così grande luce di conoscenza.[24]

In ambito monastico parole tali lasciano il segno perché, in fondo, Benedetto da Norcia non si era espresso in termini così perentori sul lavoro in generale né, tanto meno, su quello agricolo come avremo modo di vedere. Però l'ambiente e la cultura dell'ambiente appare, per usare un eufemismo, sfumata in Giovanni Cassiano il quale, come solitamente facevano i cristiani, usa belle metafore in cui protagonisti sono gli elementi naturali, ma che servono solo ad elevare l'uomo verso Dio. Non c'è mai, quindi, alcuna reale consapevolezza dell'importanza dell'ecosistema che ospita l'uomo.

Sant'Agostino, guida spirituale per il monachesimo occidentale, nel *De doctrina christiana*, affronta argomenti inerenti la natura nel suo complesso. Egli non focalizza mai, in modo esplicito, l'attenzione sugli stessi temi su cui, in precedenza, era stato posto l'accento dal mondo romano. Adesso, infatti, si tende ad osservare la caducità del mondo naturale, degli aspetti terreni dell'umanità, i suoi limiti. Il santo di Ippona tende a considerare la nostra carnalità come una sorta di prigione (in proposito si pensi al Vangelo di Matteo in cui la carnalità umana viene evidenziata e, a volte, stigmatizzata) che lega l'anima alla Terra. Il

---

22 Paul. Nol., *Preghiere*, carmen 31.425–46.
23 Cassian., *Le istituzioni cenobitiche* 3.3.1 (D'AYALA VALVA 2007).
24 Cassian., *Le istituzioni cenobitiche* 2.14.

nostro spirito però è destinato, non appena morirà il corpo, a volare via. Ad ogni modo con Agostino si capisce ciò che secoli prima aveva detto san Paolo e cioè che la parte terrena della vita umana è fondamentale, ma che la nostra aspirazione sta nell'immortalità. Tali ragionamenti, in modo, diremmo inevitabile, allontanano lo sguardo del cristiano dalle questioni pratiche, terrene, quasi fossero aliene dall'uomo. Tra San Paolo e Agostino aumenta vieppiù lo iato tra il mondo naturale, di cui pure l'uomo fa parte, e l'essere umano il quale deve perseguire il divino. D'altronde, determinati ragionamenti ricalcano ciò che viene espresso esplicitamente nella *Genesi*. Ed è sempre la *Genesi* che guida le idee cristiane su come si deve porre l'uomo nei confronti del Creato. A proposito della carne il santo di Ippona dice:

> Non c'è dunque alcuno che odi se stesso: sicché al riguardo mai c'è stata controversia con una qualche setta. E anche riguardo al corpo, nessuno lo odia, ed è vero quello che dice l'Apostolo: *Nessuno ha mai odiato la sua propria carne.* Che se alcuni dicono di preferire ad ogni costo di essere senza corpo, essi dicono una falsità: odiano infatti non il loro corpo ma la sua corruttibilità e pesantezza. Per cui non è che non vogliano avere nessun corpo ma lo vorrebbero incorruttibile e sommamente agile, deducendone però che, se il corpo fosse così, non sarebbe più corpo ma anima. Riguardo poi a coloro che sembrano quasi infierire contro il loro corpo per la continenza che praticano o le fatiche che affrontano, coloro che ciò fanno rettamente non si comportano così per non avere il corpo ma per averlo soggetto a se stessi e pronto alle opere necessarie. Combattendo faticosamente contro il proprio corpo si allenano ad estinguere le passioni che vorrebbero servirsi malamente del corpo, vale a dire tutte quelle abitudini o inclinazioni che portano l'anima a godere delle cose inferiori. Tant'è vero che costoro non si uccidono ma hanno cura della loro salute.[25]

Agostino è un romano, prima ancora di essere un cristiano, e conosce molto bene il mondo in cui vive. Di qui, nel *De doctrina christiana*, fa un riferimento significativo alle sue conoscenze botaniche che indicano un sano apprezzamento dell'ambiente.

> È facile invece capire come mai la pace permanente sia significata dal ramoscello di olivo che la colomba riportò all'arca al suo ritorno. Questo, perché sappiamo che l'olio, anche se liscio, se tocca un altro liquido non si altera e, quanto alla pianta stessa, è tutto l'anno coperta di foglie verdi. Viceversa, molti non conoscono cosa sia l'issopo e quale vigore abbia. Esso giova a liberare il polmone [dal catarro] e così pure, a quel che si racconta, riesce con le sue radici a penetrare la roccia, essendo un'erbetta bassa e piccola. Per questo non riescono a trovare il motivo per cui è detto: *Mi aspergerai con l'issopo e io sarò mondato.*[26]

Il santo di Ippona poi lancia uno sguardo sui giovani e sulle loro menti. Anche in questo caso tende a coniugare il desiderio di conoscenza con l'amore verso Dio. Il che non allontana necessariamente l'uomo da mere questioni pratiche come può considerarsi la cura dell'ambiente. Agostino esplicitamente non dirà nulla a riguardo, ma alcune sue parole possono venire interpretate come una apertura al mondo. Ora, soprattutto per quanto concerne l'ambiente crediamo opportuno sottolineare queste parole del Santo:

> Una parola sulle altre scienze che si trovano presso i pagani. Positiva è la descrizione delle cose, passate e presenti, che riguardano i sensi del corpo. Ad esse devono aggiungersi gli esperimenti e

---

[25] August., *de doct. chris.* 1.24 (Simonetti 1994).
[26] August., *de doct. chris.* 2.24.

le supposizioni delle arti utili nell'ambito della fisica. Positivo pure l'uso del metodo del raziocinio e del numero.[27]

Crediamo infatti che 'gli esperimenti e le supposizioni delle arti utili nell'ambito della fisica' si riferiscano anche alle questioni relative all'ambiente di cui i Romani, come abbiamo visto, si interessavano.

> Ai giovani appassionati del sapere, dotati di intelligenza e timorati di Dio che ricercano la sapienza si possono dare salutarmente questi precetti: non si permettano di seguire con animo tranquillo – quasi che bastassero per raggiungere la vita beata – nessuna scienza di quelle che si professano al di fuori della Chiesa di Cristo, ma le valutino con mente lucida e con diligenza. Potrà succedere che si imbattano in scienze inventate dagli uomini, diverse a causa della diversa volontà di chi le ha inventate e cadute in oblio a causa dei sospetti che suscita chi è incappato nell'errore o, soprattutto, casi in cui tali scienze contengono una società stipulata con i demoni quasi per mezzo di patti o convenzioni fondate su certi segni. In questi casi i nostri giovani le debbono radicalmente rigettare e detestare. E inoltre debbono disinteressarsi delle scienze umane superflue e di lusso. [Quanto invece alle istituzioni umane che servono alla convivenza sociale, a motivo dei rapporti che hanno con la vita presente, non le debbono trascurare. Una parola sulle altre scienze che si trovano presso i pagani.] Positiva è la descrizione delle cose, passate e presenti, che riguardano i sensi del corpo. Ad esse devono aggiungersi gli esperimenti e le supposizioni delle arti utili nell'ambito della fisica. Positivo pure l'uso del metodo del raziocinio e del numero. All'infuori di queste materie credo che altre utili non ci siano. E riguardo a tutto questo deve osservarsi la norma: *Nulla di troppo!* soprattutto riguardo a quelle cose che, avendo relazione con i sensi del corpo, sottostanno all'andare del tempo e sono contenute nello spazio.[28]

Non troppo differente è la visione dell'ambiente dei monaci orientali. Anzi essi, poiché privilegiavano l'ascetismo e l'eremitismo, vivevano la natura che li circondava in modo ancor più intenso. Nondimeno, anch'essi non dimostrarono particolari attenzioni verso l'ambiente che li circondava.

Sin dall'epoca dell'egiziano sant'Antonio (quarto secolo) il rapporto con la natura, per un cristiano, era evidentemente molto forte. Atanasio di Alessandria, nella sua *Vita di Antonio* scrive che, quando il santo andò ad abitare sulla cima di un monte (probabilmente presso l'odierna Wadi el-Arab, a una trentina di chilometri dal mar Rosso, ove si trova il monastero di sant'Antonio), per evitare disagi ad alcuni dei suoi discepoli che volevano portargli il pane, chiese loro:

> una zappa, una scure e un po' di frumento. Quando gli portarono queste cose, esplorò i dintorni della montagna e, trovato un piccolo campo adatto alla coltivazione, cominciò a lavorarlo e, dato che il fiume gli forniva acqua in abbondanza per irrigarlo, cominciò a seminare (con il sudore della sua fronte, come dice la *Genesi*). Così fece ogni anno e in questo modo si procurò il pane… In seguito, vedendo che altri ancora venivano da lui, si mise a coltivare anche alcuni ortaggi perché chi veniva a trovarlo ricevesse qualche conforto dopo la fatica di quel difficile cammino.[29]

---

[27] August., *de doct. chris.* 2.25.
[28] August., *de doct. christ.* 2.39.58.
[29] Atan. Aless., *Vita di Antonio* (CREMASCHI 1995: 172).

Dallo scritto di Atanasio si evince come Antonio 'usasse' la natura per sopravvivere senza tentare minimamente di 'modificarla' né, tanto meno, di inquinarla in qualsivoglia modalità. Egli riuscì quindi a integrarsi perfettamente nell'ecosistema, almeno secondo i dettami biblici. Non che questo fosse lo scopo della sua vita ma, *de facto*, insegnò ai suoi seguaci e quindi ai monaci di comportarsi in tal modo con l'ambiente circostante. Va d'altra parte considerato come nelle zone rurali o, come nello specifico, in montagna, il territorio offra situazioni favorevoli alla sopravvivenza dell'uomo, senza che questi debba costruire in modo sconsiderato o apporre significative modifiche all'habitat stesso. Diversa è invece la vita nelle città. Là, infatti, si deve necessariamente pensare alle tubature per le fognature, a costruire in modo anche selvaggio pur di accogliere i *cives*. Insomma la realtà cittadina ha da sempre imposto regole che a volte andavano contro gli equilibri naturali. Torniamo nell'Occidente cristiano.

Per quanto concerne la già citata *Regola* benedettina si osservi come non ci sia alcun riferimento esplicito al paesaggio circostante il monastero. Eppure Benedetto qualcosa la dice esplicitamente. Egli indica le ore in cui i confratelli devono 'essere occupati nel lavoro manuale', sottointendendo il lavoro dei campi.[30] Tale lavoro produrrà il cibo capace di saziare i monaci e questo cibo verrà distribuito secondo regole ben precise, in modo da evitare eccessi dannosi. Da sottolineare che base della dieta monastica fossero verdure cotte, mentre il consumo della carne era vietato (quindi non era previsto abbattere animali per cibarsene), se non in caso di malattia.[31] Pertanto per la refezione quotidiana '… siano sufficienti per tutte le mense due pietanze cotte, (…) sicché chi non potesse mangiarne una, si rifocilli con l'altra (…) se ci fosse modo di avere frutta o legumi freschi, si aggiunga una terza pietanza'.[32] Giudichiamo di notevole interesse l'accettazione del lavoro dei campi che si farà (con misura, *mensurate*) *si necessitas loci aut paupertas exegerit*.[33] Da qui si evince che il rapporto con l'ambiente diventa necessario ma non è auspicato. È la congiuntura economica sfavorevole che induce Benedetto a parlare e legiferare anche sul lavoro dei campi. Appare quindi evidente come egli avrebbe preferito che i suoi *monachi* si dedicassero solo alla preghiera e alla lettura di testi sacri. D'altronde anche Macario nella sua Regola invita i confratelli a dedicarsi *madefacti* ad una *laboriosam operam* e a non odiarla.[34] Al contempo, però, lo stesso Macario dice che 'all'ora della preghiera, quando è stato dato il segnale, chi non smette immediatamente il lavoro che sta facendo – poiché niente è da anteporre alla preghiera – e non sarà pronto, sia lasciato fuori affinché ne arrossisca'.[35]

Le diverse *Regulae* dei monasteri della Tarda Antichità dell'Occidente europeo dimostrano un totale disinteresse all'ambiente. E non è un disinteresse legato alla vita monastica. Anzi, come facemmo dianzi cenno, proprio i monaci avrebbero potuto mostrare una sensibilità maggiormente spiccata verso la natura. Non possiamo, d'altronde, dirci stupiti considerata anche la situazione precaria dei regni romano barbarici e della vita di quei monaci.

Poco tempo dopo le cose non sono radicalmente cambiate. Si consideri la vicenda del vescovo inglese Ceadda, di cui parla Beda il Venerabile. A lui venne affidata la regione dei

---

[30] *Regula sancti Benedicti* 48, (Pricoco 2011: 222–5).
[31] Rapetti 2013: 43; Steel 2011.
[32] *Regula sancti Benedicti* 39 (Pricoco 2011: 208–9).
[33] *Regula sancti Benedicti* 48.7 (Pricoco 2011: 224–5).
[34] *Regula Macharii* 8 (Pricoco 2011: 44–5).
[35] *Regula Macharii* 14 (Pricoco 2011: 46–7).

Merciani (l'Essex) e, alla sua morte a seguito di un'epidemia, nel 644, avvenne qualcosa di straordinario. Leggiamo le parole di Beda in proposito:

> Il luogo della sepoltura è coperto da una struttura di legno, a forma di piccola casa; nella parete c'è un'apertura attraverso la quale i pellegrini infilano la mano e asportano una manciata di terriccio; se questo terriccio viene stemperato nell'acqua, e quest'acqua viene data da bere ad animali o uomini malati, subito la malattia cessa e torna la gioia per la salute recuperata.[36]

Beda ci racconta anche di come durante il dominio di Edwin, la Britannia avesse goduto di un periodo sereno, rispetto a quelli precedenti. Il re, racconta il venerabile Beda, 'dove aveva visto sulla pubblica strada limpide fontane, fece innalzare delle pertiche con appesi boccali, perché i viandanti potessero bere'.[37] Da quel racconto si evince che, quando i tempi lo permettevano e cioè quando regnava la pace, anche in Britannia vi era una qualche attenzione per l'ambiente certificata anche dai religiosi. Si trattava però di un concentrarsi rivolto ad aspetti fondamentali ma, diremmo, basici. L'acqua è l'alimento per eccellenza senza il quale nessun essere umano può sopravvivere. All'epoca di Edwin mancava la raffinatezza dei Romani; non c'era nemmeno quel minimo di scienza dei gallo romani o degli ispano romani. Gli elementi naturali, l'acqua in particolare, vengono visti come 'mezzi' dell'uomo e, pertanto, possono essere pienamente utilizzati dall'umanità per il proprio benessere. D'altronde, l'essere umano sin dalla *Genesi* è consapevole di ciò. Il che, ovviamente, non ha mai significato uno sfruttamento della natura stessa ma un semplice suo utilizzo. I monaci, seppur maggiormente immersi nella natura rispetto al clero secolare, sembravano avere molte meno attenzioni dei pagani Romani verso la natura. Anche per il mondo cenobitico infatti, un conto era l'essere umano, un conto era il creato, quasi si trattasse di questioni differenti. Un conto era, come vedemmo, l'impegno verso la preghiera, un conto era l'impegno quotidiano nei campi. Attività nemmeno comparabili seppur fondamentali, l'una e l'altra. La prima necessaria per la vita dello spirito, la seconda per la vita materiale.

3. La cultura dell'ambiente nella Hispania visigotica

Nondimeno alcuni monaci rifletteranno maggiormente sul fatto che l'uomo sia parte di un tutto, quindi indagheranno sulla fisicità della Terra senza però arrivare alla sensibilità dei Romani circa l'ambiente. Curiosa semmai è l'ubicazione di tutti questi religiosi che nacquero e operarono nella *Hispania* dei Visigoti. Si veda, ad esempio, le *Etymologiae* di Isidoro di Siviglia il quale, in un certo qual modo, parlando del mondo 'del suo moto perpetuo: del moto del sole, della luna, dell'aria, del mare' approfondisce, insomma, temi legati alla natura. Il vescovo di Siviglia, elencherà gli elementi a seconda della loro natura di 'esseri animati' o a seconda che traggano il moto da loro stessi o subiscano l'impulso iniziale da altro.

L'aspetto essenziale del creato è l'uomo, che fa parte della natura, ma che sembra quasi non identificarsi come parte di essa. Come scrive il FONTAINE 'l'homme demeure ainsi – en un sens noveau et relatif – l'être en qui le sophiste grec Protagoras avait vu la mesure de

---

[36] Beda il Venerabile, *Hist. Eccl. Angl.* 4.2 (LAPIDGE 2010).

[37] Beda il Venerabile, *Hist. Eccl. Angl.* 2, 16.

toutes choses'.[38] Il che significa, inevitabilmente, che altri aspetti del mondo naturale, seppur ammirato per la sua bellezza dall'uomo, passano in secondo piano. Comunque anche Isidoro, come d'altronde altri padri della Chiesa, osserva come sia la natura a 'far nascere gli esseri'. Egli osserva anche come l'uomo, pur traendo origine *ab humo* (*Genesi* 2.7), alza lo sguardo al cielo per rivolgerlo al suo Creatore. D'altronde Isidoro, nella sua ultima opera, le *Sententiae*, sottolinea il cammino parallelo che fanno il mondo e l'uomo. Egli ricorda come 'la disposizione razionale del mondo la si deve considerare a partire dall'uomo. Perché se pure l'uomo tende alla sua fine percorrendo le diverse età, così il mondo tende alla medesima fine per il fatto che si distende nel tempo, la crescita apparente dell'uomo e del mondo è all'origine stessa del loro declino'.[39] E ancora 'tutte le cose sotto il cielo sono state fatte per l'uomo…'; ma l'essenza principe del pensiero di Isidoro è contenuta da tale considerazione, che 'tutte le cose sono insite nell'uomo, e in lui risiede tutta la natura delle cose. L'uomo è parte considerevole della Creazione e se il suo grado di eccellenza è superiore a quello di tutte le creature è perché lui è stato fatto ad immagine divina'.[40] Il creato è per l'uomo, ma nulla si dice circa l'atteggiamento dell'uomo verso il creato perché esso è stato fatto perché venisse sottomesso all'uomo.

Non ha torto il FONTAINE quando pensa che Isidoro in nessun altro testo abbia espresso la forza dell'immanenza del microcosmo dell'essere umano sul macrocosmo universale proprio per la sua trascendenza spirituale rispetto alla mera materialità del cosmo.[41] Isidoro di Siviglia scrisse anche il *De natura rerum* su richiesta del sovrano Sisebuto il quale sarà egli stesso autore di un poemetto di 61 esametri, di argomento astronomico.[42]

Si rende necessaria una piccola digressione solo per dire come in quel libello il re, seppure in forma ampollosa e, spesso, oscura, rivela discrete conoscenze astronomiche, che testimoniano interessi scientifici insoliti per un'epoca in cui per lo più si concentrava l'at-

---

[38] FONTAINE 2000: 306.

[39] Isid. Hisp., *Sent.* 1.8.2 (PL 83).

[40] Isid. Hisp., *Sent.*, 1.11.1.

[41] FONTAINE 2000: 308.

[42] Di seguito l'*incipit* dell'opera di Isidoro in cui, oltre alla dedica al re Sisebuto, vengono spiegati i motivi che hanno spinto il vescovo di Siviglia a cimentarsi in un impegno di tal fatta. Isid. Hisp., *de nat. rer.* (30) (FONTAINE 1960: 46): *Domino et filio Sisebuto Isidorus. 1. Dum te praestantem ingenio, facundiaque, ac vario flore litterarum non nesciam, impendis tamen amplius curam, et quaedam ex rerum natura vel causis a me tibi efflagitas suffraganda. Ego autem satisfacere studio animoque tuo decursa priorum monumenta non demoror, expediens aliqua ex parte rationem dierum ac mensium, anni quoque metas, et temporum vicissitudinem, naturam etiam elementorum, solis denique ac lunae cursus, et quorumdam causas astrorum, tempestatum, scilicet signa, atque ventorum, necnon et terrae positionem, alternos quoque maris aestus. 2. Quae omnia, secundum quod a veteribus viris, ac maxime sicut in litteris catholicorum virorum scripta sunt, proferentes, brevi tabella notavimus. Neque enim earum rerum naturam noscere superstitiosa scientia est, si tantum sana sobriaque doctrina considerentur. Quinimo si ab investigatione veri modis omnibus procul abessent, nequaquam rex ille sapiens diceret: "Ipse mihi dedit horum quae sunt scientiam veram, ut sciam dispositionem coeli, et virtutes elementorum, conversionum mutationes, et divisiones temporum, annorum cursus, et stellarum dispositiones". 3. Quapropter incipientes a die cujus prima procreatis in ordine rerum visibilium exstat, dehinc caetera de quibus opinari quosdam gentiles vel ecclesiasticos viros novimus, prosequamur, eorum in quibusdam causis, et sensus, et verba ponentes, ut ipsorum auctoritas dictorum fidem efficiat.*

tenzione su altro.[43] L'autore, poi, appare non privo di senso dell'ironia, quando per esempio sottolinea di poter dedicare ai suoi passatempi preferiti solo i ritagli di tempo, perché, da capo di stato, è chiamato ad assolvere i ben più gravi compiti di natura politica e militare, piuttosto che occuparsi di dissertazioni letterarie e scientifiche.[44] Torniamo ad Isidoro di Siviglia per aggiungere che, incoraggiato dallo stesso Sisebuto, si cimentò nella stesura del *De natura rerum,* di cui le *Origines,* sono un'epitome.[45]

Sempre nella penisola iberica, terra dove, a seguito della conversione di Recaredo al credo niceno, si avrà un notevole e repentino sviluppo di scuole monastiche, avremo altri monaci che inseriranno nelle loro usanze anche la frequentazione delle terme. Si trattava di terme romane che ancora erano funzionanti. Esse attiravano la popolazione ispano romana che fino allora non aveva perduto la gioia di immergersi in acque calde, tiepide o fredde, per ritemprare le proprie membra. Gli ispano romani infatti non erano spinti alle terme unicamente per il piacere dei bagni, ma anche dal desiderio di ottenere e mantenere un fisico atletico.[46] Come ricorda il RICHÉ, Leandro di Siviglia non solo non disdegnava andare alle terme pubbliche, ma anzi spiegava ai suoi monaci come la vista di un corpo nudo ben fatto non era qualcosa di disdicevole perché esso era *opus Dei.*[47] Dunque, e qui si può osservare come la Chiesa ispanica non si era ancora allontanata dagli ideali classici, l'ammirazione per una creazione ben fatta diventava una sorta di riconoscimento della grandezza di Dio. Semmai si doveva evitare di limitare l'uomo alla sua materialità perché *propter Deum qui eos fecit non propter pulchritudinem corporis.*[48]

Altro monaco che si dedicò agli aspetti più scientifici della natura fu Eugenio di Toledo. Anch'egli sottolineò maggiormente questioni che, seppur naturali, appartenevano al mondo degli astri.[49] Nondimeno, non ebbe timore di cimentarsi in uno studio non disprezzato nella penisola iberica ma, forse, meno apprezzato in altre zone della *pars Occidentis.*

---

[43] Sisebuto, *Carmen de eclipsibus solis et lunae* (FONTAINE 1960: 151–61, 328–35). In proposito si vd. anche FONTAINE: 1948; FONTAINE 1967: 87–147, partic. 127–8; BANNIARD 1994: 101.

[44] RICHE 1962: 304–5.

[45] Qui di seguito riporto la prefazione dell'opera isidoriana. Isid., orig. 13: *De mundo et partibus. In hoc vero libello quasi in quadam brevi tabella quasdam caeli causas situsque terrarum et maris spatia adnotavimus, ut in modico lector ea percurrat et conpendiosa brevitate etymologias eorum causasque cognoscat.* Si vd. anche l'edizione di VALASTRO-CANALE 2004, I–II.

[46] Leandr., *Regula ad virg.* (PL 72, 880): *Balneo non pro studio vel nitore corporis sed tantum pro rimedio.* Isid. Hisp., *Regula monachorum* (PL 83, 891–2): *Sub praetextu infirmitatis nihil peculiare habendum est, ne lateat libido cupiditatis sub languoris specie. Lavacra nulli monacho adeunda studio lavandi corporis, nisi tantummodo propter necessitatem languoris, et nocturnam pollutionem.*

[47] RICHE 1962: 342.

[48] Leand., *reg.* c. 3 (PL 72, 882): *Amandi sunt certe viri ut opus Dei, sed absentes … propter Deum qui eos fecit non propter pulchritudinem corporis.*

[49] Ildefons. Tolet., *De Vir. Ill.* (PL, 96, 203–4): *Eugenius discipulis Helladii, collector et consors Justi, pontifex post Justum accedit, ab infantia monachus, ab Helladio cum Justo pariter sacris in monasterio institutionibus eruditus. Hunc secum Helladius a monasterio tulit ad pontificatum tractus, qui rursus ab eo clericalibus institutus ordinibus, sedis ejus post illum tertius rector accessit. Et bonum meritum senis, qui duobus discipulis sanctisque filiis Ecclesiae Dei haereditatem meruit relinquere gubernandam. Idem Eugenius moribus incessuque gravis, ingenio callens. Nam numeros, statum, incrementa, decrementaque cursus recursusque lunarum tanta peritia novit, ut considerationes disputationis ejus auditorem et in stuporem verterent et in desiderabilem doctrinam inducerent. Vixit in sacerdotio fere undecim annis, regnantibus Chintila, Tulgane et Chindasvintho regibus.*

Isidoro, Leandro, Eugenio scrissero opere atte alla diffusione tra i religiosi di una conoscenza dell'ambiente che andasse oltre il modo di vedere dei testi sacri. La loro opera, pertanto, voleva offrire una visione, la più completa possibile della realtà, perché presupponevano che, essendo il creato e le sue leggi opera di Dio, lo studio dei fenomeni naturali fosse uno dei modi di onorare il Signore. Naturalmente, per la compilazione del *De natura rerum*, Isidoro dovette attingere a trattati di autori pagani e cristiani senza distinzione alcuna. Nello stesso periodo Gregorio Magno che, ad ogni buon conto, monaco non era, sembra trascurare l'habitat perché convinto di dover indirizzare l'attenzione del chierico verso la spiritualità e cita, come sovente accade nella letteratura cristiana, aspetti della natura solo come metafore. Da tale breve dissertazione appare chiaro come il mondo cristiano, e in particolare quello monastico, non fuggì dal mondo, ma analizzò il creato con estrema circospezione. Il retaggio culturale che i monaci avevano sull'ambiente era pur sempre quello latino. Ma essi, proprio perché dovevano occuparsi della *cura animarum*, non potevano approfondire quei temi che invece appassionarono alcuni studiosi latini. Per tacere dei versetti della *Genesi* che mettono l'uomo al centro di un Creato a lui sottoposto. Orosio, seppur con un piccolo cenno, parlò dei 'barbari che si convertirono all'aratro', con ciò lasciando intendere come anche i nuovi padroni, gioco forza, iniziarono ad avere un rapporto con il territorio che fosse anche costruttivo.[50] D'altronde quei barbari da nomadi o semi nomadi si erano trasformati in stanziali e la conoscenza delle tecniche agricole era diventata essenziale.

Ma la discriminante tra mondo antico e mondo alto medievale rispetto all'ambiente, come facemmo cenno, è l'atteggiamento della Chiesa che, più volte lo ricordammo, ha altri pensieri e usa ciò che circonda l'uomo mostrando, via via, una sempre minore preoccupazione verso appunto l'ecologia. Il mondo medievale chiedeva altro. Inevitabilmente si deve ritornare a ciò che dicemmo all'inizio e cioè che nei primi secoli medievali non esistessero strutture statali capaci di porre la giusta attenzione sul territorio.

4. La cultura dell'ambiente nei monasteri dopo il secolo ottavo

Il mondo monastico, sebbene fosse quello più immerso nella natura, sebbene fosse quello forse più attento alla cultura in senso lato, non farà propria se non in minima parte (almeno sino a Gerberto di Aurillac, cioè sino al secolo decimo), la sensibilità e il rispetto dell'ambiente dell'antichità latina. Anche Giovanni Scoto, che pure scrisse sulla natura, non affrontò con piglio deciso il tema dell'ambiente.[51] Egli, *scholasticus* e direttore della *schola palatina* alla corte di Carlo il Calvo, è senz'altro un faro della cultura occidentale del secolo nono; è, perciò, ancora più indicativo il fatto che non abbia mai posto l'accento su temi che andassero al di là del legame imprescindibile tra la natura e Dio, quasi che l'ambiente fosse 'altro' rispetto alla natura e quindi a Dio. Il suo *Periphyseon* è un'opera prettamente filosofica dove l'autore affronta anche temi inerenti la natura. Essa però è vista come mezzo per arrivare a Dio, non in se stessa dunque. Inoltre la natura di Giovanni Scoto rimane per molti versi immaginaria, provvisoria, ipotetica. Non approfondisce in modo particolare temi inerenti la natura stessa. Fa un'eccezione quando affronta argomenti di ordine astronomico e qui, primo sino al Rinascimento, esprime una teoria semi-eliocentrica quando afferma 'essi

---

[50] Oros., *Hist. contra pag.* 7.41.7 (LIPPOLD 2001).
[51] Johannes Scot., *Sulle nature dell'Universo* (DRONKE 2012).

[Giove, Marte, Venere e Mercurio] percorrono le loro orbite sempre intorno al sole, come insegna Platone nel *Timeo*'.[52] Il suo però è uno studio filosofico che non ha nulla di pratico.

Dal secolo decimo, nonostante la riforma cluniacense, inizierà una nuova fase e il pensiero di Giovanni Scoto, ormai assimilato, avrà senz'altro il suo peso. Tale riforma implicò, tra le altre cose, che i monaci di Cluny – lo scrive Cantarella – 'non hanno necessità di compromettersi con il lavoro dei campi, ma nemmeno debbono farlo; Benedetto d'Aniane, zelante riformatore del monachesimo, cent'anni prima che fosse fondata Cluny si era domandato se il lavoro agricolo dei monaci non maculasse la loro funzione presbiterale ed orante: una macchia d'ordine fisico, non morale. Se le storie raccontavano che a qualcuno capitava di praticarlo, sottolineavano l'eccezionalità dell'evento e dell'uomo'.[53] In effetti le terre dove sorge quel monastero sono di proprietà di Guglielmo, conte di Mâcon e duca di Aquitania. Di conseguenza 'il monastero non nasce affatto sfornito di beni … è dotato di tutte le pertinenze della villa: ville, cappelle, servi dei due sessi, vigne, campi, prati, boschi, acque e corsi d'acqua, mulini, vie d'accesso e d'uscita …'.[54] Insomma, seppur immersi nella natura, i monaci di Cluny, diversamente dai primi benedettini, giudicarono opportuno anche potersi affrancare dal confronto quotidiano con essa. In ciò, pertanto, essi si allontanarono anche dall'insegnamento biblico. Ciò inevitabilmente comportò un distacco dall'attenzione all'ambiente che qualche benedettino poteva coltivare. Nondimeno da Cluny scaturì una pletora di personaggi che occuparono un posto di rilievo nella cultura del tempo e che ebbero la necessità, almeno alcuni di loro, di confrontarsi con il mondo della natura. In fondo non lavorare i campi non allontanò totalmente la curiosità 'teorica' di ciò che è la natura e quindi l'ambiente. Certamente non furono tanti gli studiosi che si avvicinarono al Creato.

Uno di questi fu il monaco Gerberto di Aurillac. Egli si dedicherà, anche dopo la sua nomina a pontefice romano, all'osservazione degli astri. Ma non solo. Il suo impegno come matematico lo porterà ad osservare l'habitat con vero spirito scientifico. E la sua stessa idea teologica andrà oltre quella tipica del suo tempo. Egli, quasi fosse un antico greco, era sempre alla ricerca dell'armonia. Perciò rimaneva stupefatto di fronte agli spettacoli naturali e tentava con i suoi calcoli di capire la natura, nel pieno rispetto per essa. Gerberto di Aurillac, pertanto, non sfruttò mai la natura. La usò così come si conviene per le sue invenzioni. Si pensi agli organi idraulici o all'orologio solare. La natura è vista come razionale perché segue regole matematiche. Tali regole sono fondamento della natura poiché essa è scaturita dalla razionale mente di Dio; ecco perciò spiegato l'interesse gerbertiano per la matematica considerata, pertanto, come mezzo per poter studiare la fisica e l'astronomia. Matematica, fisica e astronomia sono discipline che permettono una comprensione approfondita dei fenomeni naturali che – è bene ribadire tale concetto – sono, per un credente, frutto della creazione dell'Eterno. Quegli studi, in ultima analisi, sono un mezzo necessario per apprezzare l'armonia dell'Universo e tentare di meglio penetrare nella mente del Creatore. L'astronomia, dunque, pur non rientrando direttamente in un discorso legato all'ambiente, fa d'altra parte riflettere su come le attenzioni dell'uomo di chiesa iniziassero a concentrarsi anche sulla natura. È solo un'attenzione maggiore verso i fenomeni naturali che porterà, molto lentamente, la Chiesa tutta e i monaci, in particolare, a interessarsi maggiormente della realtà naturale.

---

[52] Johannes Scot. *de div. nat.* 3.698a.
[53] Cantarella 1997: 21–2; Hägermann 1989: 346–51.
[54] Cantarella 1997: 20. Sulla fondazione dell'abbazia di Cluny Bernard, Bruel 1876–1903.

5. L'opera astronomica di Gerberto di Aurillac

Un *unicum* appare, in ambito alto medievale, l'interesse gerbertiano per l'astronomia che, nell'Occidente europeo, era inevitabilmente venuto meno dopo l'antichità tarda.[55] Il monaco auriliacense, oltre a dimostrare di possedere conoscenze di base diffuse all'epoca (si consideri, ad esempio l'opera di Boezio *De Astronomia*), nel suo epistolario discorre più volte, e con notevole dovizia di particolari, di tecnologia. Cioè di strumenti da lui utilizzati per individuare corpi celesti, per calcolare distanze astrali e così via.[56] Ora, la costruzione di quei marchingegni necessari per indagare la volta celeste è un qualcosa di non comune nel secolo decimo. Ma facciamoci guidare dallo stesso Gerberto in questo percorso che riflette i suoi interessi astronomici.

Egli, in alcune epistole indirizzate a Remigio, monaco di Mettlach, e allo scolastico Costantino di Fleury, fa riferimento a una sfera che, tempo prima, gli aveva consentito l'osservazione e lo studio degli astri.[57] E un riferimento specifico a questi studi è anche negli scritti del suo allievo Richerius, il quale pensò di scrivere una storia della Francia, includendo tra i personaggi celebri proprio il suo vecchio maestro.[58] La sfera, costruita, ma non in-

---

[55] Potrebbe darsi che Gerberto avesse letto da Cassiodoro nella *Varia* 1.45.4, che Boezio aveva tradotto un'opera astronomica di Tolomeo, ma non si può dire con certezza che abbia letto tale opera, quindi le sue conoscenze astronomiche, almeno in parte, potrebbe averle acquisite durante il soggiorno nella marca di Barcellona, in proposito però le fonti non ci aiutano. Cfr. LINDGREN 1985: 619–44, partic. 619. Su Gerberto di Aurillac si vd. MONTECCHIO 2011.

[56] Gerberto stesso nel 983, periodo in cui era abate a Bobbio, in una lettera indirizzata al vescovo di Reims, Adalberone, parlerà degli 8 volumi dell'opera astronomica di Boezio. Cfr. Gerberto di Aurillac, *Corr. ep.* 8 (RICHÉ, CALLU 1993: 16–9).

[57] Gerberto di Aurillac, *Corr.*, *ep.* 148 e 152 (a Remigio); *Corr.*, annessione 5, *ep.* 3.680 (a Costantino di Fleury). Le pagine seguenti sono nella sostanza riprese da MONTECCHIO 2011: 128 e sgg.

[58] Richerius monacus, *Hist. libri quattuor* (3.50–3): *In primis enim mundi speram ex solido ac rotundo ligno argumentatus, minoris similitudine, maiorem expressit. Quam cum duobus polis in orizonte obliquaret, signa septentrionalia polo erectiori dedit, australia vero deiectori adhibuit. Cuius positionem eo circulo rexit, qui a Graecis orizon, a Latinis limitans sive determinans appellatur, eo quod in eo signa quae videntur ab his quae non videntur distinguat ac limitet. Qua in orizonte sic collocata, ut et ortum et occasum signorum utiliter ac probabiliter demonstraret, rerum naturas dispositis insinuavit, instituitque in signorum comprehensione. Nam tempore nocturno ardentibus stellis operam dabat; agebatque ut eas in mundi regionibus diversis obliquatas, tam in ortu quam in occasu notarent. Circuli quoque qui a Graecis paralleli, a Latinis aequistantes dicuntur, quos etiam incorporales esse dubium non est, hac ab eo arte comprehensi noscuntur. Effecit semicirculum recta diametro divisum. Sed hanc diametrum fistulam constituit, in cuius cacuminibus duos polos boreum et austronothum notandos esse instituit. Semicirculum vero a polo ad polum XXX partibus divisit. Quarum sex a polo distinctis, fistulam adhibuit, per quam circularis linea artici signaretur. Post quas etiam quinque diductis, fistulam quoque adiecit, quae aestivalem circulationem indicaret. Abinde quoque quatuor divisis, fistulam identidem addidit, unde aequinoctialis rotunditas commendaretur. Reliquum vero spatium usque ad notium polum, eisdem dimensionibus distinxit. Cuius instrumenti ratio in tantum valuit, ut ad polum sua diametro directa, ac semicirculi productione superius versa, circulos visibus inexpertos, scientiae daret, atque alta memoria reconderet. Errantiumque siderum circuli cum intra mundum ferantur, et contra contendant, quo tamen artificio viderentur scrutanti non defuit. Inprimis enim speram circularem effecit; hoc est ex solis circulis constantem. In qua circulos duos qui a Graecis coluri, a Latinis incidentes dicuntur, eo quod in sese incidant complicavit; in quorum extremitatibus polos fixit. Alios vero quinque circulos, qui paralleli dicuntur, coluris transposuit, ita ut a polo ad polum*

ventata dal monaco di Aurillac, era uno strumento senz'altro ignorato in quell'epoca e in quei luoghi poteva essere utilizzata in tre modi. Il *modus operandi* della sfera dipendeva e variava a seconda che si volesse studiare la conformazione terrestre, individuare l'eclittica unitamente allo zodiaco, o ci si prefiggesse di osservare la volta celeste.[59] In un secondo momento quell'esemplare di sfera fu superato dallo stesso Gerberto che, per osservare con più

XXX partes, sperae medietatem dividerent; idque non vulgo neque confuse. Nam de XXX dimidiae sperae partibus a polo ad primum circulum, sex constituit; a primo ad secundum quinque; a secundo ad tertium, quatuor; a tertio ad quartum, itidem quatuor; a quarto ad quintum, quinque; a quinto usque ad polum, sex. Per hos quoque circulos eum circulum obliquavit, qui a Graecis loxos, vel zoe, a Latinis obliquus vel vitalis dicitur, eo quod animalium figuras in stellis contineat. Intra hunc obliquum, errantium circulos miro artificio suspendit. Quorum absidas, et altitudines a sese etiam distantias, efficacissime suis demonstravit. Quod quemadmodum fuerit, ob prolixitatem hic ponere commodum non est, ne nimis a proposito discedere videamur. Fecit praeter haec speram alteram circularem, intra quam circulos quidem non collocavit, sed desuper ferreis atque aereis filis signorum figuras complicavit. Axisque loco, fistulam traiecit, per quam polus coelestis notaretur, ut eo perspecto, machina coelo aptaretur. Unde et factum est, ut singulorum signorum stellae, singulis huius sperae signis clauderentur. Illud quoque in hac divinum fuit, quod cum aliquis artem ignoraret, si unum ei signum demonstratum foret, absque magistro cetera per speram cognosceret. Inde etiam suos liberaliter instruxit. Atque haec actenus de astronomia. (LATOUCHE 1930).

[59] Ecco come Gerberto, in una sua epistola, descriveva la costruzione della sfera. Gerberto di Aurillac, *Corr.*, annessione 5, *ep.* 3.680–6: *Sphaera ad coelestes circulos vel signa ostendenda componitur ex omni parte rotunda, quam dividit circumducta linea mediam aequaliter in LX partibus divisa. Ubi itaque constituis caput lineae, unum circini pedem fige, et alterum pedem e regione ibi constitue, ubi VI partes finiuntur de LX partibus praedictae lineae; et dum circinum circumduxeris, XII pertes includis. Non mutato primo pede, secundus pes extenditur usque ad locum, quo de praedicta linea undecima pars finitur; et ita circumducitur, ut XXII partes circumplectatur. Eodemque modo adhuc pes usque ad finem quintae decimae partis praedictae lineae protenditur et circumductione XXX partes habens media sphaera secatur. Tunc mutato circino in altera parte sphaerae, ubi primum pedem fixeras, attendens, ut contra statuas, praedictam rationem mensurae circumductionis et partium complexionis observabis. Nam V solummodo erunt circumductiones, quarum media aequalis est lineae in LX partibus divisae. Alterutro istorum hemisphaeriorum sumpto, interius cavato et, ubi circini alterum pedem in praedicta linea ad circumducendo fixerat, perfora, ut circumductio medium foraminis teneat. In capitibus quoque sphaerae, ubi primum pedem circini posuisti, singula foramina facis, ut medietas foraminum illorum terminet praedictum hemisphaerium. Nam ita VII erunt foramina, in quibus singulis singulas semipedales fistulas constituis; eruntque duae extremae contra se positae, ut per ultrasque, tanquam per unam, videas. Ne vero fistulae hac illacque titubent ferreo semicirculo, ad modum praefati hemisphaerii secundum suam quantitatem mensurato et perforato, utere, quo superiores extremitates fistularum coherce; quae hoc differunt a fistulis organicis, quod per omnia aequalis sunt grossitudinis, ne quid offendat aciem per eas coelestes circulos contemplantis. Semicirculus vero duorum digitorum ferme latitudinis, ut omne hemisphaerium, XXX partes habet longitudinis, srvans aequalem rationem divisionis, qua perforatus fistulas recipit. Notato itaque nostro boreo polo, descriptum hemisphaerium taliter pone sub divo, ut per utrasque fistulas, quas diximus extremas, ipsum boreum polum libero intuitu cernas. Si autem de polo dubitas, unam fistulam tali loco constitue, ut non moveatur tota nocte, et per eam stellam suspice, quam credis esse polum, et si polus est, eam tota nocte poteris suspicere, sin alia, mutando loca non occurrit visui paulo post per fistulam. Igitur preadicto modo locato hemisphaerio, ut non moveatur ullo modo, prius per inferiorem et superiorem primam fistulam boreum polum, per secundam arcticum circulum, per tertiam aestivum, per quartam aequinoctialem, per quintam hiemalem, per sextam antarticum circulos metiri poteris. Pro polo vero antarctico, quia sub terra est, nihil coeli, sed terra tantum per utrasque fistulas intulenti occurrit.*

precisione le stelle, concepì un marchingegno ancora più potente e sofisticato, paragonabile a una specie di telescopio, che gli fu utile per la scoperta delle macchie solari.[60] Volendo inoltre calcolare la posizione esatta degli astri che si stagliano nella volta celeste, il monaco di Aurillac ricorse a un ulteriore dispositivo, all'astrolabio.[61] Questo strumento, come si sa, era già stato utilizzato secoli prima da Ipparco e Tolomeo, notizia di cui Gerberto era ben consapevole, ma ora, nel decimo secolo, era caduto in disuso. Con esso si poteva calcolare la longitudine e la latitudine delle stelle che appaiono nella volta celeste e dunque, in sostanza, era possibile eseguire calcoli di geometria sferica con una certa facilità e precisione, operazioni, queste, altrimenti molto più ardue.[62] Come si può arguire, il monaco auriliacense, anche per quanto riguarda la scienza astronomica, così come per altre questioni, filosofiche, politiche, ebbe un ruolo non secondario perché concorse a far riscoprire prima, approfondire poi e, infine, a diffondere nozioni ormai perdute. Tale attenzione verso gli astri, come acutamente osserva lo Schramm, sembrava scomparsa addirittura dai tempi degli antichi egizi, mentre adesso, con questo monaco 'curioso', si riaffaccia in un mondo ormai imbarbarito, anticipando anche in questo caso quelli che saranno gli interessi degli uomini che verranno dopo, quegli uomini cioè che spenderanno tante loro energie per coronare il sogno di conquistare le stelle. Va considerato inoltre che l'astronomia, disciplina che andava di pari passo con l'astrologia, era una scienza estranea alla cultura cristiana di quell'epoca, tanto è vero che, sempre secondo lo Schramm, sarebbe stato proprio l'interesse di Gerberto per l'astronomia a far nascere tra i suoi contemporanei, subito dopo la sua morte, sospetti di magia.[63] Siffatta considerazione è da noi condivisa e temiamo si tratti di qualcosa di più di un semplice sospetto: non sarebbe possibile spiegare altrimenti come un personaggio della levatura di Gerberto d'Aurillac sia stato 'dimenticato' per secoli.

Conclusioni

Le pagine precedenti hanno voluto dimostrare quanto fosse grande la distanza tra il mondo alto medievale e quello classico rispetto al tema dell'ambiente. Laddove i Romani, soprattutto, mostrarono un atteggiamento modernissimo rispetto al rapporto tra l'uomo e la natura, il mondo medievale dimostrò una distanza inaudita. Se i Romani si posero fattivamente la questione di come convivere con la natura che, nei fatti, giudicarono un'interlocutrice, così non fece l'uomo medievale, né il monaco. Furono i Romani a bonificare laghi e paludi mentre l'uomo medievale soltanto nella seconda parte di quell'evo si cimentò in simili imprese.

Tale distanza è dovuta a due ordini di motivi. Il primo è una realtà affatto diversa tra i due mondi. Quello latino, siamo costretti a ripeterlo, era inserito in un contesto statale che

---

[60] Si trattava di 'tubi ottici' e tale tubo in arabo si chiama 'anbu'ba'. È proprio la presenza della parola 'ottico' che suggerisce l'accostamento di tali 'tubi' ai moderni telescopi. VERNET 1978: 109; cfr. anche CARRARA 1908: 19–20.

[61] SCHRAMM 1962: 67 pensa che Gerberto abbia portato con sé dalla marca di Barcellona un astrolabio e un periscopio.

[62] L'astrolabio divenne poi uno strumento più semplice usato dai naviganti: si trasformò in quadrante. CARRARA 1908: 20. Secondo POULLE 1985: 597–617, partic. 607–9, l'astrolabio gerbertiano non sarebbe altro se non un 'notturlabio', uno strumento, cioè, che serviva a far conoscere l'ora di notte.

[63] SCHRAMM 1971: 324.

solo in epoca moderna si potrà rivedere (almeno in parte). Quello medievale, a partire dalla Tarda Antichità, vivrà in tutt'altro contesto. Il secondo e, forse più incisivo motivo, riguarda il credo religioso che porterà le persone più colte a indirizzare la propria speculazione sul metafisico, prescindendo così da interessi materiali legati all'ecosistema. Le questioni inerenti all'ambiente torneranno nella seconda fase dell'evo medievale, quando cioè la Chiesa, dovrà confrontarsi con san Francesco che, dimostrando un acume eccezionale, osserverà la natura come meravigliosa creazione di Dio. Ovviamente l'uomo è l'essere per eccellenza, ma l'uomo vive immerso nella natura e quindi deve confrontarsi con lei quotidianamente. Con san Francesco, in buona sostanza si avrà una sorta di rivoluzione copernicana. Il cammino della Chiesa per avvicinarsi al Creato sarà lunghissimo. Al contrario il mondo non strettamente legato alla Chiesa potrà avvicinarvisi prima e meglio.

Va anche fatto osservare che, nonostante il riguardo del mondo romano verso l'habitat, esso, a causa del grande livello tecnologico raggiunto, procurò danni di un certo livello. Il THOMMEN a riguardo dice che 'si praticò uno sfruttamento forzato del suolo, si abbatterono foreste, si edificarono rive di laghi e di fiumi, si deviarono acque con l'ausilio della tecnologia e le si inquinarono, si sfruttarono animali e interi raccolti vennero distrutti durante la guerra'.[64] Insomma lo storico svizzero dipinge un quadro a tinte fosche della classicità latina. Ma in effetti bisogna valutare che città 'moderne', strutture statali 'moderne' significarono inevitabilmente un grande sfruttamento del territorio. Al contrario, durante la Tarda Antichità e l'Alto Medioevo, un minore sviluppo tecnologico rispetto all'evo precedente, e un livello demografico meno marcato di prima, influirono in modo meno traumatico sul rapporto dell'uomo con l'ambiente. Ciò naturalmente a prescindere da una coscienza ambientale che non poteva esistere per le cause succitate. In epoca moderna, infine, a causa di un'inusitata esplosione demografica, oltreché con la rivoluzione industriale, gli abitanti della Terra, quasi assopiti nella loro coscienza ecologica, si sono dovuti svegliare di soprassalto per porre rimedio alle tragiche conseguenze dovute al totale disprezzo nei confronti del pianeta che ci ospita e quindi della natura tutta.

Bibliografia

ALICI, L. 2001. *Agostino, La città di Dio*. Varese.

ANTONELLI, G. 1846. *M. Terentius Varro, De re rustica*. Venezia.

BANNIARD, M. 1994. *La genesi culturale dell'Europa*. Bari.

BERNARD, A. e A. BRUEL. 1876–1903. *Recueil des Chartes de l'Abbaye de Cluny*, 6 voll. Paris.

BRÜCKNER, H. 1986. 'Man's Impact on the Evolution of the Physical Environment in the Mediterranean Region in Historical Times', *GeoJournal*, 13.1, 7–17.

CANTARELLA, G. M. 1997. *I monaci di Cluny*. Torino.

CARRARA, B. 1908. *L'opera scientifica di Gerberto o sia papa Silvestro II novellamente discussa ed illustrate*. Roma.

CREMASCHI, L. 1995. *Atanasio di Alessandria, Vita di Antonio*. Milano.

d'AYALA VALVA, L. 2007. *Cassiano, Le istituzioni cenobitiche*. Magnano (BI).

DELORT, R. 2014. *La vita quotidiana nel Medioevo*. Roma-Bari.

DELORT, R., e F. WALTER. 2002. *Storia dell'ambiente europeo*. Bari.

DRONKE, P. 2012. *Giovanni Scoto Sulle nature dell'Universo*. Cles (TN).

---

[64] THOMMEN 2014: 136.

EHRSHAM, VOIGT L. 1979. 'Anglo Saxon Plant Remedies and the Anglo-Saxons', *ISIS* 70, 250–68.

FEDELI, P. 1990. *La natura violata. Ecologia e mondo romano*. Palermo.

FONTAINE, J. 1948. *La culture poétique du roi wisigoth Sisebut, Actes du Congrès de l'Association Guillaume Budé*. Grenoble.

FONTAINE, J. 1967. *Conversion et culture chez les Wisigoths d'Espagne*, in *La conversione al Cristianesimo nell'Europa dell'Alto Medioevo*, in Centro italiano di Studi sull'Alto Medioevo (Spoleto 14–19 aprile 1966). Spoleto, 87–147.

FONTAINE, J. 2000. *Isidore de Séville*. Turnhout.

GIULIANO, G. 2008. *Paolino di Nola, Preghiere*. Nola.

GRANT, E. 2007. *A History of Natural Philosophy*. Cambridge.

HÄGERMANN, D. 1989. 'Der Abtals Grundherr. Kloster und Wirtschaft im frühen Mittelalter', in F. PRINZ (ed.), *Herrschaft und Kirche. Beiträge zur Entstehung und Wirkungsweise episkopaler und monastischer Organisationsformen*. Stuttgart, 346–51.

HARRISON, P., R. L. NUMBERS, e M. H. SHANK (eds.), 2011. *Wrestling with Nature*. Chicago.

LAPIDGE, M. 2010. *Beda, Historia ecclesiastica Gentis Anglorum*. Padova.

LATOUCHE, R. 1930. *Richerius monacus, Historiarum libri quattuor*. Paris.

LINDGREN, V. 1985. 'Ptolémée chez Gerbert d'Aurillac', in M. TOSI (ed.), *Gerberto. Scienza, storia e mito*, Atti del *Gerberti Symposium* (Bobbio 25–27 luglio 1983). Bobbio, 619–44.

LIPPOLD, A. 2001. *Orosio, Historiae contra paganos*. Foggia.

MIGNE, J. P. 1844–1855. *Patrologia Latina*, Paris.

MIGOTTO, L. 1990. *Marco Vitruvio Pollione, De Architectura*. Pordenone.

MONTECCHIO, L. 2011. *Gerberto d'Aurillac. Silvestro II*. Perugia.

MONTECCHIO, L. 2012. *I Bacaudae. Tensioni sociali tra tardo antico a alto medioevo*. Roma.

POULLE, E. 1985. *L'astronomie de Gerbert (597–617)*, in M. TOSI (ed.), *Gerberto. Scienza, storia e mito*, Atti del *Gerberti Symposium*, Bobbio 25–27 luglio 1983. Bobbio, 597–617.

PRICOCO, S. 2011. *La regola di san Benedetto e le regole dei Padri*. Torino.

RACKHAM, H. 1938. *Pliny, Natural History*. Harvard.

RAPETTI, A. 2013. *Storia del monachesimo medievale*. Urbino.

RICHÉ, P. 1962. *Éducation et culture dans l'Occident barbare. VI–VIII siècles*. Paris.

SCHRAMM, P. E. 1962. *Denkmale der deutschen Könige und Kaiser*. München.

SCHRAMM, P. E. 1971. *Kaiser Könige und Päpste*, vol. 4. Stuttgart.

SIMONETTI, M. 1994. *Sant'Agostino, L'istruzione Cristiana*. Cuneo.

STEEL, K. 2011. *How to Make a Human. Animals and Violence in the Middle Ages*. Ohio SU.

THOMMEN, L. 2014. *L'ambiente nel mondo antico*. Bologna.

VENTURA, F. 2012. *Marcus Porcius Cato, Liber de agri cultura*. Reggio Calabria.

VERNET, J. 1978. *La cultura hispano-àrabe en Oriente y Occidente*. Barcellona.

WARD-PERKINS, B. 2008. *La caduta di Roma e la fine della civiltà*. Bari.

WICKHAM, C. 2014. *L'eredità di Roma. Storia d'Europa dal 400 al 1000*. Bari.

Environment, pollution, and diseases

Elizabeth Craik

Malaria and the Environment of Greece

1. Phases in Medical Historiography

A century ago it was cogently argued that malaria was prevalent in the ancient world, and at the same time suggested – with great originality but little caution – that this prevalence led to the weakening and ultimate demise of the ancient civilizations both of Greece and of Rome.[1] Subsequently, a more nuanced approach, taking account of the complex character of the disease, supported and advanced the contention that malaria was present in various forms and with a range of complications in mainland Greece of the classical period, but at the same time discarded theories of its social and political effects.[2] There is now a scholarly consensus on the prevalence of malaria in the Greco-Roman world.[3] However, debate continues on many aspects, including the nature and extent of its incidence, the circumstances and dates of its arrival in different areas of Europe, and its possible effects on human lifestyles and longevity. Much work is being done, and much remains to be done, by archaeologists and historians in collaboration with medical specialists. Archaeological investigation of cemeteries conjoined with detailed analysis of human DNA can reveal much about the ravages of the disease in particular communities in particular eras and can also illumine successive patterns of change, as human genetic mutations confer some resistance to the disease. Social and geographical study of changes in terrain, in particular changes attendant on deforestation in conjunction with introduction of new agricultural practices, can provide ancillary evidence for the development of the disease in particular regions at particular times. Study of population movements, especially those associated with warfare, can suggest the arrival of malarial disease in areas previously unaffected; here the Persian wars at the beginning of the fifth century, the Athenian expedition to Sicily towards its end and the fourth century expeditions of Alexander to India have all been canvassed as originating disease spread. There are modern parallels for such spread through 'war malaria': there was a significant outbreak of the disease among troops on the Salonika front in 1916. Analysis of different species of mosquito and the minutiae of their life cycles has led to the conclusion that malarial disease was spread to Greece and to Italy by different routes: from the Levant to Greece via the islands of the Aegean and from North Africa to Italy via Sicily and Sardinia.[4]

---

[1] See Jones 1907; Jones 1909; also the critique of van der Eijk 2014: 112–7.

[2] See Grmek 1983, Eng. tr. by Muellner 1989.

[3] See Burke 1996: 2252–81.

[4] See Sallares, Bouwman and Anderweg 2004: 311–28; Carter and Mendis 2002: 527–44.

The topic of malaria in the modern world is vast. The topic of malaria in antiquity, though certainly less vast, is nevertheless extremely complex. It can be meaningfully pursued only by selective concentration on narrow aspects.[5] In this paper, attention is focused on the Greek archipelago of the fifth and fourth centuries BC, with particular regard to environmental aspects in disease causation and to the evidence of Hippocratic physicians.[6] In a further paper, arguments advanced here will be developed with particular attention to cases of malaria in pregnancy, resulting in abortion or stillbirth with poor prospects of maternal and neonatal survival.[7]

2. Aetiology of Malaria

Malaria is not an environmental disease, but it is a disease much connected with the environment. It is classified as 'infectious', as it is caused by the spread of pathogens from one human to another. But whereas most infectious diseases have a direct bacterial or viral origin, the origin of malaria is indirect and parasitic. The malarial parasite passes from a malarial human to the biting *anopheles* mosquito and then onwards by the insect's bite to its next human victim. Thus, both parasite and *anopheles* are required for the disease, which is not directly passed from person to person and is not communicable by contagion. This process is now well documented and has been relatively well understood since fundamental research on the life cycle of mosquito parasites carried out in the late nineteenth century by two Scots: RONALD ROSS (1857–1932), a surgeon in the Indian Medical Service, and his influential predecessor PATRICK MANSON (1844–1922), commonly designated the founder of tropical medicine.

Everyone now knows that the mosquito bites can be dangerous as a cause of malaria (and indeed they can be dangerous as a cause of other diseases also, such as yellow fever and dengue fever, not of concern to us here). However, just as the nature of tuberculosis was not understood until 1882, with the discovery of the causative bacillus by ROBERT KOCH (1843–1910), though the disease had long existed in the guise of 'consumption' (as in Greece, *phthisis*), so the nature of malarial fevers (as in Greece, *puretos* or more specifically *tritaios, tetartaios* etc.) was fundamentally mysterious. Strictly, the names 'tuberculosis' and 'malaria' are anachronistic before these nineteenth century discoveries. Malarial disease had remained mysterious for some time even after a connection with mosquitoes began to be suspected by physicians in India. And for centuries before, when the basic connection of cause and effect between mosquitoes and particular diseases was quite unknown, ignorance about the aetiology of malaria was profound.

Indeed the cause could scarcely have been deduced from simple observation: the process has an intrinsic strangeness and inscrutability; also, not all mosquitoes transmit malarial disease; and further the illness does not present immediately but develops some time (typically 9 to 16 days but in certain circumstances much longer) after the precipitating mosquito bite and parasitic infection. However, it will be seen below that despite this fundamental ignorance a connection of febrile and other diseases with stagnant water, the typical

---

[5] See in general GINOUVÈS ET ALII (eds.) 1994 on environmental aspects, especially the article of CORVISIER.

[6] See CRAIK 2015 on all Hippocratic texts.

[7] CRAIK forthcoming 2017.

breeding ground of mosquitoes, was well recognised in antiquity: in this realisation, there was an oblique awareness of a connection between disease and mosquitoes and a direct consciousness of the importance of the environment. The very names for malarial disease, still current and applied before the discoveries of the late nineteenth century, indicate tellingly if indirectly that it was thought to originate in a foul and malodorous atmosphere, such as the miasmic atmosphere typical of the standing or stagnant water seen in the swampy ground, bogs, pools and marshes where mosquitoes tend to breed. English 'malaria' from Italian *mala aria* signifies 'bad air'. It was Giovanni Lancisi (1654–1720) who in the early eighteenth century linked the disease firmly with swamps. And other names for the disease are still more specific. In Modern Greek, ἑλονοσία has the connotation 'marsh-disease' and in French 'paludisme' has the same semantic connection with marshy areas.

Mosquitoes do not need much moisture to breed. Their eggs are deposited and larvae develop in clusters on the surface of stagnant water. Tracts of boggy marshland and reedy river shallows provide ideal conditions for this, but small pools and puddles, or apparently quite insignificant quantities of water, such as that gathering in the hollows of decaying tree trunks (perhaps in antiquity the ubiquitous olive stumps), are potential breeding grounds for the insects and danger areas for their victims. As mosquitoes do not fly far, the immediate breeding environment plays a continuing part in their life cycle and the original territory of the mosquito continues to be the place where it seeks a food supply.

It is evident, then, that though malaria is not an environmental disease, it can be described as an infectious disease of the environment, there being a strong environmental connection in its fundamental aetiology. That there are further environmental connections in ancillary precipitating factors is argued below with regard to ancient Greece.

3. Characteristics and Symptoms of Malaria

Malaria is a complex disease, of necessity much simplified in the following account. There are different types of malaria with distinct periodicities, engendered by different species of mosquitoes carrying and spreading different types of *Plasmodium* with different incubation periods. *Plasmodia* are the tiny parasitic organisms brought by mosquito vector to human host; once in the body they attack the erythrocytes (red blood cells) and congregate in the liver where they grow and multiply, causing progressively more and yet more damage to the red blood cells of the sufferer. Some types of malaria are relatively mild. Thus, *Plasmodium vivax* causes benign 'tertian' malaria, where the fever returns on the third day (by inclusive reckoning, that is every other day) and *Plasmodium malariae* causes benign 'quartan' malaria, where the fever returns on the fourth day (by inclusive reckoning, that is after three days). Much more serious is *Plasmodium falciparum* the cause of malignant 'tertian' malaria, sometimes given the name 'semitertian' because the fever never really abates: there is fever every day, with a stronger fever every other day. All types of malaria are liable to recur and some sufferers experience repeated attacks after the disease has been dormant for many years, or even for many decades. Different periodicity distinguishes different types of malaria and the feature of periodicity, seen in a remittent fever of regular periodic recurrence, is one salient distinction between malaria and other febrile diseases. At the same time, certain general characteristics and symptoms, especially at the first onset of the disease, make diagnosis difficult.

Even now, a firm diagnosis of malaria is dependent on microscopic examination of the erythrocytes: a test can reveal the presence of *Plasmodia* or trophozoites at a stage in the parasite's life cycle that takes place in the red blood cells and can allow identification of their different forms. The observable symptoms of uncomplicated malaria are commonly non-specific, and disease-specific symptoms are not present in every case. The fever of malaria, like the fever of many other diseases, tends to be preceded and accompanied by general malaise such as fatigue, headaches, nausea and digestive upsets. In malarial ague, the recurrent episodes of high fever are regularly prefaced by paroxysms of chills, accompanied by rigors or shivers and followed by profuse sweating. There is frequently enlargement of the spleen and frequently also yellowing of the skin or, in some cases, of the whites of the eyes. After years of malarial attacks, the skin of sufferers is sallow and darkened. As late as the discoveries of the nineteenth century, typhoid fever and malarial fever were regarded as hard to distinguish. Then, as in preceding centuries, mortality records frequently stated simply 'fever' or 'ague', without further attempt at classification or designation. Malaria is a chronic disease that increases susceptibility to other diseases. Clearly, for antiquity the situation is complex and retrospective diagnosis is, as ever, hazardous.

In its main, and certainly in its primary, manifestations malaria is not an acute disease and, perhaps because there is a degree of acquired immunity where the disease is endemic, uncomplicated early malaria is rarely the immediate cause of death. However, its repeated attacks lead to extreme bodily weakness and prostration, with weight loss, muscle atrophy and extreme fatigue. The effect is that malarial cachexia or devitalisation is as fatal as the same condition in other progressive illnesses affecting the system, such as cancer or tuberculosis. Malarial haemoglobinuria or blackwater fever (marked by the pronounced symptom of very dark urination) is a severe and frequently fatal complication in the *Plasmodium falciparum* type of malaria. The systemic effects of *Plasmodium falciparum* can be grave, leading to dysfunction of the vital organs. When the brain is affected ('cerebral malaria') coma commonly ensues and there may be lasting brain damage in the sufferer. In pregnancy and childbirth, malaria of the *Plasmodium falciparum* type carries peculiarly serious dangers.

4. Aspects of the Environment (Greece)

That marshland has long been associated with malarial conditions was observed above. Already in antiquity, some Hippocratic writers showed a remarkable awareness that to live close to marshland was undesirable: they clearly understood that homes in marshy locations were not conducive to good health. It will be seen below that the descriptions offered by these physicians of the symptoms and conditions arising in such environments are remarkably acute and accurate, though at the same time their understanding of the underlying disease causation, as of the diseases caused, was erroneous or at best very partial. One typical view regarded fevers, like illness in general, as essentially miasmic in character, thought to arise from physical pollution through breathing bad air or drinking bad water.

The short work *On Humours*, though very different in its condensed aphoristic style, is closely allied in content with the collected case histories in *Epidemics* that will feature in the next section below, where evidence for aspects of malaria is discussed. In *On Humours* (*Hum.* 12 = 5.492 L.), the causes of disease are differentiated as primarily hereditary, local or environmental; in the last category, disease is said to arise from 'muddy or marshy ex-

halations' (ἀπὸ ὀδμέων βορβορωδέων ἢ ἑλωδέων) and changing effects arising in different seasons, with different rainfall and different winds, are postulated; jaundice is regarded
an autumn occurrence and some waters are said to lead to conditions affecting the spleen
(σπληνώδεα). In *On the Nature of Man* (*Nat. Hom.* 9 = 6.54 L.) it is suggested that matter
inhaled may contain some noxious, disease-inducing element (νοσηρήν τινα ἀπόκρισιν). In
On Winds (especially at *Flat.* 6 = 6.96 L.) a visionary account is given of the part supposedly
played by air in disease causation. Here (*Flat.* 8 = 6.100 L.) the 'most common' disease, fever,
associated with all other diseases, is first to be addressed; there are two types and the narrative of the first, that widespread in a community (as opposed to the second, that restricted
and related to regimen) evinces features typical of malaria, notably the initial shivering before the coming fever and the ensuing sweating.

The views presented in *On Airs, Waters and Places* can be seen as a detailed expression
of such general theories. In that work there is clear reference to the symptoms of malaria
and to its long-term effects. Hippocrates, or rather the author of the work *On Airs, Waters
and Places* which has come down to us in the Hippocratic Corpus, has been described as
the first European environmentalist. In an elaborate, thoroughly lucid and totally coherent
argument that physical environment and human wellbeing are interdependent, the importance in medical climatology of prevailing winds and the nature of the water supply is much
stressed. It is argued that waters differ in coming from soft and marshy or from hard and
rocky ground (*Aer.* 1 = 2.12 L.). Hot winds are associated with soft regions, with plentiful
surface water and people of moist, phlegmatic bodily constitutions, whereas cold winds are
associated with hard, dry regions and people of constitutions dominated by bile. Illnesses
suffered by the former type of people include fevers of long duration (*Aer.* 3 = 2.18 L.).
Already at this early point in the work there is implicit reference to malarial conditions,
marked by recurrent and protracted fever. Water is then said to make the greatest contribution to health and further features, as well as consequences, of unhealthy waters are listed in
greater detail (*Aer.* 7 = 2.26 L.). Waters that are marsh-like, stagnant and in lakes (ἑλώδεα καὶ
στάσιμα καὶ λιμναῖα) become in summer thick and malodorous (παχέα καὶ ὀδμὴν ἔχοντα)
since they lack flow (ἅτε ἀπόρρυτα ἐόντα) and then when autumn rains come are thoroughly bad (πονηρά). People in communities with such drinking water have enlarged and
hard spleens and suffer from bodily emaciation. The men suffer from digestive disorders
and protracted quartan fevers (πυρετοὶ τερταταῖοι πολυχρόνιοι) while the women give birth
with difficulty and have sickly infants. The salient symptoms of malaria are clearly identified and described here. In addition, the statement that autumn rains make the situation
worse reflects the seasonal effect of the disease, generally rampant in late summer and early
autumn. Later in the work, enlarged spleens and quartan fevers are again connected with
stagnant water. Further, a jaundiced complexion, allied with extreme physical enervation, is
connected with the habitat of a people by a river in eastern Kolchis (*Aer.* 24 = 2.86, 88 L.).
The author's consistency in the theoretical first part and more localised ethnological second
part of *On Airs, Waters and Places* is remarkable. In the second part, both the symptoms of
malarial attacks and the effects of the disease on long-term sufferers are well described.

The two parts of the treatise are generally viewed as disparate, but it is possible that the
presence of oriental ideas is a unifying factor, present directly in the second part where the
inhabitants of 'Asia' are discussed and more indirectly in the first on the importance of wind
as a factor in health. In the Ayurvedic medicine of India, wind (*vata*) is similarly significant as a morbific power, especially in causation of fever, 'king of diseases'. The Hippocratic

works noted above all seem to contain some echoes of Ayurvedic medicine. The classic texts of Caraka, Susruta and Vagbhata contain substantial discussions of fever, all apparently with some reference to the fevers of malaria: the periodicity of 'natural' fever is stressed, in that it peaks once daily or occurs intermittently on every third or fourth day (Caraka); fever is associated with chill and skin discoloration (Susruta); the gravity of fever is noted (Vagbhata).[8]

The medical writers were not alone in their awareness that habitation close to marshy ground is to be avoided. Attempts at drainage are recorded, both in public constructions, as in building the Long Walls at Athens, and on private land, as in digging agricultural ditches, the subject of farmers' disputes in forensic oratory. And in choosing sites for military encampments, marshy ground was sedulously avoided. A narrative in Xenophon demonstrates that this choice was conditioned by the realisation, through observation of the physique and complexions of local people, that stunted growth and darkened skin were caused by an endemic disease (Xen., *Cyrop.* 1.6.16). The ideas here expressed closely parallel those of *On Airs, Waters and Places*. A different parallel to, probably a direct echo of, *On Airs, Waters and Places* can be seen in the statements in Aristotle's *Politics* on desiderata in a city's orientation and the properties of its water supply (Arist. *Pol.* 1330b). In the Aristotelian *Problemata*, connections are made between chills and fever; also climatic conditions conducive to dysenteries and chronic quartan fevers (τερταταῖοι χρόνιοι) are noted (Arist., *Problemata* 19 and 27.2). These passages may stem from direct citation of Hippocratic works but it is equally possible that an independent source of received wisdom underlies the *Problemata*.

In Greek terminology of lakes, marshes, and rivers the same terms (λίμναι, ἔλη, ποταμοί) are regularly recurrent (as Pl., *Critias* 114e; *Lg.* 824a). The first two terms are almost interchangeable, but can be differentiated in that lakes or ponds (λίμναι) are formed of water left after a river or the sea recedes, whereas marshes (ἔλη) are formed of wet and boggy land continuously fed either by rainwater or by moisture seeping up from the water table below, or by both. The term 'bog' (τέλμα) is used of low-lying land saturated by deluges of rain. Many areas of Greece were naturally marshy and rivers criss-crossed the countryside. But city as well as countryside might be waterlogged. Attica had a district with the name *Limnai*. And even in urban Athens, the river Ilissos flowed in a series of pools (ὑδάτια), creating an apparently idyllic scene (Pl., *Phdr.* 229a), but one with the reedy shallows so attractive to mosquitoes. That mud (πηλός) was a familiar hazard in the streets of Athens is clear from Aristophanic comedy (Ar., *V.* 256–7). In addition, the wells and cisterns used for domestic water supply in every community were a fertile natural breeding ground for mosquitoes. There was a general awareness that the running water of streams and springs was preferable to standing water and there was some debate on the relative merits of different types of standing water, sourced from snow or from rain. More surprisingly, there was debate on whether stored rain water was superior to fresh spring water. Wells (κρήναι) were generally preferred to cisterns (φρέατα) or reservoirs (ὑποδοχαί). Both wells, with overflowing surroundings, and cisterns, with muddy sediments of earth, were ready homes to mosquitoes and doubtless many other pests, such as rats. There were environmental opportunities for such pollution everywhere.

In areas with copious rainfall there was a sufficient supply of water to replenish many wells and meet the needs of an entire community. One such area, important here because it is the location of many case histories recorded in *Epidemics*, is Thasos. Archaeology con-

---

[8] See VALIATHAN 2003: 251–68; 2007: 420–40; 2009: 279–304; FILLIOZAT 1949.

firms that there the problem was a superfluity rather than a deficiency in the water supply.[9] In Thasos, there were many wells not only in public places but also in the courtyards, or even sometimes the interior, of private homes and in many places simple digging would give ready access to the water table, only a few metres deep. In such a community there was no need for cisterns. Most communities, however, had recourse to collecting rain water for storage and later use in times of drought in cisterns, both public and private. Cisterns varied greatly in size and sophistication.[10] In their simplest domestic form, they might resemble a water butt: a container such as a large earthenware vase would be strategically placed to catch water fed by a pipe from the house roof. In a more elaborate civic system, the cistern might be linked to an aqueduct, devised to take water some distance. Although it would have been easy to place covers of solid wood or slats over apertures to prevent contamination by leaves or other debris, it would have been impossible to cover them with mesh to exclude insects, even if the desirability of this had been recognized.

The domestic architecture of ancient Greece favoured mosquitoes. In the hours of darkness mosquitoes, which habitually attack between dusk and dawn, could readily enter by the crude apertures that served as doors and windows and that were uncovered or poorly covered and certainly not securely netted. Most houses were single storey in construction and that too gave mosquitoes, which cannot fly very far or very fast or very high, ready access to the inhabitants. In addition, as malaria-causing mosquitoes tend to be 'endophilic' and typically rest inside the house after taking a blood meal, one bite was liable to lead to another. There was even easier access to people who customarily slept in the open, either at home for coolness and relative comfort, or away from home for reasons of agricultural work: there is evidence that farmers with land in outlying districts did not return every evening to their homes in the city (Lys. 1).

It is apparent that the Greeks had a high tolerance of insects, perhaps seeing them as one of many inevitable inconveniences in life. Mosquitoes and gnats are discussed, but not differentiated, by Aristotle.[11] Aristophanes yields much evidence that fleas and other biting insects were ubiquitous in the home (Ar., *Nu.* 1 *et passim*). The buzzing of gnats, probably to be identified with mosquitoes, is a subject of humour (*Nu.* 157–8) and the tiresome buzzing of mosquitoes about the heads of people is aptly described (*Ploutos* 537–8). The 'sharp-biting insects' (ἐμπίδες ὀξύστομοι) 'at the marshy channels' (ἐλείας παρ' αὐλῶνας) 'and lovely meadow' (λειμῶνά τ' ἐρόεντα) of Marathon (*Av.* 245–9) are certainly to be regarded as mosquitoes: mosquitoes still plague the region.

5. Aspects of Malaria (Greece)

In both medical and non-medical writers 'fever' is common and viewed as a disease in its own right rather than as a symptom or syndrome. There are many terms for 'fever' and they appear in a wide range of writers (πυρετός cognate with πῦρ 'fire' – and the latter term is used as a synonym in some texts – is most common but καῦσος, καῦμα and θέρμη cognate

---

[9] See GRANDJEAN 1994: 283–95.

[10] See HELLMANN 1994: 273–82.

[11] We miss a volume of D'ARCY THOMSON on entomology, to match those on birds and fishes. But see DAVIES 1987.

with 'burning' or 'heat' are found also). All of these terms are rare before the fifth century, perhaps because medical writing became established then, or perhaps in part because malaria and its fevers gained a serious hold then. A single usage in Homer, to the effect that the dog-star (Sirius) brings 'fever', is too allusive to allow certain interpretation (*Il.* 22.31). However, there is a similar expression, surely an echo of Homer, found in doublet passages of *Epidemics* 5 and 7 (*Epid.* 5.73 = 5.246 L. and *Epid.* 7.1 = 5.364 L.), where an account of fevers with much stress on critical days, duration and alternating hot or cold conditions, is prefaced by the statement 'after the [rising of] the dog-star, fevers marked by sweating commenced' (μετὰ κύνα οἱ πυρετοὶ ἐγένοντο ἱδρώδεις). The dog-star appears in late July, at the hottest point of summer, at just the time malarial episodes would have their major onset.

Hippocratic doctors made careful detailed observations of disease onset and progression. Although they recognised symptoms and patterns of disease consistent with those peculiar to malaria, and developed terminology clearly reflecting them, they did not, even in their nomenclature, really distinguish that type of disease from other febrile conditions. Rather, various terms signifying 'fever' were loosely employed. Terminology certainly evocative of malaria centres on the periodicity of the disease: *tritaios, tetartaios, hemitritaios.* These terms are nowhere explained and it seems to be taken for granted that the reader will be familiar with them; *hemitritaios* is apparently regarded as a familiar technical term in the expression '*hemitritaios*, as it is called' (*Epid.* 1.11 = 2.672 L.).

Celsus (3.3.2) is more explicit. Fevers are introduced as a common kind of systemic disease (*morbi genus … in toto corpore*), then categorised as quotidian, tertian and quartan. Whereas quartan fevers are viewed as straightforward, tertians are differentiated as being of two types, one similar to quartans (except that there is only one free day, not two, between attacks) and the other much more dangerous (*longe perniciosius*). An account of the latter type follows: while it too recurs on the third day, it never completely remits but merely loses some of its force. This dangerous type of fever is said to be called *hemitritaios* 'semitertian' by most doctors. The use of the Greek term is revealing, surely indicative of Greek sources. Celsus' account makes it clear that there was Greek awareness of two different types of tertian, not only *Plasmodium vivax* (benign) but also *Plasmodium falciparum* (malignant) and is good independent evidence for the presence of *Plasmodium falciparum* in Greek as well as Roman lands.

Fever (καῦσος) is recognised at the start of *On Acute Diseases* as a serious disease on a par with pneumonia and others (*Acut.* 2 = 2.232 L.). In *On Affections*, the periodicity of fever (clearly malarial) is recognised and in the same context affections of the spleen are noted (*Aff.* 18 = 6.226 L.). While the 'fever' of Hippocratic descriptions often seems to refer to malaria and this is clearly so where the terms 'tertian', 'quartan' and 'semitertian' are employed, it must be allowed that other types of fever, such as might have been caused by bacterial or viral infection, were present also. Puerperal fever, or as it is now generally called puerperal sepsis, was common; and fever arising from bacterial or viral infection was surely problematic also. Such infections might range from simple digestive disorders to life-threatening gastric diseases suggestive of cholera, typically presenting in summer; and from simple respiratory complaints to serious pulmonary diseases suggestive of pneumonia, typically presenting in winter.

The cyclical periodicity that distinguishes malaria from other febrile diseases can be identified in the case histories of *Epidemics*, where the successive days and phases of illness are recorded with care. From the numerical evidence of these records, many cases seem to be clearly malarial and many more seem to have a malarial element. In a population where malaria was endemic, combinations of it with other diseases would be inevitable. Periodicity is a fundamental element in Plato's theoretical categorization of fevers in *Timaios* (Pl., *Ti.* 86a). Plato sets out a scheme where fevers (καύματα and πυρετοί) are caused by an excess of fire (πῦρ) with the refinement that quotidian fevers (ἀμφημερινοί) are due to an excess of air, tertian to an excess of water and quartan to earth. The theories of crisis and critical days, so fundamental in ancient medical texts, perhaps have their origin in the prevalence of malaria, with its regular periodicity. The circumstance that different types of malaria have different periodicities may account for the different theories propounded by different authors in different localities. It may be that this pragmatic observation is more significant than a superstitious belief in the significance of particular numbers, as has generally been imputed to the writers. Similarly, the stress on the significance of seasonal change may reflect the reality of malarial presentation at the end of summer and start of autumn.

As noted above, a marked feature of malarial fever is that shivers precede its onset; the old- fashioned term 'ague' well describes the pattern. Greek has a term specific to such 'shivers' or 'ague' (ἠπίαλος). This is defined by Suda as 'so-called chills with fever' (τὸν ῥιγοπύρετον λεγόμενον) and explained by *Etymologicum Magnum* 'signifies chill and fever'(σημαίνει τὸν ῥιγοπύρετον). Similarly, Hesychios glosses as 'chill preceding fever' (ῥῖγος πρὸ πυρετοῦ). Aristophanes uses the expression 'ague, precursor of fever' (ἠπίαλος πυρετοῦ πρόδρομος), links the terms ague and fever (ἠπίαλοι and πυρετοί) and also describes an attack by ague falling on a man as he was making his way home in a state of fatigue following over-exertion (*fr.* 315, *V.* 1038 and *Ach.* 1165). The last instance is true to the regular pattern of malarial recurrence. It is well established that latent malaria is liable to resurface when the sufferer becomes over-tired. It may be that the frequent reference in medical text to external factors, such as fatigue (κόπος) as a cause of illness is related to this phenomenon.

Peculiarly regular features of malarial attacks are swelling of the spleen and yellowing of the skin, jaundice (ἴκτερος). It would be easy to tabulate cases in *Epidemics* where these features are present in conjunction with a sequence of chills, fevers and sweating. However, although it is probable that all such cases have a malarial element, it is probable also that there are other complicating conditions. It would be easy also to list cases of malarial cachexia, but here too it is probable that a variety of conditions are at work. The complexity inherent in ancient nosological description can be readily illustrated by scrutiny of diseases identified in *On Internal Affections*. Diseases there classified in different ways seem alike in having malarial features, and perhaps simply confirm the medical reality that chronic malaria makes the sufferer susceptible to other diseases. In *On Internal Affections*, several diseases of the spleen are categorised; one in particular is marked by swollen spleen, yellow complexion and weakness (*Int.* 31 = 7.246–48 L.). Several diseases of jaundice are categorised; one in particular is marked by yellow skin and eye-whites accompanied by chills and fever (*Int.* 35 = 7.252–56 L.). Several diseases given the name *eileos* are categorised. Among these, one – a 'difficult' disease – is marked by relapses and recurrence after a period of years; sufferers are emaciated and ashen in complexion (*Int.* 44 = 7. 278 L.). Of others, one given the special name 'jaundiced *eileos*' is said to occur in marshy regions; here too face and eyes are pale and yellowed (*Int.* 45 = 7.278 L.; cf. also *Int.* 46 = 7.280 L.). In a passing remark, the author of *On Internal Affections* evinces the same environmental awareness as that seen in other Hippocratic texts.

6. Malaria in pregnancy

There are severe dangers in pregnancy of the *Plasmodium falciparum* type of malaria. These include intrauterine growth retardation resulting in miscarriage or stillbirth. There is a risk also of maternal and neonatal mortality. In a following article, cases will be examined and it will be suggested that cults of Artemis of the Marsh (Artemis being a goddess associated with pregnancy and childbirth) may reflect anxiety and prophylaxis with regard to the dangers of childbirth in malarial conditions.

Bibliography

Ancient authors and texts are cited according to the conventions set out in the lexicon of LIDDELL-SCOTT-JONES. Hippocratic passages are identified with reference to volume and page of LITTRÉ (L.), 1839–61.

BURKE, P. F. 1996. 'Malaria in the Greco-Roman World', in *Aufstieg und Niedergang der römischen Welt*, II. 37.3. Berlin-New York, 2252–81.

CARTER, R. and K. N. MENDIS. 2002. 'Evolutionary and historical Aspects of the Burden of Malaria', *Clinical Microbiology Review* 15.2, 527–44.

CORVISIER, J.-N. 1994. 'Eau, paludisme et démographie en Grèce continentale' in R. GINOUVÈS, A.-M. GUIMIER-SORBETS, J. JOUANNA and L. VILLARD (eds.) *L'Eau, la Santé et la Maladie dans le Monde Grec*, Actes du Colloque de Paris 25–27 nov. 1992 (BCH Suppl. 28). Paris, 297–319.

CRAIK, E. M. 2015. *The 'Hippocratic' Corpus*. London.

CRAIK, E. M. forthcoming, 2017. 'Malaria, Childbirth and the Cult of Artemis', in L. TOTELIN and R. FLEMMING (eds.), *Festschrift for Vivian Nutton*.

DAVIES, M. 1987. *Greek Insects*. Oxford.

FILLIOZAT, J. 1949. *La doctrine classique de la médecine indienne. Ses origines et ses parallèles grecs*. Paris.

GRANDJEAN, Y. 1994. 'L'eau dans la ville de Thasos' in R. GINOUVÈS, A.-M. GUIMIER-SORBETS, J. JOUANNA and L. VILLARD (eds.) *L'Eau, la Santé et la Maladie dans le Monde Grec*, Actes du Colloque de Paris 25–27 nov. 1992 (BCH Suppl. 28). Paris, 283–95.

GRMEK, M. D. 1983. *Les maladies au l'aube de la civilisation occidentale*. Paris, Engl. transl. by M. MUELLNER and L. MUELLNER 1989. *Diseases in the Ancient Greek World*. London.

HELLMANN, M.-C. 1994. 'L'eau des citernes et la salubrité: texts et archéologie' in R. GINOUVÈS, A.-M. GUIMIER-SORBETS, J. JOUANNA and L. VILLARD (eds.) *L'Eau, la Santé et la Maladie dans le Monde Grec*, Actes du Colloque de Paris 25–27 nov. 1992 (BCH Suppl. 28). Paris, 273–82.

JONES, W. H. S. 1907. *Malaria, a Neglected Factor in Greek History*. Cambridge.

JONES, W. H. S. 1909. *Malaria and Greek History*. Manchester.

JOUANNA, J. 1994. 'L'eau, la santé et la maladie dans le traité hippocratique des Airs, eaux, lieux' in R. GINOUVÈS, A.-M. GUIMIER-SORBETS, J. JOUANNA and L. VILLARD (eds.) *L'Eau, la Santé et la Maladie dans le Monde Grec*, Actes du Colloque de Paris 25–27 nov. 1992 (BCH Suppl. 28). Paris, 25–40.

LITTRÉ, E. 1839–61. *Oeuvres complètes d'Hippocrate*. Paris.

SALLARES, R., A. BOUWMAN, and C. ANDERUNG. 2004. 'The Spread of Malaria to Southern Europe in Antiquity: New Approaches to old Problems', *Medical History* 48.3, 311–28.

STICKER, G. 1928; 1929; 1930. 'Fieber und Entzündung bei den Hippokratikern', *Archiv für Geschichte der Medizin* 20, 150–74; 22, 313–43 and 361–81; 23, 40–67.

VALIATHAN, M. S. 2003. *The Legacy of Caraka*. Hyderabad

VALIATHAN, M. S. 2007. *The Legacy of Susruta*. Hyderabad.

VALIATHAN, M. S. 2009. *The Legacy of Vagbhata*. Hyderabad.

VAN DER EIJK, P. 2014. 'An Episode in the Historiography of Malaria in the Ancient World' in D. MICHAELIDES (ed.), *Medicine and Healing in the ancient Mediterranean World*. Oxford-Philadelphia, 112–7.

Isabella Andorlini (†)

Environmental Diseases according to Papyri from Egypt and Ancient Medical Thought

1. Environmental pollution

The aim of this paper is to examine some priorities on the notion of environmental pollution and its relationship to health. The lives of ancient Egyptians were shaped by their environment. The geographer Strabo commented on the Egyptian climate stating that 'the evaporation caused by the sun's rays […] and the air inhaled is noisome and causes pestilential fevers'. Flavius Josephus in his *Bellum Judaicum* referred to marshland in close proximity to Alexandria.[1] The land, the river Nile, and the climate were of great benefit to people affected by consumption, but in some circumstances that environment contributed to disease.

This paper is part of the Project ERC-AdG-2013-DIGMEDTEXT, Grant Agreement No. 339828 funded by the EU at the University of Parma.

[1] Jos. Fl., *Bell. Jud.* 2.386–7: τετείχισται δὲ πάντοθεν ἢ δυσβάτοις ἐρημίαις ἢ θαλάσσαις ἀλιμένοις ἢ ποταμοῖς ἢ ἕλεσιν. 'Alexandria was protected by impassable desert or harborless sea, or by rivers or marshland'. Strabo, *Geogr.* 17.1.7 observes: πληροῖ δὲ ταύτην πολλαῖς διώρυξιν ὁ Νεῖλος […] this is filled by many canals from the Nile […] γνοίη δ᾽ ἄν τις ἔν τε τῇ Ἀλεξανδρείᾳ γενόμενος if he were at Alexandria […] ὃ καὶ αὐτὸ συμβαίνει διὰ τὸ ἀμφίκλυστον καὶ τὸ εὔκαιρον τῆς ἀναβάσεως τοῦ Νείλου. αἱ μὲν γὰρ ἄλλαι πόλεις αἱ ἐπὶ λιμνῶν ἱδρυμέναι βαρεῖς καὶ πνιγώδεις ἔχουσι τοὺς ἀέρας ἐν τοῖς καύμασι τοῦ θέρους· ἐπὶ γὰρ τοῖς χείλεσιν αἱ λίμναι τελματοῦνται διὰ τὴν ἐκ τῶν ἡλίων ἀναθυμίασιν· βορβορώδους οὖν ἀναφερομένης τοσαύτης ἰκμάδος, νοσώδης ἀὴρ ἕλκεται καὶ λοιμικῶν κατάρχει παθῶν. ἐν Ἀλεξανδρείᾳ δὲ τοῦ θέρους ἀρχομένου πληρούμενος Νεῖλος πληροῖ καὶ τὴν λίμνην καὶ οὐδὲν τελματῶδες τὸ τὴν ἀναφορὰν ποιῆσον μοχθηράν. The land is washed by water on both sides and because of the timeliness of the Nile's rising; for the other cities that are situated on lakes have heavy and stifling air in the heats of summer, because the lakes then become marshy along their edges because of the evaporation caused by the sun's rays, and, accordingly, when so much filth-laden moisture rises, the inhaled air is bothersome and causes pestilential fevers, whereas in Alexandria, at the beginning of summer, the Nile is flooding and fills the lake too, leaving no marshy matter to corrupt the rising vapours. (Engl. transl. H. L. JONES, Cambridge MA 1932, slightly mod.). Malaria is transmitted by the bite of infected female mosquitos. During the autumn they could cause malaria infection in Alexandria especially at the end of the inundation period.

2. Skin diseases

The inhabitants of Alexandria, according to Galen,[2] were suffering from almost incurable skin diseases, being exposed to hot and dry air, their poor diet due to periodic malnutrition, which increased their susceptibility to infection.[3] The skin infections could be identified with the dermatological condition called *elephantiasis*.[4] The infection was revealed in test samples from the skin of a naturally desiccated body. The mummies were infected by a parasitic worm associated with irrigation ditches. There is good reason to believe that skin and mouth ulcers affected the sunburnt inhabitants of Alexandria.

The medical writer Aretaeus, who practised in the first century AD, insisted that *aphthous stomatitis* was an 'Egyptian ulcer' and this kind of mouth ulcer affected especially infant children:

ἄφθαι τοὔνομα τοῖσι ἕλκεσι. ἢν δὲ καὶ ὁ ἐπίπαγος ἴσχῃ βάθος, ἐσχάρη τὸ πάθος καὶ ἔστι καὶ καλέεται· ἐν κύκλῳ δὲ τῆς ἐσχάρης ἐρύθημα γίγνεται καρτερὸν καὶ φλεγμονὴ καὶ πόνος φλεβῶν, ὡς ἐπ' ἄνθρακος, καὶ σμικρὰ διεξανθήματα ἀραιὰ γιγνόμενα, ἔπειτα προσεπιγιγνόμενα […] διὰ τόδε παιδία μάλιστα πάσχει ἄχρις ἥβης […] χώρη δὲ τίκτει Αἴγυπτος μάλιστα· καὶ γὰρ ἐς ἀναπνοὴν ἐστὶ ξηρὴ καὶ ἐς ἐδωδὴν ποικίλη· ῥίζαι γὰρ καὶ βοτάναι καὶ λάχανα πολλὰ καὶ σπέρματα δριμέα, καὶ ποτὸν παχύ, ὕδωρ μὲν ὁ Νεῖλος, δριμὺ δὲ τὸ ἀπὸ τῶν κριθέων καὶ τὸ τῶν βρυτέων πόμα […] ὅθεν Αἰγύπτια καὶ Συριακὰ ἕλκεα τάδε κικλήσκουσι.

(Aret., Caus. acut. morb. 1.9.1–5 = CMG 2, p. 11 HUDE)

Aphtha is the name given to these ulcers. But if the concretion is deep, it is called eschar. Around the eschar there is a great redness, an inflammation, and vein trauma, as in a carbuncle; and small pustules form, at first few in number, but others appear, they coalesce, and a broad ulcer is produced […] because of that, the children are particularly affected until puberty […] The land of Egypt especially produces it, because of its dry breathing air, and its varied food, consisting of roots, many kinds of herbs, acrid seeds, and thick drinks; namely, the water of the Nile, and the sort of drink prepared from barley (i. e. beer?) […] and hence they have been named Egyptian and Syrian ulcers.

---

[2] Gal., *De alim. facult.* 1.2 (6.486.10–487.3, 487.10 Kühn) = *CMG* 5.4.2, pp. 219–220, Chapt. 8–9, Helmreich): ἐν Ἀλεξανδρείᾳ δὲ καὶ τὰ τῶν ὄνων ἐσθίουσιν, εἰσὶ δ' οἳ καὶ τὰ τῶν καμήλων […] καὶ διὰ τοῦτο νόσους χαλεπωτάτας ὕστερον οὗτοι νοσοῦσι καὶ πρὸ γήρως ἀποθνῄσκουσι […] διὰ τοῦθ' ἧττον ὑπὸ τῶν μοχθηρῶν ἐδεσμάτων βλάπτονται. In Alexandria they eat donkey meat as well, and there are also some people who eat camel […]. This is why there the people later on suffer very troublesome illnesses and die before they reach old age […]. They are consequently less injured by harmful foods.

[3] Alexandrians lived on an equally unhealthy diet of salt fish, shellfish and lentils, which produced a kind of skin disease. See Gal., *ad Glauc. de meth. med.* 11.142.3–13 (Kühn).

[4] True *elephantiasis* is the result of a parasitic infection caused by three specific kinds of round worms. An autopsy on the 3,000-year-old mummified body of Natsef-Amun, an Egyptian priest during the time of Ramses I (1113–1085 BC), revealed the presence of filarial worms. It is one of the world's most disfiguring diseases. In pharaonic times Egyptians suffered from *schistosomiasis* despite the lack of perennial irrigation and consequent lack of suitable sites for transmission. Convincing paleopathological evidence for the presence of this disease in mummies confirmed that *schistosomiasis* (also known as *bilharzia*, snail fever) is a disease caused by parasitic worms that has existed continuously in Egypt for at least 5,000 years. (See WELSH 2011). During excavations in the nineteenth century at Deir el Bahri, near Thebes, the coffin and mummy were found and transported to Europe. See DEELDER, MILLER, DE JONGE, KRIJGER 1990: 724–5.

3. Environmental health

Literary evidence suggests that a healthy site was a prior requirement in the ancient landscape. The author of the Hippocratic treatise *Breaths* (*De flatibus* 5–7) supposes that general fever, which he named *loimos*, is caused by *miasmata*, fumes that originate from organic substances and exercise a pernicious influence on bodies. The *miasmata* are a physical and natural cause of diseases. Sickness was caused by the environment or by one's poor diet.

> Ὁκόταν μὲν οὖν ὁ ἀὴρ τοιουτέοισι χρωσθῇ μιάσμασιν, ἃ τῇ ἀνθρωπίνη φύσει πολέμιά ἐστιν,
> ἄνθρωποι τότε νοσέουσιν.   (Hipp., *Flat.* 5.1: p. 108 JOUANNA)

Thus, when the air is full of *miasmata*, whose properties are hostile to human nature, this is when men are ill.

> Ἔστι δὲ δισσὰ ἔθνεα πυρετῶν, ὡς ταύτῃ διελθεῖν, ὁ μὲν κοινὸς ἅπασι ὁ καλεόμενος λοιμός· ὁ δὲ
> [διὰ πονηρὴν δίαιταν] ἰδίῃ τοῖσι πονηρῶς διαιτωμένοισι γινόμενος· Ἀμφοτέρων δὲ τουτέων αἴτιος
> ὁ ἀήρ.   (Hipp., *Flat.* 6.1, p. 109 JOUANNA)

There are two kinds of fever; one is epidemic, called pestilence, the other is sporadic, attacking those who follow a bad regimen. Both of these fevers, however, are caused by air.

> Αἳ μὲν οὖν δημόσιαί εἰσι τῶν νούσων, εἴρηται, καὶ ὅτε καὶ ὅκως, καὶ οἷσι, καὶ ἀπὸ τεῦ γίνονται· τὸν
> δὲ διὰ πονηρὴν δίαιταν […]. Πονηρὴ δέ ἐστιν ἡ τοιήδε δίαιτα, τοῦτο μὲν ὅταν τις πλέονας τροφὰς
> ἢ ὑγρὰς ἢ ξηρὰς διδῷ τῷ σώματι ἢ τὸ σῶμα δύναται φέρειν.
>   (Hipp., *Flat.* 7.1, pp. 110–1 JOUANNA)

Of epidemic diseases I have already spoken, and of their times and manners, as well as of the victims and of the cause thereof; I must now go on to describe the fever caused by bad regimen […]. By bad regimen I mean, firstly, the giving more food, moist or dry, to the body than the body can bear.

Also the treatise preserved by the *Anonymus Londiniensis* papyrus (5.35–6, second century AD), based on the Hippocratic *Breaths* 3.1, states that:

> Ἱπποκράτης δέ φ(ησιν) αἰ(τίας) (εἶναι) τῆς νόσου τὰς φύσας

Diseases are also produced by the changes in the breaths.[5]

Hippocrates' *Airs, Waters, and Places* has been considered to be one of the earliest writings on environmental health. The treatise, written in about 400 BC, and part of the *Corpus Hippocraticum*, represents a first attempt to find natural causes for the health and diseases of entire populations in different regions. *Airs, Waters, and Places* stresses the importance of environmental physicians. The Hippocratic work is a book for travelling practitioners, explaining to them everything they must check when they arrive in a new unknown place. According to the author of this manual, one of the most frequent causes of diseases is the bad quality of environmental air. A physician, who was aware of this theory, would not think that diseases could only be transmitted by contagion.

---

5   See Hipp., *Flat.* 3.1 (105.14–106.2 JOUANNA): Πνεῦμα δὲ τὸ μὲν ἐν τοῖσι σώμασιν φῦσα καλεῖται, τὰ
    δὲ ἔξω τῶν σωμάτων ἀήρ. Wind in bodies is called breath, outside bodies it is called air.

Εἰ γὰρ ταῦτα εἰδείη τις καλῶς, μάλιστα μὲν πάντα, εἰ δὲ μή, τά γε πλεῖστα, οὐκ ἂν αὐτὸν λανθάνοι
ἐς πόλιν ἀφικνεόμενον ἧς ἂν ἄπειρος ᾖ οὔτε νοσήματα ἐπιχώρια, οὔτε τῶν κοιλιῶν ἡ φύσις ὁκοίη
τίς ἐστιν.   (Hipp., *Aer.* 2.1)

> For if the physician knows these things well, by preference all of them, but at any rate most, he will
> not, on arrival in a town with which he is unfamiliar, be ignorant of the local diseases, or of the
> nature of those internal illnesses that commonly prevail.

In the opening chapter, for example, Hippocrates advises the physician to consider what effects each season and each seasonal change can produce as they all differ from one another.
Throughout the year, physicians will be able to tell which epidemic diseases might arise in the
summer and which ones in the winter, and they will be able to distinguish seasonal diseases
from those caused by an individual's lifestyle. The author tries to investigate the impact of
winds, seasonal changes with the shift from warm to cold, and the quality of food and drinking water. The author associates seasons, prevailing winds, and the quality of the air and water
with the physical condition of people and the occurrence of disease. He advises the physician
arriving in a new city to take into account the environmental factors that determine the kind
of diseases endemic to that location. He identifies the properties of water, soil, and human
behaviour and relates them to epidemiology and the maintenance of human health.

4. Egyptian ecological context

Indeed, our surviving documentary sources concerning the Egyptian ecological context,
and the co-evolution of environmental diseases, are not without interest both in themselves
and as indications of how air and water were polluted in Egyptian towns and villages. Ancient Egypt can be divided into two regions, the Nile valley, or Upper Egypt, and the Delta,
or Lower Egypt. The Fayum, a large quasi-oasis centred on a lake (Birket Qarun) is situated
west of modern Cairo. The pattern of deaths we can reconstruct from mummy labels is concentrated in a three-month period, from January to April. The mummy labels preserved in
Egypt are emblematic funerary objects from the Ptolemaic and Roman periods. Attached to
the mummy, the labels were the deceased's identification and passport to eternity. This data
can be used to reconstruct the seasonal distribution of death and diseases.

> *SB* 1.5758 (Mummy label, Hawara, January 13, Year 14 BC)
> (ἔτους) ις Καίσα(ρος), Τῦβι ιγ̄[6]

> *P.Coll.Youtie* 2.108 *verso* (Mummy label, Panopolis, March 16, Year 205 AD)
> ἔτους κε Φαμεν(ὼθ)[7] κ Πᾶσις … λα(   ) μη(τρὸς) Τιναροῦς

The treatise *Airs, Waters, and Places* correctly maintains that the inhabitants of hot regions
suffer from diarrhoea, dysentery, eye and skin infections.

---

[6] The month *Tybi* runs from December 27th/28th to January 25th/26th.
[7] The month *Phamenoth* runs from February 25th/26th to March 25th/26th.

νοσήματά τε τάδε ἐπιχώρια εἶναι [...] τοῖσί τε παιδίοισιν ἐπιπίπτειν σπασμοὺς καὶ ἄσθματα καὶ ἃ νομίζουσι τὸ παιδίον ποιεῖν καὶ ἱερὴν νοῦσον εἶναι, τοῖσι δὲ ἀνδράσι δυσεντερίας καὶ διαρροίας καὶ ἠπιάλους καὶ πυρετοὺς πολυχρονίους, χειμερινοὺς καὶ ἐπινυκτίδας πολλὰς καὶ αἱμορροῖδας ἐν τῇ ἕδρῃ. Πλευρίτιδες δὲ καὶ περιπλευμονίαι καὶ καῦσοι καὶ ὁκόσα ὀξέα νοσήματα νομίζουσι οὐκ ἐγγίγνονται τὰ πολλά.   (Hipp., *Aer.* 3.3–4, pp. 190–1 Jouanna)

The endemic diseases are these [...] Children are liable to convulsions and asthma, and to what they think causes the diseases of childhood, and to be a sacred disease. Men suffer from dysentery, diarrhoea, ague, chronic fevers in winter, many attacks of eczema, and from hemorrhoids. Cases of pleurisy, pneumonia, ardent fever, and of diseases considered acute, rarely occur.

ὀφθαλμίας ξηρὰς διὰ τὴν θερμότητα καὶ ξηρότητα τῆς σαρκός. (Hipp., *Aer.* 10.6, p. 215 Jouanna)

Dry ophthalmia appears because of the warm dryness of their flesh.

The Mediterranean climate is characterized by a seasonal pattern of mild, wet winters and hot, dry summers. The environment of the Egyptian Nile valley, as well as the regions of the Delta and Nubia, offer some very illustrative case-studies of endemic gastro-intestinal illnesses that were a serious health hazard in ancient Egypt. It is well documented that this region had one of the highest level of pollution in Egypt. Although diarrhoeas could occur in any season, hot season epidemics were very common. According to the data records of the Hippocratic *Airs, Waters, and Places,* diarrhoea and dysentery are typical of hot climates. Celsus recorded that 'diarrhoeal diseases carries off mostly children up to the age of ten'.[8] Mortality from dysentery, caused by bacterial infection, was found to peak a few weeks before the temperatures reached their maximum. The widespread outbreak of infectious diseases is further confirmed by ostraka and papyri excavated in Egypt. A papyrus of the sixth century AD, found in the Dioscoros archive from Aphroditopolis, describes a man suffering from dysenteric spasms.[9] This can cause a high temperature and painful spasms of the intestinal muscles. Ostraka found on Mons Claudianus, in the desert between the Nile and the Red Sea, are the most recent addition to our knowledge of diseases in Roman Egypt.[10] In two lists of workmen recorded as quarrymen employed at Mons Claudianus, two sick men are described as temporarily affected by dysentery, a term rendered as [δυσε]ντερικ(ός).[11]

---

[8] Cels., *med.* 2.8.30 (*CML* 1, p. 74 Marx): *Deiectionibus quoque si febris accessit, si inflammatio iocineris aut praecordiorum aut uentri,* [...] *etiam periculum mortis subest* [...]*; isque morbus maxime pueros absumit usque ad annum decimum.*

[9] *P.Berl.Brash.* 19.14 (= *SB* 14.11856) ἀνακείμενος τυγχάνω δυσεντερικὰ σπ[άσματα.

[10] See *O.Claud.* 1.83–118, and especially 191–223.

[11] The workmen are listed among the so called ἄρρωστ(οι) ('men who are temporarily ill'). See *O. Claud.* 4.708.27–28 δ[υσεν]τερικ(ὸς) α and *O.Claud.* 4.717.10 [δυσε]ντερικ(ὸς) α.

5. Evidence of leprosy

Most medical historians believe that leprosy originated in Egypt, and the leprosy bacillus, called *mycobacterium leprae*, has been found in at least two mummies that also show the typical scaly evidence of Hansen's disease that affects the skin, peripheral nerves, and upper airway. There is strong evidence of leprosy in some examined mummies. Some Coptic Christian bodies were found in a cemetery south of the temple of el-Biga, which is located on an island of the same name south of the cataract near Aswan in Nubia. A Coptic mummy, disinterred in 1907 from Cemetery No. 5, exhibits typical leprous mutilations of the hands and feet.[12] The date to which this mummified leper can be assigned probably lies between the fourth and the seventh centuries AD. Another mummified woman from Cemetery No. 6 at el-Biga exhibits pathological changes in the maxillary bone (i.e. the nasal palate) and corresponds precisely with that of an individual who suffered from advanced lepromatous leprosy. The re-discovery of this Nubian cranium, as well as the leprous mutilations of the hands and feet from the other mummy, are of great interest in the history of the disease and are strongly indicative of a *facies leprosa*.[13] In the *P.Ebers* pharaonic papyrus, dating around 1550 BC, there is evidence for tubercular leprosy in the so-called Chons' tumour in conjunction with the development of leprosy in Egypt:

> If you examine a Chons' big swelling on any limb of a man, and it is horrible when it has produced many swellings, and there arises in him something therein as if there were air therein, and it causes destruction of the swelling.   (*P.Ebers* 874, transl. EBBELL)[14]

A comarch by the name of Patermuthis, an official in charge of an Egyptian village of the Upper toparchy, was described in *P.Oxy.* 63.4356.2 as a leper:

Πατερμ[ο]ῦθις λεπρὸς κωμάρ[χης

(for the Upper toparchy) … Patermuthis the leper.[15]

Medical papyri of the Roman period usually treat leprosy and lichen with a dose of burnt papyrus, an antiseptic lotion, and then use an application of a small sheet of medicated papyrus (i.e., *chartarion*) as a bandage.

---

[12] SMITH, DERRY 1910: 1–11, with Plates I–VIII.

[13] By studying mummies we can learn about the many diseases that the ancient Egyptians suffered from, and we are able to learn much about the methods used by ancient doctors. The leprosy described in *Leviticus* could not have been true leprosy, for which the Greek medical term was *elephantiasis*.

[14] EBBELL 1937: 126 note 3, 'This seems to be a description of tubercular leprosy'.

[15] The term λεπρός is not common in documentary papyri. An analogous Maximus, the leper, is mentioned several times in *P.Mich.* 4. i 223.1189, 224.2024 (p. 195, λεπροῦ pap.), 225.1751. In the sales of Egyptian slaves, the latter are described as affected by (or free from) 'the sacred disease (epilepsy) and leprosy', a medical defect, see *P.Oxy.* 36.2777.24–5 πλὴν ἱερᾶς νόσου καὶ ἐπαφῆς, *P.Col.* 8.219.10 note, and STRAUS 2004: 153–5 and notes 282–3.

πρὸς λέπρας, ἐὰν ἐκ|δέρῃς αὐτάς, βάμμα παπύρου κεκαυμ(ένης).

(*PSI* 10.1180, Tebtunis, first-second centuries AD, A, 2.10–11)

Against leprosy: when you have scraped off these lesions, prepare a solution with burnt papyrus

τὸν λιχῆνα προεζμησάμενον κα|τάχριε καὶ ἔξωθεν γῦριν· ἐπάνω δὲ | το[ῦ] φαρμάκου χαρτάριον
ἐπίθες.   (*PSI* 10.1180, Tebtunis, A, 3.5–7)

Scrape the area affected by lichen first, smear with the finest meal externally; then cover the application with a bandage made from a papyrus sheet.[16]

There is strong evidence for several leper-hospitals in middle Egypt, located in the Hermo-
polites and Antinoopolis regions. These hospitals were called κελυφοκομ(ε)ῖα, because in Late
Antiquity κελεφός, a Greek word of semitic origin, was the usual designation for leprosy.[17]

6. λοιμός

The terms λοιμός, λοιμικός, and λοιμώδης, which occur in documentary papyri, the latter
applied to νόσος,[18] probably do not refer to plague, but to epidemic and lethal fevers.[19] It
is interesting to quote the definition of the Pseudo-Soranian question concerning this kind

---

[16] Full edition in ANDORLINI 2004: 81–118. Before chemotherapy, leprosy used to be treated with the
application of topical agents or injections in the skin lesions. The discovery of sulfonamides paved
the way to antibiotic chemotherapy to treat leprosy. *Lichen ruber* is a type of skin disease that causes
an itchy, uncomfortable rash. Lichen is characterized as bumpy lesions on arms and legs. The main
symptom of any form of lichen is a severe itching and sometimes a burning sensation. People who
suffer from a severe case of lichen may experience blistering of the skin and bleeding. Lichen is
described by Hippocrates along with leprosy: καὶ λέπραι, καὶ λειχῆνες, καὶ ἀλφοί, καὶ ἐξανθήσιες
ἑλκώδεες πλεῖσται. (Hipp., *Aph.* 3.20). Leprosy, lichens, dull-white leprosy, and eruptions turning
generally to ulcers.

[17] See GASCOU 1993: 78–9. The Hebrews fleeing Egypt during the Exodus almost certainly had been
contaminated. Leprosy is mentioned in two famous chapters of *Leviticus* (Chapt. 13–14), the very
source of stigmatization of leprosy in the Christian West (*Lev.* 13.12: ἐὰν δὲ ἐξανθοῦσα ἐξανθήσῃ ἡ
λέπρα ἐν τῷ δέρματι, καὶ καλύψῃ ἡ λέπρα πᾶν τὸ δέρμα τῆς ἁφῆς ἀπὸ κεφαλῆς ἕως ποδῶν καθ᾽
ὅλην τὴν ὅρασιν. If, however, the disease breaks out on the skin so that leprosy covers all the skin
of the person with the infection from his head to his feet).

[18] *SB* 6.9218.12 (319–320 AD) ὑπὸ νόσου λοιμώδους τινὸς ἄρδην διαφθαρῆναι.

[19] See the Hipp., *Flat.* 6. See also *P.Oxy.* 55.3817.11–15, ed. J. REA, London 1988 (see *ibidem*, p. 215
Introd., and 216 n. 11–12, dated third century AD): ἀπέθανον τῷ καταστήματι [pap. καταστέμματι].
ἐὰν γάρ τις νοσήσῃ τῶν παρ᾽ ἡμῖν ὄντων ἐν τῇ κώμῃ, οὐκ ἐγείρονται. (For if anyone among us in
the village falls ill, they do not rise [from the sick-beds]). See *P.Oxy.* 55.3816.4–10 (third-fourth
century AD): γνῶ|ναί σε θέλω ὅτι Ἀχιλλεὺς πάνυ νο|σεῖ καὶ ἐχειρίσθη ποσάκις εἰς τοὺς πό|δας καὶ
τὰ ἕως ἄρτι νο|σεῖ καὶ σχεδόν τι | προσέτι, καὶ διὰ τοῦτο οὐκ᾽ ἐδυνή|θην λαλῆσαι αὐτῷ. καὶ ἐγὼ
ἠσθένη|σα πάνυ καὶ εἰς θάνατον. (I want you to know that Achilles is very ill and has had treatment
ever so many times to the feet and has been ill right up to the present and is perhaps even more so,
and because of that I couldn't talk to him. I was very sick myself, at death's door even). The illness
on feet and legs signifies ancient epidemics which affected the extremities. See *P.Strassb.* 1.73.15
where the writer was attacked by another skin condition, called erysipelas (κατὰ τοῦ ποδός μου
ἐρυσίπελαν, ερισυπολιν pap.).

of contagion: *Quid est loimos nosema? aegritudo pestilentiosa quae de paludibus et stagnis confortatur, omnibus navigantibus maxime eveniens ex diversitate aerum, qua plurimi moriuntur.*[20] According to the Pseudo-Soranian definition, deadly contagion was caused by marshlands; disease spread because people travelled, and also because of changes in the air which contributed to lethal fevers.

Pseudo-Galen, in his *Definitiones medicae* 153 (19.391.14–392.4 KÜHN), also mentioned a disease caused by air pollution:

Λοιμός ἐστι νόσημα ἐπὶ πάντας ἢ τοὺς πλείστους παραγινόμενον ὑπὸ διαφθορᾶς ἀέρος, ὥστε τοὺς πλείστους ἀπόλλυσθαι. […] λοιμός ἐστι τροπὴ ἀέρος ὥστε μὴ τετηρηκέναι τὰς ὥρας τὴν ἰδίαν τάξιν […].

loimos is a disease which affects all or the majority of people because of the pollution of the air, so that people die […] it is a changing of air when seasonality is not observed.

A papyrus of the third century AD attests to an epidemic fever called λοιμός in Antinoopolis, and the writer begs his brother to write to him about his health:

παρακαλῶ οὖ[ν,] ἄδελφε, γράψαι μοι περὶ τῆς ὑμῶν σω|[τ]ηρίας, ἐπεὶ ἤκουσα ἐν τῇ Ἀντινόου ὅτι παρ᾽ ὑμῖν λοιμὸς | [ἐγ]ένετο. μὴ οὖν ἀμελήσῃς, ἵνα καὶ ἐγὼ περὶ ὑμῶν εὐθυ|μότερον διάξω.

(P.Oxy. 14.1666.19–22)

I beg you, brother, write to me about your well-being, as I heard at Antinoopolis that there has been a pestilential fever in your neighbourhood. So do not neglect to write, that I may feel more cheerful about you.

Another papyrus, of unknown provenance and dating back to the late second century AD (*P.Mert.* 2.82), presents an unusual feature, a self-diagnosis of an illness due to an unhealthy climate or polluted air.[21] Nike is writing to her sister Berenice protesting that:

λίαν δὲ νωθρεύομαι, πότερον | δ[ι]ὰ τὸν ἀέρα οὐκ οἶδα. ἐὰν δὲ πάλιν | ῥ[αί]σω σὺν θεοῖς, γράψω σοι.

I am very unwell; perhaps it is the climate, I don't know. But if I get better again, with the Gods, I will write to you.

The most striking item of this letter is that Nike was ill, and was uncertain whether to attribute her illness to the polluted air or to some change of the weather. When medical writers do turn their attention to such problems as air pollution, they are also able to recommend advice:

ὅταν δὲ ὑπὸ νοσήματος ἑνὸς ἐπιδημίη καθεστήκῃ, δῆλον ὅτι οὐ τὰ διαιτήματα αἴτιά ἐστιν, ἀλλ᾽ ὃ ἀναπνέομεν, τοῦτο αἴτιόν ἐστι, καὶ δῆλον ὅτι τοῦτο νοσηρήν τινα ἀπόκρισιν ἔχον ἀνίει. Τοῦτον οὖν χρὴ τὸν χρόνον τὰς παραινέσιας ποιέεσθαι τοῖσιν ἀνθρώποισι τοιάσδε· τὰ μὲν διαιτήματα μὴ μεταβάλλειν, ὅτι γε οὐκ αἴτιά ἐστι τῆς νούσου […] τοῦ δὲ πνεύματος ὅπως ἡ ῥύσις ὡς ἐλαχίστη

---

[20] Ps.-Sor., *Def. Med.*, 107 (p. 259 ROSE 1864). See *Sel.Pap.* 1.149 (Oxyrhynchus, third century AD).
[21] *P.Mert.* 2.82.14–16. Corrected by YOUTIE 1973: 1008–11. See also BAGNALL and CRIBIORE 2006: 266.

ἐς τὸ στόμα ἐσίη καὶ ὡς ξεινοτάτη προμηθεῖσθαι, τῶν τε χωρίων τοὺς τόπους μεταβάλλοντα ἐς
δύναμιν, ἐν οἷσιν ἂν ἡ νοῦσος καθεστήκη [...].

(Hipp., *De nat. hom.* 9.3, *CMG* 1.1.3, p. 190.12–192.3–5 Jouanna)

In case of epidemic disease, it is clear that its cause is not in the regimen, but in the air we breathe,
which spread any noxious exhalation it may contain. This is therefore the advice that should be
given to people: do not change regimen, because it is not responsible for the disease at all [...] but
as far as the air is concerned, here are the recommendations: breathe in as little contaminated air as
possible. For this, remove the patient as much as possible from the areas contaminated.

This treatment recorded in the Hippocratic *De natura hominis* aims to reduce the patient's
inhalation of contaminated air by removing the patient from places filled with polluted air.

7. Eye-diseases

Another Egyptian scenario involves the prevalence of eye diseases, conditions we would expect from the constant blowing of sand and dust due to the aridity of the climate. Numerous
infectious diseases may have spread in ancient Egypt due in part to their agricultural practices, in which animals and their waste tended to mix. The marshy areas of the Nile Delta
were a place where eye infections were commonplace, as the area was so heavily infested by
swarms of insects and mosquitos. The sickly season in Egypt is the spring, when hot winds
excite the malignant diseases and generate swarms of insects. Eye-salves of different consist-
encies and ointments called *collyria* were applied to the exterior of the eyes and are often
referred to in Egyptian medical papyri from the time of the Pharaohs onwards. *Collyria* are
frequently encountered in extensive *receptaria* or as isolated recipes on a single sheet of papyrus or on ostraka.[22] Honey was used in many ointments because of its anti-bacterial effect.
Grated minerals were often added, for example green copper, zinc oxide (calamine), lead
and alum. Other factors, such as malnutrition, iron deficiency and ineffective hygiene may
all have affected children, and among the ocular recipes we find eye-salves called παιδικὰ
εὔχρηστα 'useful for children'.[23] Ophthalmia, characterized by runny eyes, swellings and
thick rheum, was particularly prevalent in Egypt. The many references to recurrent eye-diseases in Egypt explain why ophthalmology was so important in this country.

Eye-diseases and their treatment in Greco-Roman Egypt are documented in a remarkable number of examples. Greek documentary papyri provide exceptional evidence for ocular disorders such as cataract, which the weaver Tryphon suffered from. Tryphon visited a
doctor in Alexandria and asked to be released from his military service, because he had lost
his ability to see well due to the opacification of his lens.[24]

---

[22] See *O.Claud.* 1.174.7 κολλυρίδια β καὶ ἐκ τῶν μικρῶν (send two sticks of collyria and small ones);
O.Claud. 2.220.5–9 ἵνα δώσι σοι τὴν κρόκον καὶ πέμψις μοι ἐπὶ ταῦτα ἅ ἔπεμψές μοι (...) κολ|λύρια
ἰατρικὰ δ̄ (so that I can send you saffron, and send me for these illnesses medicinal collyria).

[23] For this kind of eye-salve recorded as παιδικόν, i. e. 'for children', see Youtie 1976: 121–9.

[24] For cases of cataract, see *P.Coll.Youtie* 1.19 (Ptolemais Euergetis, 44 AD) ὑποκεχυμένη τὰς ὄψεις,
P.Mich. 5.321 (Tebtunis, 42 AD) Ἡρώδης Νεστηφιος ὡς ἐτῶν ἑξήκοντα πέντε εὔσημος ὀφθαλμῶι
δεξ[ιῶι] καὶ | ὑποκεχυμένος τὸν ἄλλον ἀριστερὸν ὀφθαλμόν (with a prominent right eye and with
a cataract on the left eye...,). A woman complains because she is affected by *exophthalmia*, a pro-
trusion of the eyeball from the orbit, in *P.Cair. Goodspeed* 15.22 ἐξοφθαλμιάσας. Because of ocular

ὑπο<κε>χυμένος ὀλίγον βλέπων. (*P.Oxy.* 1.39.9)

Because of cataract I am affected by a loss of sight.

Another very common ocular affliction was *leucoma*, a dense, white corneal opacity mentioned in *P.Oxy.* 31.2601.verso.32–33 (fourth century AD), which seems to have been cured by doctors in Egypt:

ἵνα θεραπευθῇ τὸ λευκωμάτιον. ἐγὼ γὰρ | εἶδον ἄλλους θεραπευθέντας.

so that he can be cured from a small *leucoma*. I have seen that others have been treated.

8. Air and water pollution

Studies on the effects of urbanism on human health have followed two basic patterns: studying urban populations and people exposed to aspects of urban environments. The air that ancient populations breathed was polluted. One of the causes may have been the manual processing of flax, which was an ancient industry in Egypt. Manual flax processing originated in Egypt in the second millennium BC. Workers who handled and processed flax were exposed to high concentrations of dust. After drying and deseeding the flax, the fibres were loosened from the woody parts of the plant by a putrefactive process. For historical and ecological reasons, the processing of flax in Egypt was concentrated in the Nile Delta, where the use of flax dates back to ancient Egyptian industrial activity, when flax was woven into linen. The workers' exposure to dust, in small workshops or at home, determines the relationship of environmental dust concentrations to the prevalence of respiratory disorders, as well as chronic bronchitis and bronchial asthma. The modern name of the disease is byssinosis. It is a disease of the lungs brought on by prolonged inhalation of dust from textile fibres, marked by coughing, wheezing, shortness of breath, and permanent lung damage. Breathing in the dust produced by raw cotton or flax could cause lung damage. Flax bundles were tossed about, scraped and pulled through hackling combs. Those movements produced a quantity of dust especially near the worker's face. An anonymous therapeutic manual, excavated in the temple area of Tebtunis in the Fayum, and copied at the end of the first century AD, preserves an original treatise on the respiratory system and related ailments. The manual mentions *dyspnea*, a shortness of breath or breathlessness, the feeling associated with impaired breathing, a rough voice that is unusually deep, characterised by the term βραγχώδης, and

---

disorders, some people were hindered from travelling (*BGU* 16.2651.8–10, Herakleopolis, 9 BC, διὰ δὲ τὴν | ὀφθαλμίαν παραπεπόδισμαι τοῦ ἀναβῆναι | πρός σε), or were compelled to ask to be exempted from public service (liturgy), as in *PSI* 10.1103.13–15 (Ptolemais Euergetis, 192–4) [ἐγ]ὼ δὲ οὐ μόνον τῇ ἡλικίᾳ, | ἀλλὰ καὶ τῷ σώ[μ]ατι ἀσθε[ν]ὴς καὶ τοῖς ὀφθαλμοῖς ἀβλυώσσων (αβλοιοσων pap.) ('I am not ill only because of my age, but also because of my weak body, and by reasons of my weak vision'). The reading ἀβλυώσσων was a correction by YOUTIE, 1973: 388. See *P.Brem.* 64.8–9 (Hermopolis, second century AD, from the dossier of Aline) τετράμενος ἀσθενῶ μου τὰ | ὄμματα (I have had a weak vision for the last four months). For further readings on diseases in Egypt, see ANDORLINI 2010: 37–44. For the *trachoma (Clamydia trachomatis)*, see *PSI* 4.299.3–11: ὡς δ' ἐκουφίσθη μοι ἡ νόσος ἐπύθετό μοι ὁ ὀφθαλμὸς καὶ τραχώματα ἔσχον.

includes symptoms of pleurisy and phthisis.[25] The background of respiratory ailments offered by this manual is perfectly consistent with the lung ailments, which affected the textile workers of ancient Egypt, and shows how lung diseases were spread and treated throughout the Egyptian environment. Based on new evidence of particulates found in the lungs of mummy tissue, ancient Egyptians may have been exposed to air pollution. This mummy of a man, who probably died around the age of 20 to 25, was discovered in Dakhleh Oasis, a remote outpost in southern Egypt.[26] The remains of the mummy show how the area around the lungs, where particulates were found, was well preserved.

Lung diseases have been detected before in Egyptian mummies, and the Egyptians did engage in cooking, metal, flax working and mining, all activities that can generate air and water pollution. Mining is another major cause of water pollution. Mining activities consume large amounts of water in processing the ores from the mines. During mining, large quantities of ore – rocks containing valued substances like metals or coal – are excavated from the generated tailings.

According to Galen's account, water became polluted once sewage or animals, fruit and vegetables, or excreta were dumped in it:

μοχθηρὸς δὲ καὶ ὃς ἂν ἔκ τινος σηπεδόνος ἢ ζῴων ἢ λαχάνων ἢ ὀσπρίων ἢ κόπρου μιαίνηται. καὶ μὴν καὶ ὅστις ὁμιχλώδης ἐστὶ διὰ ποταμὸν ἢ λίμνην γειτνιῶσαν, οὐκ ἀγαθός, ὥσπερ γε καὶ ὅστις ἂν ἐν κοίλῳ χωρίῳ πανταχόθεν ὄρεσιν ὑψηλοῖς περιεχομένῳ μηδεμίαν πνοὴν δέχηται. πνιγώδης γὰρ ὅδε καὶ σηπεδονώδης ἐστίν […].   (*De sanit. tuenda* 1.11=6.58.5–10 Kühn)

And harmful too is the air contaminated from any putrefaction of animals or vegetable or pulse or manure; and this is not good which is cloudy from a neighboring river or marshland; and likewise that which, in a hollow place, surrounded on all sides by lofty mountains, receives no breeze, for such air is stifling and foul.

Galen also notes how the muddy water of the Nile was improved by the ingenious ways in which the natives managed to filter and to purify it, through pottery vessels:

ἔστι δὲ δήπου καὶ τὸ ἰλυῶδες ὕδωρ σύνθετον ὡς καὶ τὸ τοῦ Νείλου καὶ κατ' Αἴγυπτον, ἀλλὰ διὰ τῶν κεραμίων ἀγγείων διηθούμενον ἀκριβῶς γίνεται καθαρόν.
   (*De simpl. med. fac.* 1.4=11.389.15–18 Kühn)

If the water is also so muddy as in the river Nile in Egypt, filter it with pottery vessels and it will become pure.[27]

---

[25] Full edition in Andorlini 1995: 130, 151–2. See Gal., *De comp. medic. sec. loc.* 7.1 (13.4.14–15 Kühn) καὶ βραγχώδης ἡ φωνὴ γίνεται διά τε τὰς αὐτὰς αἰτίας καὶ διὰ ψυχροῦ πνεύματος εἰσπνοήν. The term βράγχος, rendered as 'hoarseness', denotes an affliction that proceeds to travel down the throat into the chest producing throat catarrh.

[26] Jarus 2011.

[27] On the contrary, the physician Rufus of Ephesus, who travelled to Egypt, in an extract preserved by Oribasius, *Coll. med.* 5.3.16, praised the qualities of Nile water (*CMG* 6.1.1., p. 118 Raeder): καί τοι μὴ ἐπαινῶ τὰ ἐν τῇ μεσημβρίᾳ ὕδατα, τὸν Νεῖλον οὕτως ἐπαινῶ, ὥστε δοκεῖ ὀλίγοις ἂν συμβάλλεσθαι ποταμοῖς κατ' ἀρετὴν ὕδατος.

Papyri from Egypt also refer to Nile water as contaminated with oil (*P.Köln* 10.418, second-third century AD, l. 12 ὕδωρ λιπάνῃ),[28] or responsible for noxious effusions (*O.Claud.* 4.890.16–17, 150–154 AD).[29]

9. Tuberculosis

I will conclude with the examination of mummies affected by pulmonary and osseous tuberculosis. Tuberculosis certainly plagued the Nile valley and appears to have been an important cause of mortality in ancient Egypt. The specimens that have been studied come from the tombs of Dra Abu el-Naga in an Upper Egypt site, close to the temple of Hatshepsut. The best-preserved mummy was that of an approximately five-year-old child. The child suffered from osseous and pulmonary tuberculosis. Fibrotic areas of the lung showed deposits of iron pigments. The finding of childhood tuberculosis is indirect evidence of a multigenerational disease, which affected the Egyptians since the early dynastic period, as has been demonstrated by the pathologist MICHAEL ZIMMERMAN of the University of Michigan at Ann Arbor, who studied this mummy.[30]

Bibliography

ANDORLINI, I. 1995. *Trattato di medicina su papiro*, Istituto Papirologico 'G. Vitelli'. Firenze.

ANDORLINI, I. 2004. 'Un ricettario da Tebtynis: parti inedite di PSI 1180,' in I. ANDORLINI (ed.), *Testi medici su papiro. Atti del Seminario di Studio (Firenze, 3–4 giugno 2002)*, Istituto Papirologico 'G. Vitelli'. Firenze, 81–118.

ANDORLINI, I. 2010. 'Segni di malattia nelle lettere dei papiri,' in *Actes du 26e Congrès International de Papyrologie.* Genève, 37–44.

BAGNALL, R. S. and R. CRIBIORE. 2006. *Women's Letters from Ancient Egypt*, 300 BC–AD 800. Ann Arbor.

DEELDER, A. M., R. L. MILLER, N. DE JONGE, and F. W. KRIJGER (eds.). 1990. 'Detection of schistosome antigen in mummies,' *The Lancet* 335, 724–5.

DILLER, H. 1999[2]. *Hippocratis De aere aquis locis, CMG* 1.1.2. Berlin.

EBBELL, B. 1937. *The Papyrus Ebers. The Greatest Egyptian Medical Document*. Copenhagen.

GASCOU, J. 1993. *Un Codex Fiscal Hermopolite (P. Sorb. II 69)*. Atlanta Ge. (*ASP* 32)

HELMREICH, G. (ed.). 1923. *Galeni De alimentorum facultatibus, CMG* 5.4.2. Leipzig-Berlin.

HUDE C. (ed.). 1958[2]. *Aretaeus. Editio altera lucis ope expressa, nonnullis locis correcta, indicibus nominum verborumque addendis et corrigendis aucta, CMG* 2. Berlin.

JARUS, O. 2011. 'Egyptian Mummies Hold Clues of Ancient Air Pollution', *Live Science*, June 3, see online at http://www.livescience.com

JONES, W. H. S. 1931. *Hippocrates IV. Nature of Man. Regimen in Health. Humours. Aphorisms. Regimen 1–3. Dreams*. Cambridge MA.

JOUANNA, J. 2002. *Hippocratis De natura hominis, CMG* 1.1.3. Berlin.

JOUANNA, J. 2003. *Hippocrate II.2. Airs, eaux, lieux*, CUF. Paris.

JOUANNA, J. 2003. *Hippocrate V.1. Des Vents*, CUF. Paris.

---

[28] The verb λιπαίνω describes rivers with thick waters due to plant and animal oil residues.

[29] Θέρμουθις ὄνους β ὕδατος. ἵνα μὴ | ὄζῃ τὸ ὕδωρ. Thermouthis ordered two donkeys for the water, so that the water does not produce noxious effusions.

[30] ZIMMERMAN 1979: 604–8.

KOCH, K. (ed.). 1923. *Galeni De sanitate tuenda libri vi.*, CMG 5.4.2. Leipzig-Berlin.

KÜHN C. G. (ed.). 1821–1833. *Claudii Galeni opera omnia*, 20 vols. Leipzig (repr. Hildesheim 1964–1965).

LITTRÉ, É. 1839–1861. *Oeuvres complètes d'Hippocrate*, 10 vols. Paris (repr. Paris 2012).

MANETTI, D. 2011. *Anonymus Londiniensis. De medicina*. Berlin-New York.

MARX F. (ed.), 1915. *A. Cornelii Celsi quae supersunt. CML* 1. Leipzig-Berlin (repr. Hildesheim 2002).

RAEDER, J. (ed.) 1926–1933. *Oribasii Collectionum medicarum reliquiae I–L, Libri incerti, Eclogae medicamentorum*, CMG 6.1.1–6.2.2. Leipzig-Berlin.

ROSE, V. (ed.). 1864. Pseudo-Soranian, *Definitiones medicae, Anecdota Graeca et Graecolatina*. Berlin.

SMITH, G. E. and D. E. DERRY. 1910. 'Report of human remains,' *Bulletin No. 6 Archaeol. Survey of Nubia*, Cairo: National Printing Department, 1–11, with Plates I–VIII.

STRAUS, J. A. 2004. *L'achat et la vente des esclaves dans l'Égypte romaine: Contribution papyrologique à l'étude de l'esclavage dans une province orientale de l'empire romain* (*Archiv für Papyrusforschung und verwandte Gebiete*, Beih. 14). Munich-Leipzig.

WELSH, J. 2011. 'Nubian Mummies Had Modern Disease', *LiveScience,* June 08, *online.*

YOUTIE, H. 1973. *Scriptiunculae 2.* Amsterdam.

YOUTIE, L. C. 1976. 'A Medical Prescription for an Eye-Salve (P.Princ. 3.155 R),' in *ZPE* 23, 121–9.

ZIMMERMAN, M. R. 1979. 'Pulmonary and osseous tuberculosis in an Egyptian mummy,' *Bulletin of the New York Academy of Medicine* 55, 604–8.

Pollution and the Environment in Ancient life:
Material Evidence

Alain Bresson

Anthropogenic Pollution in Greece and Rome

1. Definitions

How can we define pollution? The *American Heritage Science Dictionary* defines pollution this way: 'The contamination of air, water, or soil by substances that are harmful to living organisms. Pollution can occur naturally, for example through volcanic eruptions, or as the result of human activities, such as the spilling of oil or the spilling of industrial waste'.[1] The entry goes on to mention that light, noise and heat can also be sources of pollution.

The key word in this definition is the 'harm': the harm that can be done to 'living organisms'. The reference to harm implies a value judgment that presupposes the existence of a set of values. For anyone studying pollution in the ancient world, the first question is thus to determine whether judgments will be made in accord with the ancient world's values, which implies an analysis of the ideological constructs of that time, or with those of our own world, as applied to the ancient world. This is not a minor question, since any steps taken by ancient communities to prevent anthropogenic pollution can only have been made according to their values, not ours. Linked to this question is that of the extension of the definition quoted above, which includes all living organisms. It makes perfect sense in the framework of a modern scientific definition, but even in modern times it is problematic in legal terms. Legal rights are commonly described with reference to the rights of the human beings. Even in modern legal systems, animals or plants have been only exceptionally defined as having rights of their own. It is only very recently that this situation has changed.

The starting point for the definition of pollution, the notion of contamination, supposes a modification of the environment that proves to be harmful for living organisms. If light, noise and heat are included, one may end up studying all the modifications of the natural milieu by human activity, insofar as they inevitably impact the environment. However, this study will focus only on modifications of the environment produced by human activity that could impact both directly and negatively the health and welfare of human populations.

2. Domestic pollution: burning wood, charcoal and coal

Wood and charcoal were the only sources of heat production in most of the ancient world. From references in extant ancient sources and parallels from modern situations, we may infer that in some regions of the ancient world the use of domestic fires in densely populated

---

[1] *The American Heritage Science Dictionary.* Boston, 2005, s. v. 'Pollution', p. 495.

areas and the intensification of 'industrial production' must have already have been a major factor in air pollution.[2] Two sources of this type of pollution can be identified.

The first and most common source of pollution was the domestic burning of wood or charcoal, primarily for cooking (which used chiefly twigs and small branches), and secondarily for house heating. It is difficult to test directly the level of this type of pollution, so that we must resort mainly although not only to comparative evidence to evaluate the possible level of contamination and its impact on the health of ancient populations. In association with a number of other wood combustion products (particulate matter, nitrogen dioxides and sulfur dioxides), carbon monoxide is responsible for respiratory diseases, and also for cardiac diseases and cancer.[3]

In today's world, an extreme case of household carbon monoxide contamination is provided by the case of villages of Ladakh, a high altitude region of northern India.[4] There, most houses are not equipped with chimneys. Unsurprisingly, in houses without chimneys individuals inhale much more carbon monoxide than those who live in houses equipped with chimneys. In the former, the concentration of carbon monoxide, 102 ppm (part per million), is more than twice the recommended upper limit of the time-weighted eight-hour level of 50 ppm, whereas in houses with chimneys the level is just below this level (47 ppm).[5] There is also a very distinct difference between summer and winter: it is in the winter, when families live indoors because of the bitter cold outside, that the inhalation of carbon monoxide is maximal. There is also a significant gender difference: in winter, women exhale more carbon monoxide (17.3 ppm) than non-cigarette-smoking men (13.9 ppm). Women's cooking activities and the fact that they remain longer at home can easily explain this difference. Domestic pollution is a major contributor to chronic respiratory illness. The prevalence of coughing with chronic phlegm (production of mucus provoked by an infectious inflammation of the lungs) remains moderate for young women (10–15 % for women aged ca. 20–35) but reaches ca. 65 % for women over fifty-five. Men are on average slightly less impacted, but their curve of respiratory difficulties increases in parallel to women's.

A more moderate but still significant case of domestic air pollution is that of the native populations of the Highlands of New Guinea, where temperature commonly drops below 10°C during the night and where for lack of proper clothing people use small fires to heat their huts. Rates of carbon monoxide of 1.08 ppm and 21.3 ppm, but with peak values of 3.8 ppm and 150 ppm, have been reported. The levels of contamination were thus in general lower than those in Ladakh. But they have been assessed as contributing to the development of the (nontuberculous) lung disease prevalent among the New Guinea highlanders.[6] Lung infection, in the form of lobar pneumonia or bronchopneumonia, is the major cause of death among these populations (67 %). Although individuals of all ages are affected, the damage increases with age, and women are comparatively more affected. The parallel with Ladakh is striking.

---

[2] For general views about air pollution in the ancient world, see WEEBER 1990: 119–21; BORSOS ET ALII 2003; MAKRA and BRIMBLECOMBE 2004; MAKRA 2015.

[3] COULTAS and LAMBERT 1991; MARBURY 1991; LAMBERT 1997.

[4] NORBOO ET ALII 1991.

[5] In this study, the results given by some authors in μg/g are systematically expressed in ppm, as 1 μg/g = 1 ppm.

[6] CLEARY and BLACKBURN 1968; MASTER 1974.

An interesting (although isolated) prehistoric parallel is provided by the case of the 'Iceman' ('Ötzi'), whose body was found in an Alpine glacier in 1991 and who lived around 3300–3100 BC. Analysis of his lung revealed the presence of large quantities of carbon particles, which can be attributed to inhalation from indoor open fires.[7] The question is thus to determine how to situate the ancient world on the scale provided by the Ladakh and New Guinea cases.

From the Archaic to the Imperial period, many ancient Greek houses had permanent hearths in the rooms that performed the role of kitchen, but many others did not have such hearths, and we must assume that braziers were used instead.[8] Things do not seem to have been fundamentally different in the Roman world, at least if we can extrapolate from the case of Pompeii. In this city, some houses also had permanent hearths, while others did not, and the locations of the hearths suggest that they had a culinary, not a heating function.[9] There also we find braziers (mostly terracotta ones). In large houses, there were spaces for kitchens but they were poorly ventilated, if at all.[10] The same characteristics can be observed in the Roman houses of the Rhineland region in the Imperial period.[11] There, however, the real innovation is the presence even in more modest houses of underfloor or wall heating systems, the warm air coming from a furnace (*praefurnium*) located beneath the house: this provided an efficient heating without the inconvenience of the smoke in the house.[12]

The question of the presence or absence of chimneys in the houses is thus obviously a crucial one. In the ancient Mediterranean world, most houses seem to have lacked the chimneys that have been characteristic of western Europe since the medieval period. We see from various authors that there might have been some ways of evacuating smoke in Greek houses. But they took the form of a hole in the roof (κάπνη, καπνοδόκη) that could be closed by a trapdoor (τηλία) rather than that of a properly built chimney flue.[13] This supposes that much more smoke remained in the house than with the use of a properly built chimney. Things do not seem to have been much different in the Roman world.

It is true that the use of braziers limited the volume of smoke within the house, because they could easily be taken to the indoor courtyards that were common in Greek houses, where, under a porch (παστάς) or the roof of a peristyle for wealthier houses, it was possible to cook conveniently even during the rainy season. This was certainly a way to limit the volume of smoke in the house. Besides, it is also true that, given the absence of windowpanes, these houses may seem to have been well ventilated. In the Roman world, windowpanes were introduced in the homes of the wealthy in the early imperial period. Windowpanes and casement windows appear to have spread rapidly through the northern provinces of the Empire, which allowed better control of ventilation, but still did not make up for the lack of chimneys.[14]

---

[7] Pabst and Hofer 1998.

[8] Tsakirgis 2007.

[9] Boman 2005.

[10] Kastenmeier 2007: 79–85; Roberts 2013: 248–9.

[11] Jansen 1999: 833–5.

[12] Jansen 1999: 847.

[13] Κάπνη: see for instance Ar., *vesp.* 143; καπνοδόκη: Hdt. 4.103 and 8.137; τηλία, trapdoor: Ar., *vesp.* 147.

[14] Windowpanes: Dell'Acqua Boyvadaoğlu 2008; Jansen 1999: 846–7 (for the German provinces).

But beyond cooking, there remains the question of heating the houses. Indeed, in the Mediterranean area, temperatures rarely fall to the levels that can be observed in Ladakh. Home heating is thus comparatively less necessary. However, it would be impossible to deny that heating was necessary in many regions, especially in mountainous areas and in the northern part of the 'Mediterranean area,' especially when in the Imperial period the 'classical way of life' extended up to central and northern Europe. Heating was also necessary, at least temporarily, in the core of winter even in more privileged regions. There, the tradition of the *focus*, the hearth that was used for both cooking and heating the room, and was also the symbolic center of the house, was certainly maintained. Tibullus describes the Italian farmer who for the country festival, when all activities have stopped in the house, 'will heap huge logs upon his blazing hearth'.[15]

Also, windows might have been less common in ordinary people's houses than is usually thought. If, during the coldest winter nights, fires were maintained and doors were closed, air pollution peaked. If charcoal was used, air pollution was minimized, because it produces less fumes and carbon monoxide than burning wood. But charcoal, which is a more efficient source of energy, was much more expensive, and the poor could buy it only in small quantities. Lighting the house with oil-lamps was probably less of a problem, because they seem to have been efficient and produced little smoke.[16] We might also add that in most poor families, again for reasons of cost, the use of lamps must have been strictly limited.

But cooking and secondarily heating were not the only sources of domestic pollution. As early as the Archaic period, in the case of Sardis, we know that houses were commonly used for production purposes, including metal or glass working.[17] Later examples confirm these observations, as has been proven for various sites.[18] With poorly ventilated houses, wood and charcoal smoke as well as the fumes produced by the processing of various items must inevitably have affected the health not only of the workers themselves, but also of members of the household.

Beyond wood and charcoal, Britain represents a special case. In this region, coal was in common use in the Imperial period for both heating and cooking, and also for production processes, principally in metalworking. This was true both in military sites and in the residential areas, especially in the towns.[19] This must inevitably have been a source of specific pollution, if once again the houses were not equipped with chimneys.

Gauging the impact of domestic pollution remains mainly, although now not solely, a qualitative enquiry. Promising results have been obtained for Britain between the Bronze Age and the medieval period by recording the lesions left on skulls by severe chronic sinusitis as a marker of pollution.[20] Analysis shows an absence of incidence of this disease in

---

[15] Tib., *el.* 2.1.22: *ingeret ardenti grandia ligna foco.* (Engl. transl. G. P. GOOLD 1913[2], London-Cambridge MA). See also FOSS 1994, vol. 1: 63–5, for other references from various authors.

[16] BOMAN 2005.

[17] CAHILL 2005.

[18] See below and n. 29–30.

[19] DEARNE and BRANIGAN 1995.

[20] WELLS 1977; WALDRON 2009: 113–5, for the methodology, and MUSHRIF-TRIPATHY 2014 (with rich comparative ethnographical material) for the link with domestic pollution and the increase of mortality by inflammation and infection of the respiratory tract, impacting particularly girls and women occupied at cooking). On the pioneering work of C. WELLS, see ROBERTS and MANCHESTER 2012. Of course more research should now also be made for Britain itself.

pre-Roman Britain (0 % recorded), a significant one in the Roman period (2.8 %), followed by a further increase in the Anglo-Saxon period 6.8 %) and a decline in medieval period (3.6), but with levels still higher than in the Roman period.[21] Similar longue durée research has also been done for India (though with samples of much more limited size).[22] What is still lacking is new and large-scale similar studies for the core of the ancient world.

By comparison with the Ladakh domestic spaces, with their heavy concentrations of smoke from burning yak dung during the long winter season, it is unlikely that contamination by domestic smoke and fumes in the Greek and Roman world reached the high levels that are observed there. But the common absence of proper ventilation by chimney flues remains a useful indication of the negative impact on the health of the household members. The skeletons found at *Herculaneum*, which like Pompeii was destroyed by the explosion of Mount Vesuvius in 79 AD, provide proof of the prevalent pollution in this city. These skeletons show traces of newly formed subperiosteal bone in the visceral surfaces of the ribs in about 11.6 % of the individuals examined. These types of lesions are undoubtedly related to pleural inflammation, which in all likelihood correspond to a high degree of particulate pollution and with indoor fires as the most likely cause of the pathology. There was no prevalence by sex or age and children were also affected.[23]

The case of Ladakh suggests, however, that contamination might have different impacts on the several members of the household. In the ancient world, some sort of gender differentiation is thus also most likely. Because like their Ladakh counterparts they spent their time and worked mostly inside the houses, Greek and Roman women must have been more exposed to winter carbon monoxide pollution than men. Even cooking outside does not eliminate the risk of pollution, as is proved by contemporary experience.[24] In late Hellenistic and Imperial urban contexts men seem often to have bought their food in 'bars' along the streets, which must have limited their exposure to smoke at the time of cooking.[25] There must also have been a social distinction in this regard. A social differentiation can be observed in bones or teeth. There we see that individuals who were probably slaves have poorer teeth and their bones often bear the traces of heavy muscular efforts begun at a young age.[26] In any case, it seems logical to think that slaves, and especially female slaves, must have been the most exposed to domestic air pollution. Further analysis is thus needed on the question of the social and gender differentiation. But indoor pollution from domestic fires in the Greco-Roman world should, however, be considered as an established fact.

3. General sources of pollution

Beyond domestic pollution proper, there were also other sources of pollution in the Greek and Roman cities. This meant that one could be affected both by 'home-produced pollution' and by the pollution coming from other houses (for instance by the smoke produced for

---

[21] WELLS 1977, with BRIMBLECOMBE 1999: 5–8.
[22] MUSHRIF-TRIPATHY 2014.
[23] CAPASSO 2000.
[24] MUSHRIF-TRIPATHY 2014: 22–5.
[25] See below on the 'bars'.
[26] BISEL and BISEL 2002: 460–73.

domestic purposes by other houses) or from craft workshops (as we saw, homes could also be workshops), while in the towns there was also a significant level of noise pollution.

Indeed noise pollution should not be neglected. Craft activities, like metal-hammering and grain-grinding were sources of noise, both for the inhabitants of the house (if the workshop was also a residence) and for people living nearby.[27] Long-term exposure to noise must have had negative effects on the health of the inhabitants. Horace (*odes* 3.29.12) encouraged Maecenas to leave the noise of the city. Juvenal may have exaggerated the impact of noise pollution, but he does not fail to underscore the difference between rich and poor in this regard: 'Here at Rome very many invalids die from insomnia, although it's food undigested and clinging to the fevered stomach that induces the malaise in the first place. Which lodgings allow you to rest, after all? You have to be very rich to get sleep in Rome. That's the source of the sickness. The continual traffic of carriages in the narrow twisting streets and the swearing of the drover when his herd has come to a halt would deprive a Drusus or the seals of sleep'.[28]

But air pollution was even more damaging for the health of the inhabitants. Indeed the Greek and Roman towns are now increasingly recognized as full-scale production centers.[29] The presence of craft activities such as the production of purple dye, glass, bronze, lead, ceramics or perfume is well attested in urban and often, as observed above, in fully domestic contexts in Classical Olynthos, Kassope (in Epirus) and Athens and in Hellenistic Rhodes and Delos.[30] The picture is no different at *Herculaneum*.[31] The development of bakeries in late Hellenistic and imperial-period urban contexts added its share to urban pollution.[32] Many people also bought their food at 'bars' along the streets (no less than 158 have been identified in Pompeii).[33] In all likelihood it is because they had no proper kitchen installation at home that they resorted to these 'bars'.[34] These 'bars' had their own ovens, with rudimentary systems of smoke evacuation, although apparently they usually had no proper chimneys.[35]

Even before the Imperial period, from the Archaic period onwards, baths were common in the Greek cities, first in the gymnasiums, normally located outside the walls of the city, but then later, in the Hellenistic period, both *palestrae* and gymnasiums were located within the walls.[36] In the Imperial period, both in the East and in the West, the number of baths

---

[27] HARRINGTON 2015 for this type of noise pollution.

[28] Juv., *sat.* 3.232–8: *Plurimus hic aeger moritur vigilando (sed ipsum / languorem peperit cibus inperfectus et haerens / ardenti stomacho); nam quae meritoria somnum / admittunt? Magnis opibus dormitur in urbe. / inde caput morbi. Raedarum transitus arto / vicorum in flexu et stantis convicia mandrae / eripient somnum Druso vitulisque marinis.* Drusus and the seals had a reputation of being heavy sleepers. (Engl. transl. S. M. BRAUND 2004, London-Cambridge MA).

[29] See WILSON and FLOHR 2016 for the world of the Imperial period.

[30] ZIMMER 1999 for the Classical period examples. For Hellenistic Rhodes: KAKAVOYIANNIS 1984; Delos: KARVONIS 2008: 170–9, with also KARVONIS and MALMARY 2009 for the notion of houses as multi-purpose spaces.

[31] MONTEIX 2013: 169–217.

[32] See MONTEIX 2013: 133–67 for *Herculaneum*; MONTEIX 2016 for Pompeii.

[33] MONTEIX 2013: 89–132, and 2016.

[34] MAYER 2012: 30.

[35] ELLIS 2002.

[36] HOFFMANN 1999: 46–82 and 86–7.

increased dramatically. In the Rome of the fourth century AD (that is, in a period when the city had already lost some of its former population (at its maximum Rome might have had between 450,000, as a low estimate, and 600,000–800,000 or 900,000 inhabitants), there were 11 (large) *thermae* and 952 (smaller) *balneolae*.[37] Heating the baths also contributed to general air pollution.[38]

We must consider lime kilns as well. They have been acknowledged as a major source of air pollution in the Middle Ages.[39] Lime kilns must also have been major contributors to air pollution in the Late Republican and Imperial periods. This was a period of spectacular development in the use of concrete in Roman architecture, implying a large increase of lime production, which in turn must have contributed to increasing air pollution.[40]

Horace's reference to Rome's 'statues stained with black smoke' (*foeda nigro simulacra fumo*, *Odes* 3.6.4) as a product of time is no more than a rhetorical *topos*, but his advice to Maecenas to leave 'Rome and its smoke, riches, and noise' (*fumum et opes strepitumque Romae*, *Odes* 3.29.12) sounds more realistic. In a letter to his friend Lucilius, Seneca explicitly states that he is leaving Rome for his villa at *Nomentum* to escape the fever that was endangering him (*Epist.* 104.1). Then he can boast (104.6): 'As soon as I escaped from the oppressive atmosphere of the city, and from that awful odor of reeking kitchens which, when in use pour forth a ruinous mess of stem and soot, I perceived at once that my health was mending'.[41] But not everyone owned a villa at *Nomentum*.

This should not mean that the countryside was immune from air pollution. There a major source of pollution must have been the burning of charcoal. Charcoal was the only fuel that produced the high temperatures required for large sectors of metal production, especially, but not only, for metal production. The aim of charcoal burning is to obtain pure carbon that will reach very high temperatures when burnt. In antiquity charcoal was produced using large piles of wood. With traditional methods, charcoal is produced at a temperature ranging between 200°C and 400°C. The efficiency depends on the dryness of the wood (with fresh, wet wood it is significantly lower). Normally an efficiency of 15 % can be seen as a satisfactory, meaning that 7 kg of wood will produce 1 kg of charcoal.[42] Burning charcoal generates large quantities of highly toxic methane. Charcoal burning was certainly a major factor in deforestation and had a massive impact on the environment.[43] But it inevitably also contributed to large-scale air pollution in the countryside. Admittedly the impact on the health of the local population must have varied depending on the distance between the charcoal kilns and the human habitat and on the number of kilns. In regions where charcoal

---

[37] Boëthius and Ward-Perkins 1970: 164 and 561, quoting the *Libellus de regionibus Urbis Romae* (ed. Nordh 1949; the aggregate numbers correspond to the summation of the numbers of baths in the various districts). Population of Rome: low count by Storey 1997, and methodological discussion in de Ligt 2008: 149–52; 2012: 277 (who suggests a population of 900,000 inhabitants in 27 BC).

[38] Miliaresis forthcoming, on the question of the presence of windowpanes for the windows of baths.

[39] Brimblecombe 1987: 6–7, and 11–12 for pollution by lime kilns in British towns in the Middle Ages.

[40] Lechtman and Hobbs 1986; Lancaster 2008: 260–3.

[41] *Ut primum gravitatem urbis excessi et illum odorem culinarum fumantium, quae motae quicquid pestiferi vaporis obferunt, cum pulvere effundunt, protinus mutatam valetudinem sensi.* (Engl. transl. R. M. Gummere 1925, London-Cambridge MA).

[42] See Olson 1991, with FAO Forestry Paper 41, 1983 and Veal 2013.

[43] Harris 2011a and b; 2013.

burning was a large-scale occupation, this impact may have seriously damaged the health of the local population.

The sites of metal production were also a major source of pollution. Metal production required large quantities both of wood and charcoal to process a ton of ore. It has been calculated that 100,000 tons of rock dug out resulted in 1,500 tons of slag, one ton of silver, and 400 tons of lead. Producing these amounts required 500 to 2,000 tons of charcoal, obtained from 10,000 tons of wood.[44] Naturally, the quantities at stake varied with the ore richness. If in the first half of the fifth century Athens produced (?) an average 1,000 talents of silver per year, viz. ca. 26 tons, this production alone required an aggregate total of 260,000 tons of wood per year, burnt partly on the sites where charcoal was produced, and partly in the Laurion. Pollution linked to mining was thus not limited to 'industrial sites'.

Today satellite photos give us a macro-view of pollution prevailing in very large regions, like Northern India. Admittedly part of this pollution proceeds from modern industry, but small fires are the other major contributor. Owing to the much lower level of population density, no such massive level of pollution can have existed in the ancient world. But locally there must have been a large number of small pockets of dense air pollution that had a massive impact on the health of the neighboring populations.

Finally, it should be observed that in some privileged cases it is possible to highlight the professional diseases, the existence of which could be predicted from what we know of the prevailing work conditions of the time. Thus, as shown by ISABELLA ANDORLINI (this volume), analysis of a mummy of the Dakhleh Oasis illustrates the impact of flax processing. Particulates of flax have been found in its lungs (as is well known Egypt in general was the major producer of flax and linen of the ancient world).

4. Special cases (1 and 2): fulling mills and mines

Two special cases deserve special attention: fulling mills (fulleries) and mines. Their impact was of different nature. Fulling mills were to be found in all urban contexts, which means that the pollution they created was universally felt by the inhabitants of ancient Greek and Roman cities. Mines were ordinarily located far from cities, but their impact on their environment could be highly destructive.

The large development of the fulling mills is a characteristic feature of ancient Greek and Roman towns. While fulling mills are mentioned by written sources or known from archaeological sources in most regions of the ancient world, the Pompeii fulling mills are the best known, due to the circumstances of their preservation.[45] A fulling mill was basically a workshop that refined new woolen cloth, or washed it if it had been worn. For lack of soap, the basic agent for cleaning was human or animal urine. Urea in urine breaks down into ammonia, or more precisely, into an ammonia solution, i.e. ammonium hydroxide. This is why urine was considered a valuable product. Collected by contractors operating for the state, it was sold to fullers.

---

[44] PATTERSON 1972: 231 (followed by PICARD 2001: 4–5). For CONOPHAGOS 1980: 352, see 275–6, the Laurion mines consumed 600 tons of wood and 6,500 tons of charcoal annually. On the massive quantities of wood also necessary for copper production, see DOMERGUE 2008: 45.

[45] FLOHR 2006; 2007; 2009; 2013.

To refine or clean woolen cloth, three operations were needed. First came the cleaning, done in 'fulling stalls', i.e. in tubs filled with a mixture of water and urine. The workers trampled the cloth or wrung it out with their hands. Then came the rinsing. Finally the cloth was dried, brushed, and was given the finishing touches. It is above all in the first phase, and to a lesser degree in the second (although not only there, as sulphur could also be used for bleaching in the first phase) that fulling could damage the workers' health. The United States Department of Labor states that 'Ammonia is considered a high health hazard because it is corrosive to the skin, eyes, and lungs. Exposure to 300 ppm is immediately dangerous to life and health'.[46] Constant exposure to ammonia gas must have severely damaged the health of the fulling mill workers. The smell of the fulling mills, which could be located in residential areas, was also a source of nuisance and health damage for the neighbors. Besides, even though it was discharged to the sanitary sewers, in the end the wastewater from the fulling mills inevitably polluted the ground water.

Mines were also a specific and major source of pollution, although their impact was geographically more limited. It has already been noted that metallurgical sites produced large quantities of carbon monoxide. But they were also sources of metal contamination. The global impact of the pollution linked to ancient metal industry, specifically for copper and for lead, has famously been observed in Greenland ice. Such concentrations can be found again only in the early modern period.[47]

Mining was itself a major source of industrial pollution. First of all, it should be underscored that if there is one ancient production activity that was similar, in terms of space occupation and pollution, to what can be observed in the modern world, it is certainly the mining sector. The Laurion silver mines in southern Attica and the Las Médulas mines in Spain are famous examples of this activity and of the pollution they can cause, but of course they were not the only ones. The miners themselves were the first victims of mining operations. This is made clear by Strabo (12.3.40) concerning the mercury mines of Mt. Sandarakourgion, near Pompeiopolis, in northern Asia Minor. Mortality caused by the extreme toxicity of mercury was so high that maintaining the profitability of the mine proved difficult.

The conditions of the workers extracting ore in the Spanish mines, as described by Pliny (*nat. hist.* 33.21 [4]), were also extremely hard.[48] In all underground mines, the dust from the smoke of the lamps and poor air ventilation must have seriously damaged the workers' health, leading to silicosis. Using vinegar meant additional acid smoke. Finally, mining mercury, lead, or other metals involved working in a toxic environment, and directly caused irremediable health damage. This was also true for the workers occupied in washing and especially separating the metal for slag by cupellation (for galena), cementation (for gold) or grilling (for copper and iron), all of which created toxic fumes.[49]

---

[46] https://www.osha.gov/SLTC/ammoniarefrigeration/index.html

[47] ROSMAN 1993; HONG ET ALII 1994; HONG ET ALII 1996; DE CALLATAŸ 2005; SCHEIDEL 2009a. See also SHOTYK ET ALII 1998 for the results of the analyses from peat bogs, which confirm the results of ice core.

[48] On Pliny and pollution, see O. D. CORDOVANA, this volume.

[49] DOMERGUE 2008: 146–55 for washing operations, and 158–78 for separating metal from slag.

Ancient writers do not fail to report the fact that mining districts were insalubrious environments. Xenophon stressed that the Laurion was an unhealthy place.[50] Strabo advised that the installations for processing silver should be located on hilltops so that the fumes could be dispersed more easily.[51] Pliny stressed the danger of the silver mines' fumes, the processing of lead being especially dangerous for dogs (this makes sense since dogs breathe closer to the ground, where the concentration of fumes is higher, but of course this does not mean that there was little or no impact on humans).[52] So far, no analysis has been performed on bones coming from mining districts. Analyses of this type might reveal the actual level of contamination and the differential impact on the health of the populations of the mining districts (especially the average age of death).[53]

The ancient slag heaps may have had a limited impact on the landscape. They were and remain sterile, but their extent was comparatively limited and in general they were located far from densely inhabited areas.[54] But although they were limited in size, the slag heaps seem to have been more extensive than hitherto envisaged. Measuring residual pollution, as has been done at the large ancient copper mine of Khirbet Faynan in the desert of southern Jordan, proves that the adjacent terrains were polluted by copper, but also by other metals associated with copper.[55] The copper mine, which made heavy use of water in a half-desert environment, was thus also itself a large source of pollution in addition to the slag heaps proper, and this mine cannot have been an exception.

Thus the pollution produced by the mines was twofold. On the one hand, it had a geographically limited but massive polluting impact. On the other hand, it was a source of new and major contamination for the whole planet, foreshadowing the levels of pollution of the industrial era.

5. Special case (3): lead poisoning

Lead is now commonly acknowledged as having a very negative impact on human health. Even with a very low dosage, the brain, kidney, and blood are famously affected by lead absorption. But virtually every organ of the human body also is. Among the common symptoms of lead contamination are gastrointestinal pain, vomiting and severe anemia. It is especially dangerous for children, in whom it may cause severe intellectual disability.[56]

Whether or not lead pollution had a major impact on the human population has been a hotly debated question for decades. In an essay of 1909, RUDOLF KOBERT collected the ancient sources that testify to a clear awareness of the danger of lead for human health.[57]

---

[50] Xen., *Mem.* 3.6.12: λέγεται βαρὺ τὸ χωρίον. The passage shows that wealthy people from Athens even never visited the place.

[51] Strabo 3.2.8.

[52] Plin., *nat. hist.* 33.98 and 34.167.

[53] DOMERGUE 2008: 45–8.

[54] DOMERGUE 2008: 42–4 and 46–8, with also modern comparanda.

[55] GRATTAN ET ALII 2013.

[56] NEEDLEMAN 2004; AMITAI ET ALII 2010: 20–31.

[57] KOBERT 1909.

The debate was revived by an article by SEABURY COLUM GILFILLAN in 1965.[58] Taking a macro-demographic approach, he concluded that massive lead poisoning was at the origin of the fall of the Roman Empire. In his monograph of 1983, JEROME OKU NRIAGU gathered the various sources relating to lead poisoning and also the possible symptoms of undetected lead poisoning in the medical writings of the time.[59] For NRIAGU, it was wine consumption that was the major vector of contamination, since lead vessels were commonly used in the preparation of *defrutum* (*sapa*), wine juice reduced by boiling, the consumption of which was very popular in ancient Rome. As massive wine drinkers, the Roman aristocrats would have been seriously impaired by lead poisoning, which in turn would have had a massive impact on their fertility.

These views provoked intense debate, but on the whole they were strongly rejected. CHARLES ROBERT PHILLIPS rightfully stressed the lack of skeleton evidence and the weakness of the link established between the consumption of *defrutum*, lead poisoning and gout, not to mention the connection established between the decay of Roman aristocracy and the fall of the Roman Empire.[60] JOHN SCARBOROUGH severely criticized the lack of accuracy in NRIAGU's references to ancient sources. Analyzing in more detail the ancient literary evidence, he concluded that the unfermented juice ('must') boiled in bronze or lead cauldrons did not remain in contact with the metal for a long time, and that in any case the Romans must have preferred bronze cauldrons. From this and similar analyses he concluded that 'our ancient evidence (literary, archaeological, and even the limited osteological studies) shows exposure to lead in various forms in food, water, and wine, but does not support the theory that lead poisoning was endemic or pandemic in the Roman Empire, much less the cause of the "Fall of Rome"'.[61] He even added in a note that atomic absorption analyses showed that 'ancient exposure to lead generally was one-tenth that of modern man'.[62]

In a different vein, TREVOR HODGE, an expert on ancient aqueducts, admitted that Vitruvius (*de arch.*, 8.6.10–11) points to lead poisoning but stressed that (according to HODGE) he was careless in giving as an argument the contamination by lead smoke. For HODGE, Vitruvius should have emphasized that lead pipes were used only for the siphons or other water-pipe connecting devices and in houses, while there were no lead pipes in the aqueducts. In addition HODGE stressed that at least with water containing calcium carbonate the quick calcification of the pipes was sufficient to insulate the lead, which in consequence was not in contact with water. Finally, the system of constantly flowing water (even though taps were perfectly known) meant that water did not remain a long time in contact with lead in the pipes. HODGE thus concluded: 'As a general principle it is plain that the Romans did not poison themselves though their use of lead pipes, the water they drank from them was harmless, and in suggesting the contrary Vitruvius was wrong'.[63]

It should, however, be pointed out that although in his review HARRY ARTHUR WALDRON was also very critical of NRIAGU's methods, on the basis of his own experimental

---

[58] GILFILLAN 1965: 53–60.

[59] NRIAGU 1983a. See also the digest of his views in NRIAGU 1983b.

[60] PHILLIPS 1984.

[61] SCARBOROUGH 1984: 475.

[62] SCARBOROUGH 1984: 475, n. 23.

[63] HODGE 1981: 491. See also HODGE 1992: 308, the argument being repeated unchanged in the second edition of the book, HODGE 2002: 308.

research on lead accumulation in bones he insisted that lead contamination in Roman times had reached very high levels.[64]

The question of lead contamination is an exemplary case of a dramatic turnabout in the understanding of a historical phenomenon. A clear answer, based especially though not exclusively on the analysis of lead in bones, can now be given to the question of the level of lead contamination in the ancient world. Far from confirming the view that there was no lead contamination, new analyses show that it could be very high, far above the level that is considered harmful for human health. This in turns leads to a new analysis of the impact lead contamination had on the ancient world. This does not at all mean that we should return to the view that lead contamination was the 'cause' of the fall of the Roman Empire. First, there is no specific proof that contamination affected predominantly aristocrats, as NRIAGU argued. Moreover, it seems that we are for now unable to correlate in any manner lead contamination with social level. Finally, establishing a causal relationship between lead contamination and the fall of the Roman Empire is a bold jump, to say the least. It is more important to observe that lead contamination was in fact much more widespread than previously thought.

A direct result of mining activity, and especially of silver production, was the massive increase in the production of lead. Silver is commonly and for good reason associated with lead in galena. Silver mines were in fact first of all lead mines. But a survey of lead mines in the ancient world shows that lead production was not confined to the most famous silver-producing regions like the Laurion, Thrace, and Spain.[65] In addition, lead was also a massive object of trade.[66]

As is now well known, the production of lead and copper was so massive in the ancient world that these metals can be shown to be present in ice cores from the Greenland ice sheet. Production peaked in the late Hellenistic and early Roman Imperial periods, and did not reach such levels again before the late medieval and early modern periods.[67] On the basis of the historical level of air pollution, the yearly production of lead has been estimated to be ca. 160 tons between 2000 BC and the silver coinage revolution of the seventh century BC. From then on, worldwide production is supposed to have first climbed to 10,000 tons and then reached 80,000 tons in the early Imperial period, mostly produced and consumed in the Mediterranean region. After a marked decrease in the early medieval period, production is supposed to have increased again up to 100,000 tons around 1700. As a matter of comparison, it reached one million tons around 1900 and was around three million tons in 2000.[68] In the last decade it has suddenly and massively increased and reached five million tons.

For the ancient world, it is precisely in the period in which lead production peaks that the concentration of lead accumulated in human bones also peaks. A large-scale study of lead exposure in Italy between 800 BC and 700 BC shows convincingly how lead exposure parallels the production of lead as it can be traced from the Greenland ice.[69] The results are

---

[64] WALDRON in *Isis* 76.1, 1985, 118–20 (*Review* of NRIAGU 1983a).

[65] MEIER 1995: 14–136, with map p. 14, and DOMERGUE 2008, maps pp. 16–22. For Spain, DOMERGUE 1990: 71–5. For the German provinces in the Imperial period, BODE ET ALII 2009.

[66] DOMERGUE 2004; MONTEIX 2004; TRINCHERINI ET ALII 2009.

[67] DE CALLATAŸ 2005; SCHEIDEL 2009.

[68] SETTLE and PATTERSON 1980: 1170.

[69] AUFDERHEIDE ET ALII 1992.

fully convincing because the same method has been used for analyzing all bones (atomic absorption spectrometry) and because the samples are comparable (mid-diaphysis of tibia or femur of adults of comparable age and sex distributions). Special care was taken to determine whether soil contamination might have distorted the result, but it was also proved that this was not the case.

Table 1 – Lead Accumulation in Bones in Italy 9th Century BC – 8th Century AD (in ppm) (from AUFDERHEIDE ET AL. 1992, combined selected data from tables 1 and 2)

No.	Archaeological Site	Province	Region	Date/Cent.	No. Adults	Mean Bone Lead
1	Castiglione	Rome	Latium	9 BC	19	8.8
2	Castiglio di Gabii	Rome	Latium	8–7 BC	2	0
3	Via Bracciano	Rome	Latium	7–6 BC	3	6.8
4	Siena	Siena	Tuscany	6 BC	15	1.5
5	Castiglione Fosso S. Giuliano	Rome	Latium	4 BC	6	0.8
6	–	–	–	4 BC	1	24
7	Tarquinia	Viterbo	Latium	3 BC	47	9.1
8	Sulmona	L'Aquila	Abruzzi	3–2 BC	27	5.8
9	Castiglione Fosso S. Giuliano	Rome	Latium	Aug. per.	2	42
10	Lunghezza	Rome	Latium	Aug. per.	10	43.1
11	Grottaperfetta	Rome	Latium	1 AD	7	93.3
12	Civita Castellana	Viterbo	Latium	1–2 AD	1	77.7
13	Alba	Cuneo	Piedmont	1–2 AD	4	18.8
14	Via Gragnano	Rome	Latium	2 AD	10	72.2
15	Lunghezza	Rome	Latium	2 AD	2	18.3
16	Via Vigna Fabbri	Rome	Latium	2 AD	4	70.1
17	Via Latina	Rome	Latium	2 AD	2	1.8
18	San Paolo	Rome	Latium	4–5 AD	27	81.7
19	Villa Carcina	Brescia	Lombardy	7 AD	31	2.6
20	Malvito	Cosenza	Calabria	7–8 AD	20	18.6

From negligible levels in the Early Iron Age, the accumulation of lead in bones climbs abruptly between the mid-third century BC and the Common Era. This is the period when Italy began to have access to Spanish silver and lead, as well as to other sources of metal, such as Magdalensberg. The increase is spectacular. The 'lag' as compared to the curve of world lead production may be explained by the fact that in the eastern Mediterranean the massive extraction of silver and lead had begun even earlier – especially, although not only, with the exploitation of the Laurion and of the Thracian mines. The mean level of contamination (82.7 ppm) observed at *Herculaneum* in a separate study and with a very large sample of

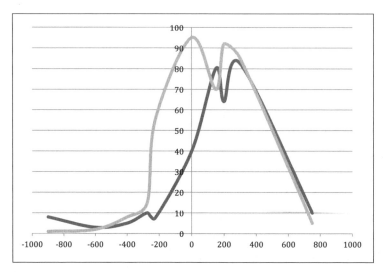

Fig. 1: Lead Production in Thousand Tons Per Year (light grey) and Accumulation of Lead in Bones in Italy, in ppm (dark grey). (Based on Aufderheide et al. 1992, Fig. 1. Lead Production from Scheidel 2009, Fig. 1).

individuals (92) is perfectly in accord with the highest figures observed for Rome in the same period.[70]

In detail, the analysis of Italian bones shows that the increase in lead contamination was not confined to Rome. The figures for Rome show a massive increase between the fourth century BC (no. 5: 0.8 ppm) and the later period (no. 9–11 and 14–18, from 42 ppm up to 93.3 ppm, with only one site, no. 15, at 18.3 ppm, but the sample consists of only two individuals). A parallel evolution can be observed in Etruria: in the third century BC, the sample from Tarquinia (no. 7) has only 9.1 ppm, but in the early Imperial period the site of Civita Castellana (no. 12, in inland Etruria, with admittedly only one individual in the sample) shows an accumulation of 77.7 ppm. If the observations from *Herculaneum* are brought into the picture, and even if more analyses should be made, it is clear that lead contamination affected not only the populations of the capital but also those of the smaller towns. *Herculaneum* and Pompeii were comparatively rich towns in an affluent region, but it does not mean that they were exceptional. However, whether we can extrapolate a massive lead contamination to the whole population of Italy, at a time when a large majority of the people lived in the countryside, remains to be proved. In the early Imperial period the site of Alba in Piedmont (no. 13) presents much lower levels of contamination (18.8 ppm). But, conversely, the mean levels of accumulation of lead in bones at the Romano-British sites of

---

[70] Bisel and Bisel 2002: 459–60. We keep here the figure of 82.7 ppm in order to be able to compare it with the numbers provided by Drasch 1982. To compare the *Herculaneum* result with the results of analyses on contemporary individuals, the figure should be divided by 1.2 to account for the organic material lost from the bones during post-mortem decomposition, as pointed out by Keenan-Jones et alii 2011: 144, basing themselves on Waldron et alii 1976: 226. The corrected result is 68.9 ppm (see below fig. 2).

Henley Wood in Somerset (65.7 ppm) and Poundbury in Dorset (112.7 ppm) are consistent with the high levels of the Italian towns. In any case, it is clear that Romanization brought about a huge and unprecedented lead contamination of human populations.

After the collapse of the empire, the site of Villa Carcina, in seventh-century Lombardy, which corresponds to a Lombard population, shows an accumulation of lead in bones of only 2.6 ppm. The lead supply had decreased and so had lead contamination. The very low levels observed even earlier (in the fourth to fifth centuries AD) at the site of Augsburg in Southern Germany justify a similar analysis. With a comparatively large number of femurs in the sample (121), the level of contamination is remarkably low, from 2.36 ppm for children younger than six to 6.19 ppm for adults over sixty. It should be observed that Raetia Secunda, of which *Augusta Vindelicorum* had been the capital, was evacuated by Rome under the pressure of the Alamanni in the course of the fourth century.[71] These low levels of contamination thus probably reflect the process of de-Romanization that was taking place at that time in this region.

The high levels of human lead contamination in the Classical world thus correspond perfectly well to the levels of lead production. It is also interesting that the per capita production of lead was far above that of our modern world. With an annual lead production of 80,000 tons in the first century AD and on the basis of a population of 60 million inhabitants for the Roman Empire (which was by far the main lead-producing and lead-consuming area in the world of the time), the per capita yearly additional available quantity of lead in the Mediterranean world was 1.33 kg. With an annual world lead production of 3 million tons, a world population of six billion inhabitants in 2000 and a universal use of lead, the quantity was recently only 0.5 kg. But contamination is not directly a matter of the per capita proportion of lead available.

Indeed, an overview of human lead contamination shows a more complex history (see fig. 2, which provides samples illustrating this evolution). Societies of hunter-gatherers or non-Mediterranean Neolithic societies had in general very low levels of contamination, below 1 ppm.[72] Interestingly, the lead accumulation (3.3 ppm) in the bones from the Neolithic period at the Franchti Cave (Southeastern Argolis, Greece) cannot be natural, but testify to a first, if limited, contact with lead. After the huge increase in lead production corresponding to the ancient Classical world's growth, the very low levels of contamination in the Regensburg samples of the early or central Middle Ages reflect the overall decrease in lead supply in this region during this period. But from the late Middle Ages onwards, lead contamination again reaches high or very high levels. The case of a farm site in Virginia shows that the poorer population (the laborers) was less contaminated than the 'richer' planters, who probably had access to more metal objects.[73]

A global survey of lead accumulation in children's tooth enamel in England from the pre-Roman period until the nineteenth century confirms this general pattern.[74] Pre-Roman samples are systematically below 0.87 ppm. Roman-period samples are between 1 and 10 ppm. Early medieval samples can be very low (below 0.87 ppm) or much higher (up to c. 30 ppm). But all post-medieval samples have high accumulation levels, up to 100 ppm. Given

---

[71] BURNS 1994: 122–3.
[72] SETTLE and PATTERSON 1980; DRASCH 1982 for Peru.
[73] AUFDERHEIDE ET ALII 1981.
[74] MILLARD ET ALII 2014.

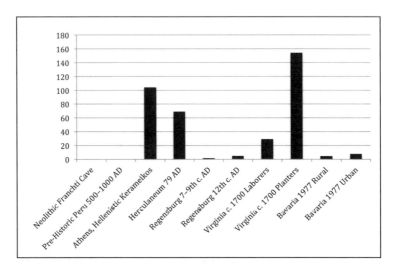

Fig. 2: Historical Accumulation of Lead in Bones (in ppm ash). Franchti Cave and *Herculaneum*: Bisel and Bisel 2002. Pre-Historic Peru, Regensburg 1–2 and Bavaria 1977 1–2: Drasch 1982 (the numbers are those for adults 40–60 years old). Virginia 1–2: Aufderheide et al. 1981 (see above n. 68 for the reduction of 1.2 applied to all pre-contemporary results).

that this corresponds to the contamination of children, it is clear that at least in Britain, starting in the late sixteenth century and during the transition to the 'Industrial Revolution,' the levels of lead contamination were very high, even higher than those found in the city of Rome in the Imperial period.[75] These levels are consistent with what can be observed for English colonists in Virginia.

In terms of lead contamination, the levels of lead contamination in the ancient world, like those in early modern Europe, were so high that public health was badly impaired. There were probably various levels of contamination. The countryside must have been significantly less contaminated than the towns (as in modern Bavaria). We know from *Herculaneum* that men were more impacted than women (the accumulation of lead in bones was 93.8 ppm for men, and 70.1 ppm for women).[76] Whether that was because men drank more *defrutum* or because in their occupations they had more contact with lead, or both, cannot be determined. Whether rich people were more or less affected than the poor also remains uncertain, and the answer might be more complex than previously imagined. The parallel of early modern Virginia shows that very poor laborers, probably because they could not buy lead plates or similar objects, were less affected than the richer planters.[77] But in general in England in the early modern period the poor were affected more than the rich.[78] Finally, it should be underscored that in terms of lead contamination early modern Europe did not fare better than the ancient Classical world, which in this regard is not an exception.

[75] Montgomery et alii 2010.
[76] Bisel and Bisel 2002: 460.
[77] Drasch 1982.
[78] Millard et alii 2014.

The reasons why lead contamination was massive in the ancient world are not difficult to identify. Lead was both cheap and abundant. It was also easy to process (its melting point is 327°C, and it thus requires less energy than other metals), cast and solder.[79] As a material, lead played in many respects the role played by plastics today. Thus, for example, it was used to make water pipes and taps in the Roman West. These pipes, or *fistulae*, were produced in large quantities by specialized workers, the pipe-producers (*fistularii*), or lead-workers (*plumbarii*).[80] The question is to gauge the level of contamination brought about by the lead water pipes. As observed by HODGE, the aqueducts, which brought water from distant sources, were not equipped with lead pipes. But as is well demonstrated by the case of Pompeii, in Roman towns the systems of water distribution of water within the town used lead pipes, lead taps and lead junctions.[81] Indeed, also as mentioned by HODGE, the deposit of a layer of calcium carbonate on the pipes can be observed, which partly insulates the water from the lead. But the sinter itself has now been proved to be contaminated by lead, and in any case the inevitable repairs to the lead piping system inevitably created peaks of contamination. The result is that the lead accumulation in samples from various locations of the water system within the town of Pompeii, reaching 1171 ppm in one case, are usually far higher than the rates of lead accumulation in the samples from the *Aqua Augusta*, the aqueduct that brought water to Pompeii, which were in the range of 25–101 ppm.[82] Thus the case of Pompeii shows that it should not be doubted that lead piping was a major contributor, although certainly not the only one, in the lead contamination of human population in Roman times.

Greek towns preferred stone and above all ceramic pipes, and this remained true in the Imperial period. Even siphons could be made of stone, although there were exceptions, as was the case with the large siphon of Pergamon built by Eumenes II, which was made of lead.[83] However, even if in the Roman West lead pipes were commonly employed to supply water to towns, they were not in universal use. Thus, although it produced lead in large quantities, Britain also used wood pipes, oak trees being in great abundance and certainly cheaper than lead itself. The same was true in Germany.[84] The high level of lead accumulation in bones at the Kerameikos cemetery in Athens in the Hellenistic period, 125 ppm, is also a warning against concluding that lead pipes were the only cause of contamination.[85]

In fact, the population of the ancient Greek and Roman world lived in an environment that was heavily polluted by lead. All kinds of everyday objects were made of lead. This was the case for plates, drinking vessels, coins (when it was alloyed with bronze from the late Hellenistic period onwards), tokens for various uses, market weights, lamps, writing tablets, dice, knucklebones (jacks), toys, gym weights, cosmetics, medical instruments, water heaters, drug containers, pumps, loom weights, fishing net weights, hooks, anchors, sound-

---

[79] HODGE 2002: 308.
[80] HODGE 2002: 307–15; PAGANO 2004; ROTHENHOEFER and HANEL 2013: 278–9, on the technique of fabrication of pipes.
[81] KEENAN-JONES ET ALII 2011; OLSSON 2015.
[82] See in detail KEENAN-JONES ET ALII 2011.
[83] HODGE 2002: 24–41, with a series of examples of stone siphons, and 43–5 for the lead siphon of Pergamon.
[84] HODGE 2002: 307–8, with 106–15 for a general discussion of the types of pipes.
[85] BISEL and BISEL 2002: 459.

ing-weights, sling-bullets, etc.[86] Roman-period shipwrecks off the coast of Israel have been the object of a detailed study.[87] The hull sheathing, anchors, fishing gear, braziers, and cooking equipment were all made of lead. This means that these ships were a traveling source of pollution. We can assume that a ship of that time carried a total of between 385 and 535 kg of lead. The lead inevitably poisoned the sailors and also, to a certain extent, the population of the ports visited by these ships.

The ubiquitous presence of lead in the daily life of the Greeks and the Romans explains the major traces of lead contamination that have been recorded in the ports of ancient towns. In contrast to the pre-harbor Tiber valley, and later-period harbor deposits, the analysis of the sediments at the Trajanic harbor at *Portus*, which was connected to the Tiber, show a peak in lead contamination corresponding to the early Imperial period.[88] There is no doubt that the water system of Rome was a contributor to this contamination, although, for the reasons previously mentioned, probably not the only one. The low levels of lead accumulation in the sediments in the port of Ephesos, a city that used terracotta lead pipes rather than lead ones, seem indeed to tip the balance in favor of lead water piping as a major factor of contamination.[89] But in the port of Alexandria, a city that does not seem to have resorted to lead piping, very high levels of lead accumulation have been observed in the sediments of the period corresponding to the early Imperial period, even twice as high as those found in contemporary industrialized estuaries.[90]

Today, at least in developed countries, levels of contamination are back to far lower levels as compared to the early modern period. The level of lead accumulation among the North-American population is commonly below 10 ppm.[91] The levels of accumulation in Bavaria are significantly below 10 ppm, with a clear gap between urban and rural populations (which show less contamination). It has been proved that what may seem a detail, such as the lead soldering of food cans, was in fact a major contributor to lead ingestion by human populations.[92] Public health regulations like the ban on lead-based paints have had a major impact.[93] The ban on leaded gasoline produced an immediate and spectacular drop in blood lead levels (by a factor of eight between 1974 and 2000 in the United States).[94] This shows that contamination by lead is not a fatality, even in a society that produces much lead.

---

[86] See in general BOULAKIA 1972: 144; MEIER 1995; RETIEF, CILLIERS 2005. Coins: FAUCHER 2013: 36–80, for Ptolemaic Alexandria; SCHEIDEL 2009b: 173, for Rome. Tokens: CROSBY in LANG and CROSBY 1964, Part II for Athens; ROSTOVTZEFF 1905 for Rome (with the *caveat* of VIRLOUVET 1995: 309–62, for the *tesserae frumentariae*). Market-weights: LANG in LANG and CROSBY 1964, Part I: 13–18; GATIER and OLIVIER 2015. Gym weights: Sen., *epist.* 56 (*plumbo graves* hurled out by athletes). Cosmetics: KEENAN-JONES ET ALII 2011: 144. Medical instruments: HEALY 1978: 248. Water heaters: MONTEIX 2010: 97–102, and 2013. Pumps: DOMERGUE 2008: 122–3 and fig. 66. Sounding-weights: OLESON 2008 and GALILI ET ALII 2010.

[87] ROSEN, GALILI 2007.

[88] DELILE ET ALII 2014.

[89] DELILLE ET ALII 2015.

[90] VÉRON ET ALII 2006.

[91] KEENAN-JONES 2011: 145, although older people and some minority population may have higher lead bone accumulation.

[92] SETTLE and PATTERSON 1980.

[93] AMITAI ET ALII: 38–41.

[94] AMITAI ET ALII: 36–8.

But strict public health measures have to be taken against lead contamination. In this matter the very high levels of contamination observed in the ancient world easily find a counterpart in the similarly high levels of contamination in early modern Europe. It was only when contamination could be scientifically measured, when public health became a major concern and, crucially, when intentions could be translated into political action, that effective measures could be taken against lead contamination. In the history of Western Europe, this did not happen before the nineteenth and, above all, before the twentieth century. In the ancient world, lead was too convenient and too cheap. The awareness of the danger of lead among doctors and other scientists of the time was never translated into systematic public policies.[95] For this reason, lead contamination soared dramatically.

Until a very recent period, the ancient world was perceived as a mainly rural civilization, a face-to-face society living on its local productions and which could be easily imagined as a world immune to pollution. In fact, just like Britain in the late medieval or early modern period, this ancient world we have lost was heavily polluted. Two major factors of the contamination of the human population have been identified in this study. The first one is the domestically produced carbon monoxide smoke produced in houses that had no proper chimneys, to which should be added the toxic emissions of urban activities such as metal workshops, lime kilns, and fulling mills. In a rural context, charcoal burning, mining and metal processing were major factors of contamination. Although not easy to trace archaeologically, these smokes must have induced endemic respiratory diseases among both urban and rural populations. The second factor is contamination by lead, which is now fully confirmed by the chemical analysis of bones or tooth enamel. It is no exaggeration to say that Greece and Rome (Rome even more than Greece) were 'lead civilizations'. There was a form of universal poisoning that inevitably had a serious negative impact on the health of populations, especially urban ones. Lead piping in the western part of the Mediterranean was certainly a major factor of contamination, although not the only one. Specific populations, like the workers involved in mining or metal processing, were even more severely impacted. It should, however, be acknowledged that, at least as far as lead is concerned, this high level of pollution was even surpassed in certain regions of late medieval or early modern Europe. The ancient awareness of anthropogenic pollution and the policies to reduce it, if any (two topics deliberately left aside in this study), are thus obviously crucial questions.[96]

Bibliography

AMITAI, Y., ET ALII 2010. *Childhood Lead Poisoning*. World Health Publications. Geneva.

AUFDERHEIDE, A. C., F. D. NEIMAN, L. E. JR. WITTMERS and G. RAPP. 1981. 'Lead in Bone II: Skeletal Lead Content as an Indicator of Lifetime Lead Ingestion and the Social Correlates in an Archaeological Population', *American Journal of Physical Anthropology* 55, 285–91.

AUFDERHEIDE, A. C., G. JR. RAPP, L. E. JR. WITTMERS, J. E. WALLGREN, R. MACCHIARELLI, G. FORNACIARI, F. MALLEGNI and R. S. CORRUCCINI. 1992. 'Lead Exposure in Italy: 800 BC–700 AD', *International Journal of Anthropology* 7.2, 9–15.

---

[95] RETIEF and CILLIERS 2005 for the perception of the danger of lead.

[96] See already the other papers of this volume.

BISEL, S.C. and J.F. BISEL. 2002. 'Health and Nutrition at Herculaneum: An Examination of Human Skeletal Remains', in W.F. JASHEMSKI and F.G. MEYER (eds.). *The Natural History of Pompeii*. Cambridge, 451–75.

BODE, M., A. HAUPTMANN and K. MEZGER. 2009. 'Tracing Roman Lead Sources Using Lead Isotope Analyses in Conjunction with Archaeological and Epigraphic Evidence – A Case Study from Augustan/Tiberian Germania', *Archaeological and Anthropological Sciences* 1.3, 177–94.

BOËTHIUS, A. and J.B. WARD-PERKINS. 1970. *Etruscan and Roman Architecture*. Harmondsworth.

BOMAN, H. 2005. 'White Light – White Heat. The Use of Fire as a Light and Heat Source in an Atrium House in Roman Pompeii', *Current Swedish Archaeology* 13, 59–75.

BORSOS, E., L. MAKRA, R. BÉCZI, B. VITÁNYI and M. SZENTPÉTERI. 2003. 'Anthropogenic Air Pollution in the Ancient Times', *Acta Climatologica et Chorologica* 36, 5–15.

BOULAKIA, J.D.C. 1972. 'Lead in the Roman World', *American Journal of Archaeology* 76, 139–44.

BRIMBLECOMBE, P. 1987. *The Big Smoke: a History of Air Pollution in London since Medieval Times*. London-New York.

BRIMBLECOMBE, P. 1999. 'Air Pollution and Health History', in S.T. COLGATE, J.M. SAMET, H.S. KOREN and R.L. MAYNARD (eds.), *Air Pollution and Health*. London, 5–18.

BURNS, T.S. 1994. *Barbarians within the Gates of Rome*. Bloomington.

CAHILL, N. 2005. 'Household Industry in Anatolia and Greece', in B.A. AULT and L.C. NEVETT (eds.), *Chronological, Regional, and Social Diversity*. Philadelphia, 54–66.

CALLATAŸ, F. de. 2005. 'The Graeco-Roman Economy in the Super Long-Run: Lead, Copper, and Shipwrecks', *Journal of Roman Archaeology* 18, 361–72.

CAPASSO, L. 2000. 'Indoor Pollution and Respiratory Diseases in Ancient Rome', *The Lancet* 356 (9243), 1774.

CLEARY, G.J., and C.R.B. BLACKBURN. 1968. 'Air Pollution in Native Huts in the Highlands of New Guinea', *Archives of Environmental Health* 17, 785–94.

CONOPHAGOS, C. 1980. *Le Laurium antique et la technique grecque de la production de l'argent*. Athens.

COULTAS, D.B. and W.E. LAMBERT. 1991. 'Carbon Monoxide', in J.M. SAMET and J.D. SPENGLER (eds.), *Indoor Air Pollution: A Health Perspective*. Baltimore-London, 187–222.

DE LIGT, L. 2008. 'The Population of Cisalpine Gaul in the Time of Augustus', in L. DE LIGT (ed.), *People, Land, and Politics. Demographic Developments and the Transformation of Roman Italy 300 BC-AD 14*. Leiden-Boston, 139–83.

DE LIGT, L. 2012. *Peasants, Citizens and Soldiers: Studies in the Demographic History of Roman Italy 225 BC – AD 100*. Cambridge-New York.

DEARNE, M.J., and K. BRANIGAN. 1995. 'The Use of Coal in Roman Britain', *Antiquaries Journal* 75, 71–105.

DELILE, H., J. BLICHERT-TOFT, J.-P. GOIRAND, S. KEAYE and F. ALBARÈDE. 2014. 'Lead in Ancient Rome's City Waters', *Proceedings of the National Academy of Sciences of the United States of America* 111.18, 6594–9.

DELILE, H., J. BLICHERT-TOFT, J.-P. GOIRAN, F. STOCK, H. BRÜCKNER and F. ALBARÈDE. 2015. 'The Geochemistry of the Harbor Sediment at Ephesus Provides a Record of Disturbances in the Mediterranean in Late Antiquity and the Medieval Period', *Journal of Archaeological Science* 53, 202–15.

DELL'ACQUA BOYVADAOĞLU, F. 2008. 'Between Nature and Artifice: "Transparent Streams of New Liquid"', *RES: Anthropology and Aesthetics* 53/54, 93–103.

DOMERGUE, C. 1990. *Les mines de la péninsule ibérique dans l'Antiquité romaine*. Rome.

DOMERGUE, C. 2004. 'Les mines et la production des métaux dans le monde méditerranéen au Ier millénaire avant notre ère: du producteur au consommateur', in A. LEHOËRFF (ed.), *L'artisanat métallurgique dans les sociétés anciennes en Méditerranée occidentale: techniques, lieux et formes de production*. Rome, 129–60.

DOMERGUE, C. 2008. *Les mines antiques. La production des métaux aux époques grecque et romaine*. Paris.

DRASCH, G.A. 1982. 'Lead Burden in Prehistorical, Historical and Modern Human Bones', *The Science of the Total Environment* 24, 199–231.

ELLIS, S. J. R. 2004. 'The Pompeian Bar: Archaeology and the Role of Food and Drink Outlets in an Ancient Community', *Food & History* 2, 41–58.

FAO Forestry Paper 41. 1983. *Simple Technologies for Charcoal Making*. FAO Forestry Paper 41 (reprint 1987). Rome Food and Agricultural Organization of the United Nations.

FAUCHER, T. 2013. *Frapper monnaie: la fabrication des monnaies de bronze à Alexandrie sous les Ptolémées*. Alexandria.

FLOHR, M. 2006. 'Organizing the Workshop. Water Management in Roman Fullonicae', in G. WIPLINGER (ed.), *Cura Aquarum in Ephesus. Proceedings of the 12th International Congress on the History of Water Management and Hydraulic Engineering in the Mediterranean Region, Ephesus/Selçuk, October 2–10, 2004*. Leuven, 193–200.

FLOHR, M. 2007. '*Nec quicquam ingenuum habere potest officina?* Spatial Contexts of Urban Production at Pompeii, AD 79', *Babesch* 82, 129–48.

FLOHR, M. 2009. 'The Social World of Roman Fullonicae', in M. DRIESSEN, S. HEEREN, J. HENDRIKS, F. KEMMERS and R. VISSER (eds.), *Proceedings of the Eighteenth Annual Theoretical Roman Archaeology Conference*. Oxford, 173–85.

FLOHR, M. 2013. *The World of the Fullo. Work, Economy and Society in Roman Italy*. Oxford.

FOSS, P. W. 1994. *Kitchens and Dining Rooms at Pompeii: the Spatial and Social Relationship of Cooking to Eating in the Roman Household*. 2 vols. Diss. Ann Arbor.

GALILI, E., J. P. OLESON, and B. ROSEN. 2010. 'A Group of Exceptionally Heavy Ancient Sounding Leads: New Data Concerning Deep-Water Navigation in the Roman Mediterranean', *Mariner's Mirror* 96, 136–48.

GATIER, P.-L. and J. OLIVIER. 2015. 'Les poids des cités du Proche-Orient hellénistique et romain au Cabinet des Médailles', *Bulletin de la Société française de Numismatique* 70.5, 111–6.

GILFILLAN, S. C. 1965. 'Lead Poisoning and the Fall of Rome', *Journal of Occupational Medicine* 7, 53–60.

GRATTAN J. P., D. D. GILBERTSON and M. KENT. 2013. 'Sedimentary Metal-Pollution Signatures Adjacent to the Ancient Centre of Copper Metallurgy at Khirbet Faynan in the Desert of Southern Jordan', *Journal of Archaeological Science* 40.11, 3834–53.

HARRINGTON, K. 2015. 'Privacy and Production: Sensory Aspects of Household Industry in Classical and Hellenistic Greece', in M. DALTON, G. PETERS and A. TAVARES (eds.), *Seen & Unseen Spaces*. ARC 30.1, 143–9.

HARRIS, W. V. 2011a. 'Bois et déboisement dans la Méditerranée antique', *Annales. Histoire, Sciences Sociales* 66, 105–40.

HARRIS, W. V. 2011b 'Plato and the Deforestation of Attica', *Athenaeum* 99, 479–82.

HARRIS, W. V. 2013. 'Defining and Detecting Mediterranean Deforestation 800 BCE to 700 CE', in W. V. HARRIS (ed.), *The Ancient Mediterranean Environment between Science and History*. Leiden-Boston, 173–94.

HEALY, J. J. 1978. *Mining and Metallurgy in the Greek and Roman World*. London.

HODGE, A. T. 1981. 'Vitruvius, Lead Pipes, and Lead Poisoning', *American Journal of Archaeology* 85, 486–91.

HODGE, A. T. 2002². *Roman Aqueducts & Water Supply*. (1st ed. 1992). London.

HOFFMANN, M. 1999. *Griechische Bäder*. Munich.

HONG, S., J. P. CANDELONE, C. C. PATTERSON, and C. F. BOUTRON. 1994. 'Greenland Ice Evidence of Hemispheric Pollution for Lead Two Millennia ago by Greek and Roman Civilizations', *Science* 265, no. 5180, 1841–3.

HONG, S., J. P. CANDELONE, C. C. PATTERSON, and C. F. BOUTRON. 1996. 'History of Ancient Copper Smelting Pollution during Roman and Medieval Times Recorded in Greenland Ice', *Science* 272, no. 5259, 246–9.

JANSEN, B. 1999. '"Wo der Römer siegt, da wohnt er". Wohnen in den nordwestlichen römischen Provinzen', in W. HOEPFNER (ed.), *Geschichte des Wohnens*. Vol. I. *5000 v. Chr. – 500 n. Chr.: Vorgeschichte – Frühgeschichte – Antike*. Stuttgart, 785–854.

KAKAVOYIANNIS, E. 1984. 'Production of Lead from Litharge in Hellenistic Rhodes', *Athens Annuals of Archaeology* 17.1–2, 124–40 (in Greek).

KARVONIS, P. 2008. 'Les installations commerciales dans la ville de Délos à l'époque hellénistique', *Bulletin de Correspondance Hellénique* 132, 153–219.

KARVONIS, P. and J.-J. MALMARY. 2009. 'Étude architecturale de quatre pièces polyvalentes du Quartier du Théâtre à Délos', *Bulletin de Correspondance Hellénique* 133, 195–226.

KASTENMEIER, P. 2007. *I luoghi del lavoro domestico nella casa pompeiana*. Roma.

KEENAN-JONES, D., J. HELLSTROM and R. DRYSDALE. 2011. 'Lead Contamination in the Drinking Water of Pompeii', in E. POEHLER, M. FLOHR, and K. COLE (eds.), *Pompeii: Art, Industry and Infrastructure*. Oxford, 131–48.

KOBERT, R. 1909. 'Chronische Bleivergiftung im klassischen Altertum', in P. Diergart (ed.), *Beiträge aus der Geschichte der Chemie dem Gedächtnis von Georg W. A. Kahlbaum*. Leipzig-Wien, 103–19.

LAMBERT, W. E. 1997. 'Combustion Pollution in Indoor Environments', in E. J. BARDANA and A. MONTANARO (eds.), *Indoor Air Pollution and Health*. New York, 83–103.

LANCASTER, L. 2008. 'Roman Engineering and Construction', in J. P. OLESON (ed.), *The Oxford Handbook of Engineering and Technology in the Classical World*. Oxford, 256–84.

LANG, M. and M. CROSBY. 1964. *The Athenian Agora*. Vol. X. *Weights, Measures and Tokens*. Princeton.

LECHTMAN, H. N. and L. W. HOBBS. 1986. 'Roman Concrete and the Roman Architectural Revolution', in W. D. KINGERY and E. LENSE (eds.), *High-Technology Ceramics: Past, Present, and Future: the Nature of Innovation and Change in Ceramic Technology*. Westerville OH, 81–128.

MAKRA, L. 2015. 'Anthropogenic Air Pollution in Ancient Times', in P. WEXLER (ed.), *History of Toxicology and Environmental Health: Toxicology in Antiquity*, vol. 2, 2015. London-San Diego, 21–41.

MAKRA, L. and P. BRIMBLECOMBE. 2004. 'Selections from the History of Environmental Pollution, with Special Attention to Air Pollution', Part 1. *International Journal of Environment and Pollution* 22.6, 641–56.

MARBURY, M. C. 1991. 'Wood Smoke', in J. M. SAMET and J. D. SPENGLER (eds.), *Indoor Air Pollution: A Health Perspective*. Baltimore-London, 209–22.

MASTER, K. M. 1974. 'Air Pollution in New Guinea. Cause of Chronic Pulmonary Disease Among Stone-Age Natives in the Highlands', *JAMA* 228.13, 1653-5.

MAYER, E. 2012. *The Ancient Middle Classes: Urban Life and Aesthetics in the Roman Empire, 100 BCE – 250 CE*. Cambridge MA.

MEIER, S. M. 1995. *Blei in der Antike: Bergbau, Verhüttung, Fernhandel*. Diss. Zürich.

MILIARESIS, I. Forthcoming. 'Throwing Money out the Window: Fuel in the Terme del Foro at Ostia', in R. VEAL and V. LEITCH (eds.), *Fuel and Fire in the Ancient Roman World: Towards an Integrated Economic Approach*. Cambridge.

MILLARD, A. R., J. MONTGOMERY, M. TRICKETT, J. BEAUMONT, J. EVANS and S. CHENERY. 2014. 'Childhood Lead Exposure in the British Isles during the Industrial Revolution', in M. K. ZUCKERMAN (ed.), *Modern Environments and Human Health: Revisiting the Second Epidemiological Transition*. Hoboken NJ, 279–300.

MONTEIX, N. 2004. 'Les lingots de plomb de l'atelier VI, 12 d'Herculanum et leur usage: aspects épigraphiques et techniques', in A. LEHOËRFF (ed.), *L'artisanat métallurgique dans les sociétés anciennes en Méditerranée occidentale: techniques, lieux et formes de production*. Rome, 365–78.

MONTEIX 2010. *Les lieux de métier. Boutiques et ateliers d'Herculanum*. Rome.

MONTEIX 2013. 'Cuisiner pour les autres. Les espaces commerciaux de production alimentaire à Pompéi', *Gallia* 70.1, 9–26.

MONTEIX 2016. 'Contextualizing the Operational Sequence: Pompeian Bakeries as a Case Study', in A. WILSON and M. FLOHR (eds.), 2016. *Urban Craftsmen and Traders in the Roman World*. Oxford, 153–82.

MONTGOMERY, J., J. EVANS, S. CHENERY, V. PASHLEY and K. KILLGROVE 2010. '"Gleaming, White, and Deadly": Using Lead to Track Human Exposure and Geographic Origins in the Roman Period in Britain', in H. ECKARDT (ed.), *Roman Diasporas. Archaeological Approaches to Mobility and Diversity in the Roman Empire* (Journal of Roman Archaeology Suppl 78), 199–226.

MUSHRIF-TRIPATHY, V. 2014. 'Maxillary Sinusitis from India: A Bio-Cultural Approach', *Korean Journal of Physical Anthropology* 27.1, 11–28.

NEEDLEMAN, H. 2004. 'Lead Poisoning', *Annual Review of Medicine* 55, 209–22.

NORBOO, T., M. YAHYA, N. G. BRUCE, J. A. HEADY and K. P. BALL. 1991. 'Domestic Pollution and Respiratory Illness in a Himalayan Village', *International Journal of Epidemiology* 20.3, 749–57.

NORDH, A. 1949. *Libellus de Regionibus Urbis Romae*. Skrifter utgivna av svenska Institutet i Rom III. Lund.

NRIAGU, J. O. 1983a. *Lead and Lead Poisoning in Antiquity*. New York.

NRIAGU, J. O. 1983b. 'Saturnine Gout Among Roman Aristocrats: Did Lead Poisoning Contribute to the Fall of the Empire?', *New England Journal of Medicine* 308, 660–3.

OLESON, J. P. 2008. 'Testing the Waters: The Role of Sounding-Weights in Ancient Mediterranean Navigation', in R. L. HOHLFELDER (ed.), *The Maritime World of Ancient Rome* (Memoirs of the American Academy in Rome, Suppl. 6). Ann Arbor, 119–76.

OLSON, D. S. 1991. 'Firewood and Charcoal in Classical Athens', *Hesperia* 60, 411–20.

OLSSON, R. 2015. *The Water-Supply System in Roman Pompeii*. Lund.

PABST, M. A., and F. HOFER. 1998. 'Deposits of Different Origin in the Lung of the 5300-Year-Old Tyrolean Ice Man', *American Journal of Physical Anthropology* 107, 1–12.

PAGANO, M. 2004. 'Un'officina di plumbarius a Ercolano', in A. Lehoërff (ed.), *L'artisanat métallurgique dans les sociétés anciennes en Méditerranée occidentale: techniques, lieux et formes de production*. Rome, 353–63.

PATTERSON, C. C. 1972. 'Silver Stocks and Losses in Ancient and Medieval Times', *Economic History Review* 25.2, 205–35.

PHILLIPS, C. R. 1984. 'Old Wine in Old Lead Bottles: Nriagu on the Fall of Rome', *Classical World* 78, 29–33.

PICARD, O. 2001. 'La découverte des gisements du Laurion et les débuts de la chouette', *Revue Belge de Numismatique* 147, 1–10.

RETIEF, F. P. and L. CILLIERS. 2005. 'Lead Poisoning in Ancient Rome', *Acta Theologica* 26.2, Suppl. 7, 147–64.

ROBERTS, C. A. and K. MANCHESTER, 2012. 'Calvin Percival Bamfylde Wells', in J. E. BUIKSTRA and C. A. ROBERTS (eds.), *The Global History of Paleopathology: Pioneers and Prospects*. Oxford, 141–5.

ROBERTS, P. 2013. *Life and Death in Pompeii and Herculaneum*. Oxford.

ROSEN, B. and E. GALILI. 2007. 'Lead Use on Roman Ships and its Environmental Effects', *International Journal of Nautical Archaeology* 36.2, 300–7.

ROSMAN, K. J. R., W. CHISHOLM, C. F. BOUTRON, J. P. CANDELONE and U. GORLACH. 1993. 'Isotopic Evidence for the Source of Lead in Greenland Snows since the Late l960's', *Nature* 362 (25 March 1993), 333–5.

ROSTOVTZEFF, M. I. 1905. *Bleitesserae. Ein Beitrag zur Social- und Wirtschaftsgeschichte der römischen Kaiserzeit*. Leipzig.

ROTHENHOEFER, P. and N. HANEL. 2013. 'The Romans and Their Lead – Tracing Innovations in the Production, Distribution, and Secondary Processing of an Ancient Metal', in S. BURMEISTER, S. HANSEN, M. KUNST and N. MÜLLER-SCHEESSEL (eds.), *Metal Matters. Innovative Technologies and Social Change in Prehistory and Antiquity*. Rahden, 273–82.

SANIDAS, G. M. 2013. *La production artisanale en Grèce: Une approche spatiale et topographique à partir des exemples de l'Attique et du Péloponnèse du VIIe au Ier siècle avant J-C*. Paris.

SCARBOROUGH, J. 1984. 'The Myth of Lead Poisoning Among the Romans: An Essay Review', *Journal of the History of Medicine* 39, 469–75.

SCHEIDEL, W. 2009a. 'In Search of Roman Economic Growth', *Journal of Roman Archaeology* 22, 46–70.

SCHEIDEL, W. 2009b. 'The Monetary Systems of the Han and Roman Empires', in W. SCHEIDEL, W. (ed.) *Rome and China. Comparative Perspectives on Ancient World Empires*, Oxford, 137–207.

SETTLE, D. M. and C. C. PATTERSON. 1980. 'Lead in Albacore: Guide to Lead Pollution in Americans', *Science* 207, no. 4436, 1167–76.

SHOTYK, W., D. WEISS, P. G. APPLEBY, A. K. CHERBRUKIN, R. FREI, M. GLOOR, J. D. KRAMERS, S. REESE and W. O. VAN DER KNAAP. 1998. 'History of Atmospheric Lead Deposition Since 12,730, (14)C. yr. BP from a Peat Bog, Jura Mountains, Switzerland', *Science* 281, no. 5383, 1635–40.

STOREY, G. R. 1997. 'The Population of Ancient Rome', *Antiquity* 71, 966–78.

TRINCHERINI, P. R., C. DOMERGUE, I. MANTECA, A. NESTA and P. QUARATI. 2009. 'The Identification of Lead Ingots from the Roman Mines of Cartagena (Murcia, Spain): the Role of Lead Isotope Analysis', *Journal of Roman Archaeology* 22, 123–45.

TSAKIRGIS, B. 2007. 'Fire and Smoke: Hearths, Braziers and Chimneys in the Greek House', in R. WESTGATE, N. FISHER and J. WHITLEY (eds.), *Building Communities: House, Settlement and Society in the Aegean and Beyond.* London, 225–31.

VEAL, R. J. 2013. 'Fuelling Ancient Mediterranean Cities: a Framework for Charcoal Research', in W. V. HARRIS (ed.), *The Ancient Mediterranean Environment between Science and History.* Leiden-Boston, 37–58.

VÉRON, A., J. P. GOIRAN, C. MORHANGE, N. MARRINER and J.-Y. EMPEREUR. 2006. 'Pollutant Lead Reveals the Pre-Hellenistic Occupation and Ancient Growth of Alexandria, Egypt', *Geophysical Research Letter* 33, L06409, doi:10.1029/ 2006GL025824.

VIRLOUVET, C. 1995. *Tessera frumentaria: les procédures de distribution de blé public à Rome à la fin de la République et au début de l'Empire.* Rome.

WALDRON, H. A., A. MACKIE and A. TOWNSHEND. 1976. 'The Lead Content of some Romano-British Bones', *Archaeometry* 18, 221–7.

WALDRON, T. 2009. *Palaeopathology.* Cambridge-New York.

WEEBER, K.-W. 1990: *Smog über Attika: Umweltverhalten im Altertum.* Zürich.

WELLS, C. 1977. 'Disease of the Maxillary Sinus in Antiquity', *Medical and Biological Illustration* 27, 173–8.

WILSON, A. and M. FLOHR (eds.), 2016. *Urban Craftsmen and Traders in the Roman World.* Oxford.

ZIMMER, G. 1999. 'Handwerkliche Arbeit in Umfeld des Wohnens', in W. HOEPFNER (ed.). *Geschichte des Wohnens.* Vol. I. *5000 v. Chr. – 500 n. Chr.: Vorgeschichte – Frühgeschichte – Antike.* Stuttgart, 561–75.

J. Donald Hughes

Deforestation and Forest Protection in the Ancient World

Introduction

Sustainability is a much-debated term in contemporary economics and ecology. One of the first instances where it appeared was in forest science as 'sustained-yield forestry'. A reason that forest management can be sustainable is that forests are a renewable resource, that is, once cut they can in many cases grow again through natural regeneration or replanting. These are ideas that originated in the nineteenth and twentieth centuries, but to a surprising extent they had antecedents in ancient times, particularly among the classical Mediterranean civilizations.

Deforestation was a major environmental trend in the ancient Mediterranean world in the period from the Bronze Age through the Roman Empire.[1] The result of continual use for construction and dependence on wood and charcoal for fuel among other activities, the degradation of the landscape through deforestation became lasting where the rate of use exceeded the rate of forest regeneration and growth. To supply the heat necessary for metallurgy, ceramic production, and heating of homes and baths, local supplies were exhausted and it became necessary to import forest products by sea.

Anthropogenic impacts on the environment were issues even in antiquity, and ancient people found some ways of dealing with them.

1. Extent and causes of deforestation

Deforestation was a widespread disaster caused by human economic activities in the ancient world from the Bronze Age through the period of the Roman Empire.[2] Writings of the time, archaeological discoveries, pollen studies, and the vegetation history of the Mediterranean area all attest to a more or less permanent removal of forests in districts accessible to urban and imperial centers. Forests are renewable through natural reproduction and replanting, but deforestation results from depletion of forest resources at a rate higher than growth. The demands for use exceeded the rate of replacement in many districts of the Mediterranean during periods of the rise, flourishing, and decline of ancient societies.

Use for fuel constituted the most consumptive demand for wood and its partially oxidized product, charcoal. Fuel was necessary in the refining of metallic ores for the production of copper, iron, gold, and silver. For example, Athens produced its fine silver coinage

---

[1] HUGHES 1984: 331–43.
[2] HUGHES 2011: 43–57.

Fig. 1: Athenian silver mines at Laurion: ancient ore washing facilities. (Photo by author, 2011).

from mines at Laurion, and the demand for fuel there exceeded the annual forest growth of the entire homeland of Attica.[3] This meant that additional wood and charcoal resources needed to be imported from elsewhere in the lands around the Aegean, Black, and Mediterranean seas. Where did the wood come from? Overland transportation was slow and expensive, so inland forests, especially in high mountains, were spared the worst exploitation. Some would have come from islands and coastlands in the Athenian-dominated Aegean. The furnaces of Laurion were relocated near the coast, and a port for barges was constructed. JOHN ELLIS JONES noted that 'The coastal position of these furnaces reflects the deforestation of Laurion by then and dependence on foreign fuel'.[4] Theophrastus cites sources for timber in Macedonia, Thrace, Asia Minor, and the south shore of the Black Sea.[5] Wood came from Cyprus, Sicily and southern Italy. These places were accessible through ports and by rivers that could float rafts of logs. So the ripple of deforestation that began at Laurion spread out like a tsunami to the Mediterranean coasts. The result was a timber shortage reflected in rising prices and a decline in Athens' power. As PAUL GILDING exclaimed, 'If you cut down more trees than you grow, you run out of trees'.[6] Since silver ore

---

[3] HUGHES 2014: 136–42.

[4] JONES 1982: 169–83.

[5] Theophr., *hist. plant.* 4.5.

[6] GILDING 2011.

was lead for the most part, its production resulted in lead pollution of the environment and resultant danger to health.

Even so, metallurgy was not the only drain on forest resources. Ceramic production was a major industry that required firing. Homes were usually heated by charcoal fires, and water heating was necessary for cooking and bathing. The Roman Empire's large and numerous baths and the hypocausts that warmed floors in mansions are cases in point. Construction required wood for scaffolding and beams for roof supports, and bricks had to be baked at high temperatures, from 900°C to 1200°C. Shipbuilding required not only specialized wooden parts such as ribs and keels, but also masts demanding tall, straight trees that frequently had to be transported from distant mountains, so that materials to build ships had to be transported by sea. Few forests escaped exploitation to one extent or another, except those located far inland and away from rivers that were navigable or at least large enough to float logs.

In calculating forest loss, factors such as wildfires and losses to diseases and insects must not be forgotten. Not only were there natural fires from lightning and volcanoes, but also herders, hunters, and armies commonly set forest fires. Ancient writers knew that the destruction attendant upon pastoralism included fire to clear brush and forests. Vergil said:

Just as, in summer, when the winds he wished for
Awake at last, a shepherd scatters fires
Across the forests; suddenly the space
Between the kindled woods takes fire, too.[7]

Wildfires usually burned until they reached a barrier or were put out by rains; they were not fought unless they threatened a settlement. Fires during the long, dry Mediterranean summer are often catastrophic and bare the slopes to erosion, though many plants are adapted to fire and show remarkable powers of recovery if not prevented by grazing.

Regrowth of vegetation after such disasters, or after clear-cutting due to the activities mentioned above, usually does not restore a forest similar to the one that was removed. If a forest reappears, it will contain species more tolerant of sun and aridity than before. For example, evergreen oak may replace deciduous oak. In a typical Mediterranean forest zone, a high forest is initially replaced by a bushy plant association called *maquis*, or after repeated degradation, to a scrubland of low-growing, drought-resistant plants termed *garigue*.

The extension of agriculture into forested areas was extensive and represented a more or less permanent change. Lucretius saw one aspect of what was going on:

The opulence of the Earth
Led folk to clear its wealth, convert the woods
To open harvest-fields, kill the wild beasts…
[Woodcutters] made the woods climb higher up the mountains,
Yielding the foothills to be tilled and tended.[8]

---

[7] Verg., *Aen.*10.405–9. (Engl. transl. by Mandelbaum 2007: 267).
[8] Lucr., *de re. nat.* 5.1247–9, 1370–1. (Engl. transl. by Rouse 1924).

Fig. 2: Scene of wood cutting and several kinds of wood use by Roman soldiers in Dacia, Trajan's Column, Rome. (Photo in public domain, from *Reliefs of Trajan's Column by Conrad Cichorius*, Plate No. 88, Verlag von Georg Reimer, 1900).

The expansion of farmland and pasture at the expense of the forest led to a change in habitat over large swaths of land. The result was more or less permanent, because the open landscape was maintained by those whose livelihood depended on it.

The damage done to forests during Roman warfare is prominently portrayed in the great spiral relief of Trajan's Column. That monument in Rome celebrates Trajan's conquest of Dacia, a territory in modern Romania, and is regarded by experts as a principal source of information about Roman military equipment and operations.[9] More than two hundred trees are represented in the relief. Many are shown being chopped down vigorously by axe-wielding Romans. Sometimes the military axmen are clearing roads through thick woodland to allow passage for the legions. They can also be seen carrying away logs and using them to make siege terraces, catapults, battering rams, and beacon fires. One such beacon, not yet ablaze, is made of 144 logs.[10] There are many structures that demanded timber in their construction: camps, forts, palisades and other defense works, warships, boats, and barges loaded with barrels. Then there are bridges of boats, huge assemblages of wood. Sections near the upper end of the relief suggest deforestation of areas of Dacia.

---

[9] Rossi 1971.
[10] Lepper and Frere 1988, Plate IV.

2. Varying interpretations of deforestation

Scholarly unity is rarely if ever attained on major economic issues concerning ancient life, and deforestation is no exception to this observation. Indeed, Robert Sallares remarked that it is 'the most controversial question of the environmental history of the Mediterranean', although one might argue that other questions such as population, biodiversity, climate, and connectivity are equally controversial.[11] Earlier views, from ancient times through the early modern period, regarded deforestation as positive because it cleared the way and provided materials and fuels for agriculture and industry. Cicero approved of it among other human activities, saying 'by means of our hands we endeavor to create as it were a second world within the world of nature'.[12]

The controversial question being investigated, however, may be summarized: Did widespread deforestation occur in the ancient Mediterranean world, and if so did its environmental effects damage the welfare of ancient societies? Such a question was first formulated, and answered in the affirmative, in the late nineteenth and early twentieth centuries. George Perkins Marsh was a classicist and politician who served U. S. embassies in Istanbul and Athens, and was ambassador to Italy for 21 years (1861–1882). In his book *Man and Nature* (1864), he maintained that the originally advantageous territory of the Roman Empire had been reduced to various degrees of desolation. Forests had disappeared, and as a result rich soils washed away, water supplies were disrupted, streams dried up in summer and become torrents in winter, and harbors were blocked by erosive deposits. Much of what remained, he added, was 'no longer capable of affording sustenance to civilized man'.[13] While there were some natural causes of this, he concludes that for the most part it was 'the result of man's ignorant disregard of the laws of nature', along with war, tyranny and misrule, which contributed to that disregard.[14] Similar views were voiced by the ecologist Paul Sears and the soil conservationist Walter Lowdermilk, who published reports on his surveys of the Mediterranean area before and after the Second World War.[15]

The present author, based on classical texts, history, archaeological data, scientific studies such as palynology and dendrochronology, arrived at the views presented in *Pan's Travail*, and summarized in this paper.[16] These are that there was deforestation in the ancient Mediterranean zone, particularly during periods of high civilization and in areas close to cities and/or accessible by water. Curtis Runnels, using archaeology and geological stratigraphy, examined the relationship between human settlement and landscape through time, concluding that 'The evidence mounts for episodes of deforestation and catastrophic soil erosion over the past 8,000 years. Many scholars believe they resulted from a long history of human land use and abuse'.[17]

Some writers have doubted this account. Alfred Thomas Grove and Oliver Rackham have a different answer for the question of ancient deforestation, which follows in

---

[11] Sallares 2007: 15–37, quoted on p. 21.
[12] Cic., *de nat. deo.* 2.60 (152). (Engl. transl. by Rackham 1933).
[13] Marsh 1864 (1965): 10.
[14] Marsh 1864 (1965): 11.
[15] Sears 1935: 27–30; Lowdermilk 1944a; 1944b: 413–27.
[16] Hughes 1994: 73–90.
[17] Runnels 1995: 96

paraphrase: the landscape is not ruined.[18] Deforestation exists mainly in the imagination of those who mistakenly assume that cutting down trees destroys forests, whereas trees grow again. Grazing by goats and sheep is not bad, since it renders vegetation less vulnerable to wildfires. But fires do not destroy forests. Deforestation does not make erosion worse. Badlands, where erosional features dominate, are stable landscapes. They see no deserts on the march. The state of the Mediterranean lands in 1950 AD was little worse than it was at the end of the Bronze Age. These opinions are not only counterintuitive, but go against the conclusions of students of historical land use who have been studying these questions. Grove and Rackham are hostile to the use of ancient texts as evidence, and whenever these sources seem to argue for deforestation caused by human action, they exhaust themselves in a search for ways to explain it away, and demand unassailable proof from those who judge that the evidence shows that humans cause serious forest removal.

Peregrine Horden and Nicholas Purcell do not go so far. While admitting that forests were destroyed by factors such as overgrazing and mining, they opine that such damage was rare and localized, and that deforestation was seen as a 'Good Thing', because it improved the landscape for agriculture.[19]

Alain Bresson makes a statement describing the Mediterranean economy that is certainly applicable to the trade in wood products:

> To sum up, thanks to the sea, connectivity in the Mediterranean zone greatly accelerated the movement of history. In the first millennium BC, with the unification of the Mediterranean, the concentration of wealth and the movement of ideas reached levels and forms that were then completely unknown anywhere else on the globe. But the specific market that emerged at that time was in fact heavily dependent on the ecological milieu: this milieu gave the market its best opportunity, but it also set definite limits on its development.[20]

This can be understood to imply that in the classical period, Mediterranean communities moved beyond local autarchy to provision of resources such as timber by sea throughout the zone, but that supplies from overseas forests were subject to depletion.

A study by William V. Harris offers an interpretation of many types of evidence both textual and proxy archives interpreted by scientific analyses.[21] His tentative conclusions, summarized, are as follows: In classical Greece, mild deforestation was widespread, but almost total deforestation was limited to the supply areas of Athens and a few other cities. Crises of fuel supply occurred where smelting of metals was intense, such as the silver mines of Laurion in Attica during the fifth and fourth centuries BC. Hellenistic and Roman shipbuilding and fuel needs seriously deforested some areas, especially in Italy, but this was ameliorated to some extent by the trading network and forest management. High demand increased through the time of greatest imperial expansion ('perhaps down to Severan times').[22] In the late Empire, some areas recovered forest, but due to climate change or poor management, others did not. I agree with these conclusions, although my quantitative estimate of deforestation is possibly greater than that of Harris.

---

[18] Grove and Rackham 2001.
[19] Horden and Purcell 2000: 334. The expression, 'a Good Thing', is offered in the index: 743.
[20] Bresson 2005: 94–116, quoted on 113–4.
[21] Harris 2013: 173–96. See also Harris 2011: 105–40.
[22] Harris 2013: 194.

Fig. 3: Laphistos, Greece: Boulders brought down by torrents from deforested land upstream near Pieria. (Courtesy of the Goulandris Museum of Natural History, Kifissia, Athens, Greece, 1988).

3. Erosion and pollution as results of deforestation

Plato in the *Critias* observed that the mountains near his native city of Athens had been forested not long before his own time, but that the cutting of timber and subsequent grazing had robbed them of trees. The result was serious erosion that had washed away the rich, deep soil and consequently dried up springs and streams that formerly existed there, so that 'what now remains compared with what then existed is like the skeleton of a sick man, all the fat and soft soil having wasted away, and only the bare framework of the land being left'.[23]

The roots of a forest hold the soil together and encourage the growth of soil organisms, giving the soil the consistency of a sponge that absorbs water from rainfall and releases it at a slower rate. The branches and leaves of a forest resist the force of rainfall in a storm, protecting the soil from washing away. The combined effect of these factors reduces erosion and provides springs and streams with a moderate, more dependable flow. The removal of a forest allows runoff to occur with greater force, causing rapid erosion of soil and increasing the load of material in streams. The runoff occurs in a shorter period, resulting in intermittent streams that are dry for much of the year in the Mediterranean climate regime. As Plato put it:

---

[23] Pl., *Criti.* 111b.

Trees grew tall and strong,
and the soil produced plentiful pasturage for flocks.
It was enriched by yearly rains from Zeus,
and did not lose it, as now, by flowing from the bare ground into the sea;
but the soil it had was deep, and it received the water,
storing it up in the retentive loamy soil;
and let water flow down from high ground to the low ground of every district,
providing abundant springs to feed streams and rivers.
Even now there are still shrines,
left over from the old days at sites of former springs,
as evidence of the truth of this account of the land.[24]

Some commentators tend to dismiss the evidence of this oft-quoted passage because it is part of a description of the mythical lost continent of Atlantis in the *Critias*, but they miss the fact that Plato is making a comparison with his own day (i. e. 'as now', 'even now') and processes that he himself had observed.[25] Studies of watersheds in Italy have shown that erosion after forest removal is fifty times as rapid as before. The danger to agriculture is evident, and farmers countered erosion by building terraces and check dams. Further, the sudden excess of runoff caused flash floods that swept riverbanks and settlements. Due to the deforestation of the Tiber watershed, Rome suffered floods that covered the lower parts of the city and backed up the sewers with their contents. At such times, the drains in buildings such as the Pantheon could become fountains of polluted water.[26]

Water pollution by silt and mud interferes with the water supplies of cities, requiring the construction of aqueducts from sources at higher elevations and provision of settling tanks at points where water was distributed in urban centers. Pollution harmed fish, including those such as salmon that return to their ancestral watercourses to spawn. Erosion also increases the salt content of streams and rivers, adding to the danger of salinization in lowland agricultural landscapes where evaporation is high, as it is in the warm Mediterranean climatic zone. Along the coast, erosional material is deposited in shallow areas, extending river deltas and creating marshes and other wetlands. These new areas may provide habitat for fish and birds, but also for mosquitoes that are vectors for malaria and other diseases. The spread of malaria in the ancient Mediterranean lands may be correlated with increasing deforestation. The deposition of silt, sand, gravel, and boulders filled harbors near river mouths, a barrier to shipping that required dredging and the excavation of new basins for commerce if former ports were not to be left isolated from the sea.

Air pollution is also connected to deforestation because much of the wood removed was combusted for fuel, contributing to the turbidity of the atmosphere, especially over and near cities. Roman poets complained of the smoke that hung over the city, and those that could afford it built villas in the countryside where they could find fresher air. Countless cooking and baking fires, smoky lamps, charcoal fires to heat rooms, the smoke pouring from furnaces in the baths, from metal-working, and from kilns for firing of pottery, not to mention the ubiquitous dust, meant that a city could be seen a long way off because of its polluted air.

---

[24] Pl., *Criti.* 111b–d. (Engl. transl. by WATERFIELD 2008).
[25] HARDY and TOTELIN 2016: 28.
[26] BALL PLATNER 1911: 75; Plin., *nat. hist.* 36.105.

To Homer, smoke was the first sign of human habitation.[27] Horace remarked on the thousands of wood-burning fires in Rome.[28] Temperature inversions, which were natural occurrences as common in ancient times as they are in the Mediterranean basin today, held smoke and dust in suspension over cities. Air pollution was familiar to the Romans, who termed it *gravius caelum* ('heavy heaven') or *infamis aer* ('infamous air').[29] Air pollution resulted not only from wood and charcoal smoke, but also from the fumes of various noxious substances that were heated or burnt. There was other damage noticed by ancient writers, the pollution of air and water by poisons released in extractive activities and industrial processes. Because wood and charcoal were the usual fuels, and fossil fuels were almost unknown, forests suffered proportionately from ancient industry. In addition to its ritual and moral meaning, pollution was also given something like its modern sense of contaminating water, air, and earth with the waste products of human activities, including industrial processes. Silver was refined from ores consisting mainly of lead. Noting the threat of lead poisoning from the smoke of silver smelters, Strabo recommended the construction of high chimneys, but this spread the pollution further.[30] That air pollution from smelting was not minor or merely local may be indicated by the fact that measurements of the lead content of arctic snow preserved in glaciers in Greenland show a marked increase in concentration at the time when the Romans discovered a more efficient method of smelting in the second century BC.[31]

Air flowing through forests is cleansed of some of the pollution, and wind is reduced in force. Both of these effects are favorable to agriculture. When forests were removed, agriculture had to compensate for increased evaporation and wind damage by planting windbreaks.

4. Forest protection and restoration

Granted that deforestation-caused environmental problems were happening, were the Greeks and Romans aware of them? Did they reflect on them and attempt to deal with them? Interestingly, they made a start in these respects. Theophrastus in *Historia Plantarum* and *De Causis Plantarum* made a start in understanding what would today be called forest ecology and sylviculture. He maintained that a tree grows best when it is in an 'appropriate' (*oikêôs*) environment: this, along with a similar use of the word by Aristotle about animals, may be the root of our word 'ecology'. The importance of timber supply and the effects of deforestation and erosion were evident to ancient observers, who often lamented them, especially rising prices due to scarcity. Therefore it is not surprising that governments as well as private landowners exercised care in assuring a continued supply of wood from the forests under their control. Unfortunately such efforts were far from universal, not always effective, and were diminished by other policies that encouraged exploitation and destruction of forests. Plato recognized that deforestation was a cause of interference with water supply, including the disappearance of springs, and advised the conservation and plantation of trees

---

[27] Hom., *Il.* 1.317; 8.183; 9.243; 18.207; 21.522.
[28] Hor., *car.* 3.29.12; 2.15.12.
[29] Front., *de aquae duc.* 88; Tac., *hist.* 2.94.
[30] Vitr. *de arch.* 8.6.11; Str. 3.2.8, C146.
[31] Hong, Candelone, Patterson and Boutron 1994: 1841–3.

to ameliorate that process.[32] As he observed, water that rushed unimpeded down mountain-sides was no longer available to feed springs. Perhaps for this reason, he portrayed his ideal Atlantis as having springs surrounded by plantations of appropriate trees.[33]

Agriculture included some forestry; Greek and Roman farmers had reason not to clear all their land. Cato, listing amenities for a good farm, included a woodlot as the seventh of nine.[34] Large estates contained forests that supplied timber, nuts, berries, honey, and fo-liage for fodder. Plantations for timber were common, and in addition trees were planted to line roads, shelter fields, and mark boundaries. Landowners had their own nurseries to propagate trees. Pliny the Younger remarked that the mountain slopes around his villa were 'covered with plantations of timber'.[35] Cultivated trees added so much to property value that when Crassus would not sell some large trees with his estate, Domitius refused to buy it, even though he had previously offered a princely sum.[36] Columella condemned neighbors who cut down trees near property lines, since it might depreciate the value of adjoining property as well as their own.[37]

Because timber was of great economic and military importance, governments consid-ered forests a proper area of their concern. The sovereign power, whether city or empire, usually asserted its ownership of all unoccupied forestland within its territory. Government supervision of forests and watersheds included regulation of the forest products trade, of the timber harvest, and the construction of works to provide or control water supply, drain-age, and erosion. Responsibility for these matters was regularly delegated to designated offi-cials; in some cities the timber trade was under *agoranomoi* (overseers of commerce), while forestland in the countryside was supervised by *hyloroi* (custodians of forests) who, says Aristotle, had 'guard-posts and mess-rooms for patrol duty'.[38]

It was a recurrent practice of governments to encourage private exploitation of forests by leasing the right to cut trees on public land as a means of revenue, or by sale or grant of public forestland to private entrepreneurs or consortiums, but their policy was fortunately not always directed to encourage deforestation. Conscious of the danger of a diminishing supply of wood, the state in many cases regulated private land so as to encourage conserva-tion. Plato's recommendation that landowners be fined if fire spread from their property to timber on a neighbor's land doubtless represented actual Greek law.[39] Land leases elsewhere also contained restrictions on timber cutting and stipulations for replanting.

Authorities protected and managed public forestlands as well. Large tracts remained in government hands, and measures were taken to prevent encroachment and assure their use for the needs of the state. When Scipio Africanus needed fir trees to make masts for the fleet he raised against Carthage in 205 BC, he found them in 'forests belonging to the state'.[40] Wise administrators limited timber harvest; Theophrastus said that in Cyprus, 'the kings

[32] Pl., *Criti.* 111b–d; *leg.* 6.761b–c.
[33] Pl., *Criti.* 117a–b.
[34] Cato, *de agr.* 1.7.
[35] Plin. Iun., *epist.* 5.67.7–13.
[36] Plin., *nat. hist.* 17.1.
[37] Col., *de re rus.* 1.3.7.
[38] Arist., *pol.* 6.5.4. (1321b); 7.11.4 (1331b).
[39] Pl., *leg.* 8.843E.
[40] Livy 28.45.18 (*ex publicis silvis*).

Fig. 4: High forest near Pertouli, Macedonia. Forests such as this were common in moderate elevations in the Mediterranean region in pre-classical times, but became rare. (Courtesy of the Goulandris Museum of Natural History, Kifissia, Athens, Greece, 1988).

used not to cut the trees … because they took great care of them and managed them'.[41]
He added that later rulers of that island reaped the benefit of their predecessors' restraint;
Demetrius Poliorcetes cut timber of prodigious length there for his ships. Some magistrates
were foresighted enough to protect public lands against greed-motivated exploitation, and
found political support for their efforts. When the Roman tribune Servilius Rullus proposed
that Rome sell some state forestland to raise money for other programs, Cicero appealed to
popular sentiment to keep the forests in the ownership of the Republic. Knowing that there
were profiteers in high position behind Rullus' proposal, Cicero attacked them:

> What they need now is money, money that cannot be questioned, money that can be counted. I
> wonder what this watchful and shrewd tribune has in mind? 'The Scantian Forest is to be sold', he
> says. Did you discover the forest in the list of abandoned land holdings? … Would you dare to sell
> the Scantian Forest in my consulship? … Would you rob the Roman people of what gives them
> strength in war, and in peace a more easy life?[42]

Decrees of Ptolemy Euergetes, Macedonian ruler of Egypt, prohibited unauthorized cutting
of wood by private individuals on their own land and required planting trees.[43] The Ptolemies
fostered tree plantations, since in that part of the Mediterranean world the need for local wood
was acute. A nationwide tree-planting project operated in wasteland, private land, royal estates, and the banks of rivers and canals.[44] Trees were started in government nurseries. Laws
protected these plantations by regulating the felling of older trees, the lopping of branches, and
the removal of fallen trees. Sheep and goats were excluded from areas where young trees had
been planted. In addition, governments created and protected parks and sacred groves of trees.
    The latter category of land use had a major role in protecting forests. Sacred space (*hie-
ron temenos* in Greek and *templum* in Latin) consisted of areas set aside for worship and
dedicated to gods and goddesses. These reserves could be located physically in towns, fields,
or woodlands, but sacred groves are in principle as untouched as wilderness because economic activities including hunting and wood gathering were forbidden. The presence of
deity was recognized in the quality of the environment itself. As Seneca remarked:

> If you come upon a grove of old trees that have lifted up their crowns above the common height
> and shut out the light of the sky by the darkness of their interlacing boughs, you feel that there is a
> spirit in the place, so lofty is the wood, so lone the spot, so wondrous the thick unbroken shade.[45]

Or Ovid, 'Here stands a silent grove black with the shade of oaks; at the sight of it, anyone
could say, "there is a spirit here!"'[46] Virgil had his character Evander remark that when they
saw the old tree-covered Capitoline Hill, the rural folk who lived around it exclaimed, 'Some
god has this grove for dwelling!'[47] This primal attitude long persisted in popular feeling, as
Pliny the Elder indicated:

---

[41] Theophr., *hist. plant.* 5.8.1.
[42] Cic., *de l. agr.* 1.3. (Engl. transl. by FREESE 1930).
[43] HUNT and EDGAR 1934, vol. 2, No. 210.
[44] ROSTOVTZEFF 1941. 1: 299, 480–1; 3: 1169–70.
[45] Sen., *ep.* 4.12.3. (Engl. transl. by GUMMERE 1917).
[46] Ov., *fas.* 3.295–6: *Numen inest.*
[47] Verg., *Aen.* 8.351–2.

Trees were the first temples of the gods, and even now simple country people dedicate a tree of exceptional height to a god with the ritual of olden times, and we … worship forests and the very silences they contain.[48]

Virgil noted that the gods favor wild trees 'unsown by mortal hand'.[49] They were preserved from changes wrought by humans, but in use for worship. Temples never lost their connection with trees; it was felt that every one needed to have trees around it, and if there were none, they were planted. The Parthenon, where pits were excavated in the limestone of the Acropolis to provide for the planting of cypress trees in an artificial grove, is an example of this.

Forms of pollution, in the case of sacred space understood as ritual pollution or *miasma*, were strictly forbidden. Guards and priests, who were often the same individuals, often lived in the groves to watch them and report violations to local authorities that enforced severe penalties. Fines in amounts ranging from fifty to a thousand drachmas were exacted for cutting down a tree or carrying away leaves, costs high enough to discourage attempts to profit by theft from the precincts. Expensive mandatory sacrifices could be assessed. Slaves were whipped or imprisoned for the same offenses. To these heavy punishments, ritual curses were joined. Ancient people never laughed at these, since the gods might hear them, and tradition was full of examples of divine retribution for infractions, stories that the common people took seriously.

Conclusion

In conclusion, it is evident that deforestation, one of the most important environmental trends in the ancient Mediterranean world, caused widespread erosion and both water and air pollution. It was also regarded as a form of ritual pollution of sacred groves.

Causes of deforestation included dependence on wood and charcoal for fuel, use for construction, and other forms of consumption. Degradation of the landscape through deforestation became lasting where the rate of use exceeded the rate of forest regeneration and growth. To supply the energy necessary for metallurgy, ceramic production, and heating of homes and baths, local supplies were exhausted and it became necessary to import forest products by sea.

Forests protected the soil by moderating the force of storms and regulating the release of water into springs and streams. Forest removal left the land open to erosion by rainfall. The products of erosion – silt, sand, gravel, and boulders – are forms of pollution that are the results of human industry that demanded forest resources. The effects of this pollution include disturbance of watercourses, infilling of ports, reduction of the quality of water supply, and damage to aquatic organisms such as anadromous fish.

Ancient societies took measures to ameliorate these problems through forest plantations, the reservation of areas as sacred groves, and oversight of royal and public lands. Government officials were appointed to regulate forest contracts, and guards with stations and regular patrols limited illegal cutting. Water supervisors directed construction of means of purification such as settling tanks for urban water supplies. Even with these measures,

---

[48] Plin., *nat. hist.* 12.2. (Engl. transl. by RACKHAM 1945).
[49] Verg., *geor.* 1.21–2.

powerful private consortia had political influence, and in many districts the demand for fuel and wood products exceeded the supply that Mediterranean forests could provide.

Bibliography

BRESSON, A. 2005. 'Ecology and Beyond', in W. V. HARRIS (ed.), *Rethinking the Mediterranean*. Oxford.

FREESE J. H. 1930. *Cicero, Orations*. English translation. Cambridge MA, 94–114.

GILDING, P. 2011. *The Great Disruption: Why the Climate Crisis Will Bring On the End of Shopping and the Birth of a New World*. New York.

GROVE, A. T., and O. RACKHAM. 2001. *The Nature of Mediterranean Europe: An Ecological History*. New Haven.

GUMMERE R. M. 1917. *Seneca, Epistles volume I: Epistles 1–65*. English translation. Cambridge MA.

HARDY, G. and L. TOTELIN, 2016, *Ancient Botany*. London.

HARRIS, W. V. 2011. 'Bois et déboisement dans la Méditerranée antique', *Annales HSS*, 105–40.

HARRIS, W. V. 2013. 'Defining and Detecting Mediterranean Deforestation, 800 BCE to 700 CE', in W. V. HARRIS (ed.), *The Ancient Mediterranean Environment between Science and History* (Columbia Studies in the Classical Tradition 39). Boston-Leiden, 173–96.

HONG, S., J.-P. CANDELONE, C. C. PATTERSON and C. F. BOUTRON. 1994. 'Greenland Ice Evidence of Hemispheric Lead Pollution Two Millennia Ago by Greek and Roman Civilizations', *Science* 265, No. 5180, 1841–3.

HORDEN, P. and N. PURCELL. 2000. *The Corrupting Sea: A Study of Mediterranean History*. Oxford.

HUGHES, J. D. 1984. 'Sacred Groves: The Gods, Forest Protection, and Sustained Yield in the Ancient World', in H. K. STEEN (ed.), *History of Sustained-Yield Forestry: A Symposium*. Durham NC, 331–43.

HUGHES, J. D. 1994, *Pan's Travail: Environmental Problems of the Ancient Greeks and Romans*. Baltimore.

HUGHES, J. D. 2011. 'Ancient Deforestation Revisited', *Journal of the History of Biology* 44.1, 43–57.

HUGHES, J. D. 2014. *Environmental Problems of the Greeks and Romans: Ecology in the Ancient Mediterranea*. Baltimore.

HUNT, A. S. and C. C. EDGAR. 1934. *Select Papyri*, 2 vols. London

JONES, J. E. 1982. 'The Laurion Silver Mines: A Review of Recent Researches and Results', *Greece & Rome* 29.2, 169–83.

LEPPER, F. and S. FRERE. 1988. *Trajan's Column: A New Edition of the Cichorius Plates*. Gloucester.

LOWDERMILK, W. C. 1944a. *Conquest of the Land through 7,000 Years* (U. S. Department of Agriculture Information Bulletin no. 99). Washington DC.

LOWDERMILK, W. C.1944b. 'Lessons from the Old World to the Americas in Land Use', *Smithsonian Report for 1943*, 413–27.

MANDELBAUM, A. 2007. *The Aeneid of Virgil: A Verse Translation*. Berkeley-Los Angeles-London.

MARSH, G. P. 1864. *Man and Nature*. New York (Reprint edited by David Lowenthal. Cambridge, MA 1965).

PLATNER, S. B. 1911. *Topography and Monuments of Ancient Rome*. Boston.

RACKHAM H. 1933. *Cicero. On the Nature of the Gods. Academics*. English translation. Cambridge MA.

RACKHAM H. 1945. *Pliny. Natural History, Volume IV: Books 12–16*. English translation. Cambridge MA.

ROSSI, L. 1971. *Trajan's Column and the Dacian Wars*. Ithaca NY.

ROSTOVTZEFF, M. 1941. *The Social and Economic History of the Hellenistic World*, 3 vols. Oxford.

ROUSE W. H. D. 1924. *Lucretius. On the Nature of Things*. English translation. Cambridge MA.

RUNNELS, C. N. 1995. 'Environmental Degradation in Ancient Greece', *Scientific American* 272.3, 96.

SALLARES, R. 2007. 'Ecology', in W. SCHEIDEL, I. MORRIS, and R. SALLER (eds.), *The Cambridge Economic History of the Greco-Roman World*. Cambridge, 15–37.

SEARS, P. B. 1935. *Deserts on the March*. Norman.

WATERFIELD, R. 2008. *Plato: Timaeus and Critias*, English translation. Oxford.

Jocelyne Nelis-Clément

Roman Spectacles: exploring their environmental implications

Spectacles can reveal many aspects of a society's sense of identity and of its vision of its place in the world. They offer a useful perspective from which to consider important questions about values, ideologies, and codes of behaviour. The nature of the various types of spectacles, the way they are staged, their evolution, and the interactions of the different groups involved (sponsors and overseers, performers and spectators) can all offer valuable insights into the organization, the self-image, and the aspirations of a community. Why, how often, and in what contexts a group or community decides to organize and participate in a mass spectacle, what kind of shows it chooses to be entertained by, what political, cultural, economical and social discourses are highlighted and displayed in the performances, these are some of the aspects which merit consideration. Studying them can lead to a better understanding of a community and of its interactions with the social and natural contexts and environments within which it functions.

The extravagant spectacles of the ancient Roman world provide a particularly rich terrain for such an enquiry.[1] Their shows and games and the remarkable buildings in which they took place still represent today one of the main emblems of the civilization of the Romans.[2] When travelling all around the Mediterranean and far beyond, from England to Syria, it is difficult to ignore the material trace left by the Roman culture of spectacles on our contemporary landscapes: the remains of numerous theatres, amphitheatres, stadia, circuses-hippodromes are still impressively visible in the sites of many ancient cities. Each of these monuments is in itself a *lieu de mémoire*, testimony to the various spectacles and

---

* For assistance of various kinds, I express my warmest thanks to Sinclair Bell, Alain Bresson, Marialetizia Buonfiglio, Kathleen Coleman, Hazel Dodge, Daria Lanzuolo, Damien Nelis, Hugh Possingham, Anne de Pury-Gysel, Gianpaolo Urso, Andreas Wacke, and the editors.

[1] By 'spectacles' I mean, accepting the definition of the *OED*, a 'specially prepared or arranged display of a more or less public nature (esp. one on a large scale), forming an impressive or interesting show or entertainment for those viewing it.' In the Roman context these include sports and games of many kinds, dramatic performances, triumphal parades and so on. I emphasize the question of scale, since the evidence shows that in many cases several hundreds or thousands of people would assemble to watch such events, and that the occasions in question were planned and organized in ways that involved the participation of very considerable numbers of people, both those implicated in the actual preparations and those taking part in the peformances. The events could be held in streets, in buildings built specifically to contain them, in temporary structures or in suitable natural settings, and they took place in the city of Rome and in cities across the Roman world. I will focus mainly on chariot races, staged hunts and gladiatorial combats.

[2] See, for example, PURCELL 2013: 441, in a splendid treatment of Rome as the 'City of the Games': 'In the modern imagination, the ancient city of Rome is tied inextricably to Games'.

performances held there. Each of them, on a more general level, plays a part in one of the most striking examples of the cultural impact of Roman civilization on both the ancient Mediterranean and the contemporary world. They have over the years inspired many writers and artists, who have seen their ruins as symbols of the fall of Rome itself and of the collapse of her civilization.[3] Some of these monuments, such as the amphitheatres of Verona, Nîmes, Arles and Avenches, in Italy, France and Switzerland, the theatre of Orange, in France, or the circus of Jerash, in Jordan, are still used as venues for various types of spectacles, musical events, and performances re-enacting ancient games, such as chariot races, gladiatorial fights, and even the movements of legionary troops. Those which survive today represent of course only a small percentage of all the types of temporary and permanent structures in which spectacles were mounted during the Roman period. In the entire empire each city of a certain importance disposed at least of one building in which public shows and entertainments could be presented to the population, and the presence of several entertainment buildings was not exceptional in most cities of reasonable size. They usually occupied a prominent position in the landscape, often in proximity to the sacred buildings or monumental complexes dedicated to the gods. The scale and durability of many Roman entertainment buildings stand in evident contrast to the ephemeral character of the shows they housed and for the mounting of which impressive resources were often expended, only to disappear almost without leaving any trace *in situ*.[4] The total number of spectacles and performances must have been nevertheless extremely high, if we consider their ordered recurrence throughout the year and the long duration of Roman rule across its vast empire.

The importance of spectacles in the eyes of the Romans, the place they occupied in their lives, and the fascination they so obviously exercised, are all reflected in a large amount of surviving evidence, both written (literary texts, inscriptions and papyri) and iconographical (mosaics, frescoes, reliefs, coins, plates, cups, glasses or knives, made out of various materials such as stone, bronze, silver, ivory, glass, ceramic, bone). All such artefacts offered their owners and all those who saw or handled them a way of recreating a connection to ephemeral spectacles. For the sponsors and organizers their very materiality was a way of memorializing the expensive shows they had offered to and shared with their community. As for the spectators, these products could have the effect of reminding them of the strong and constrasting emotions they had felt during the many days and hours spent watching the events. The mass of these different archaeological, iconographical, and textual testimonies, coming from all over the Roman world, constitutes a body of evidence with which we can build up a picture of Roman spectacle culture.

As an expression of Roman power, spectacle culture is tied up with notions such as the political control of the *oikumene*, human domination of the animal world, the virtues of domestication of the wilderness, the celebration of urbanisation and of advances in technological capacities. Many aspects of this topic have of course attracted detailed attention. But only relatively recently have scholars begun to pose and try to answer questions about the

---

[3] On the Colosseum, see for example the Venerable Bede's famous remark: *Quandiu stat Colyseus, stat et Roma; / Quando cadet Colyseus, cadet et Roma; / Quando cadet Roma, cadet et Mundus*, with Byron's translation, cited by WOODWARD 2001: 11, 'While stands the Coliseum, Rome shall stand / When falls the Coliseum Rome shall fall / And when Rome falls – the world.' See also EDWARDS 2015: 217, on GIBBON and Rome.

[4] For some examples of material evidence see ALDRETE 2014: 447–8.

organization and mounting of spectacles, and this is an important aspect that still requires further study, with many questions remaining open. Awareness of the energy deployed by public authorities and by private benefactors in putting on games and festivals of various kinds, in Rome, in Italy, and across the entire Roman world, several times each year and during hundreds of years, invites us to consider the extraordinary extent of the human, natural, and financial resources involved. And so within the framework of a colloquium and a subsequent volume entitled 'Pollution and the Environment in Ancient Life and Thought', it seems legitimate to raise the question of the environmental implications of Roman spectacles. In this paper, therefore, what I intend to do first is to offer a very broad introductory survey of some of the ancient evidence that allows us to get some sense of the scale of the activity that went into planning, organizing, and putting on spectacles in the Roman world.[5] I will consider some contemporary Roman reactions to these processes and suggest how modern scientific methods of measuring environmental impact could be 'good to think with' in relation to the ancient evidence. In other words, I will attempt to illustrate with some examples the number and diversity of the questions that would have to be tackled, if a historian of the ancient Roman world, an expert in the field of mathematics and statistics in ecology, an archeozoologist and an archeobotanist were to collaborate in order to try to apply modern scientific methods to ancient evidence. What follows will, of course, involve many more questions than answers. Considering the types of evidence we have at our disposal, it is obvious that the observations and the analyses depend largely on more or less rough estimations, fragile hypotheses, and broad generalizations. But to begin with, one question is worth posing: what precisely do we know today about the data to be considered and about the methodology to be applied in order to assess environmental impact? It is only in recent decades that this subject has become a matter of intense discussion, as ecological concerns about environmental sustainability have come to constitute the key-words of the new rhetoric applied to the organisation of global sporting events. In spite of the manifest differences between the contemporary world and a pre-industrial society, a quick look at the environmental issues surrounding one recent mass sporting spectacle event can offer much to think about.

Despite its designation as the 'Green Cup' and all the efforts made to put on an ecologically sustainable tournament and to minimise negative environmental impact, Brazil's FIFA Soccer World Cup of 2014 has come in for much criticism. Both FIFA and the Brazilian government have been widely accused of failing to reach their environmental goals.[6] Some

---

[5] For the different concepts used in relation to environmental issues and their meanings, see THOMMEN 2012: 3–9. Pointing out that 'environment' and 'ecology' did not exist as such in Antiquity, he notes that the concept of environment used in English since the beginning of the nineteenth century as a translation of the German word *Umwelt* (attested since 1800) appears as an 'anthropocentric concept' to qualify the 'sum of all phenomena which influence the life situation of a human community', but that its meaning has degenerated since the 1970s to become an 'empty phrase, a shell' (p. 4–5). He reminds us also that the term 'ecology', derivated from the ancient Greek *logos* and *oikos,* referring to a house or household and in turn its budget, describes a 'kind of budget science of nature'. The notion of sustainability in today's terminology emphasizes the capacity of endurance of an ecosystem, its ability to remain productive and diverse over time. For selections of literary texts concerning ecology, see FEDELI 1990 and VOISIN 2014.

[6] SPANNE 2014a and 2014b; http://www.fifa.com/sustainability/football-for-planet.html. All the on-line papers and reports cited in this chapter were accessed for the last time during the month of April 2016.

positive results have been acknowledged, mostly in relation to the fact that several buildings constructed or restored for these games have been awarded LEED (Leadership in Energy and Environmental Design) certification. Among these buildings figures the legendary Maracanã Stadium, built for the 1950 World Cup in the middle of Río de Janeiro, which at that time was the biggest football stadium in the world, with an original capacity of approximatively 220,000 spectators, exceeding by 43,000 the capacity of Hampden Park in Glasgow. Today, after its renovation for the 2014 tournament, this stadium, with a capacity of almost 79,000, is still the largest in Brazil, but now only the second largest in South America. It has been provided with all the technological equipment required by new international standards and regulations.[7] As largely relayed in the media, thanks to its 2,500 m^2 rooftop of photovoltaic panels, for example, it produces around 400,000 kilowatt hours of solar energy, in other words 9 % of its own power requirements, and enough capacity to supply electricity to 240 single-family homes.[8] Besides, a rainwater capture system provides water for the football pitch and for the flushing systems of 292 toilets, saving water consumption by between 40 and 50 %.

On the other hand, the new Amazon Arena in Manaus, built after the demolition of the old Vivaldão stadium, offers a striking counter-example in terms of environmental success, despite the fact that it too has been awarded a LEED certification label. The original plan to provide all its power needs by means of solar energy was abandoned, due mainly to construction delays. More generally, the decision to build such a huge and expensive stadium in a remote area deep in Amazonia raises further questions, in particular regarding its maintenance and the uncertainty of its reuse for future events.[9] And so, despite all the setting of ambitious targets and all the efforts deployed in getting across a positive message about environmental sustainability, numerous questions still remain to be answered concerning the green gain and the environmental impact of the Soccer World Cup of 2014, and more generally in connection with the sustainability in ecological terms of such huge sporting events.[10] The environmental impact of the Winter Olympics of 2014 in Sochi, for example, is almost certainly even worse than that of any of its predecessors.[11]

---

[7] The Brazilians refused to see their stadium replaced by a new structure. In order to qualify for the LEED certification, the renovated Maracanã stadium was assessed in relation to the following seven areas: 'sustainable space, water, energy and atmospheric efficiency, materials and resources, internal environmental quality, innovation processes and regional priority credits': see http://www.supatank.com.au/rainwater-harvesting-sustainable-stadiums/. See also http://www.copa2014.gov.br/en/noticia/maracana-use-solar-energy-and-savings-lighting-and-water, and http://www.fifa.com/classicfootball/stadiums/stadium=214/.

[8] This system will thereby prevent each year the release into the atmosphere of over 2,560 tons of carbon dioxide: http://www.yinglisolar.com/assets/uploads/projects/downloads/Yingli%20Case%20Study_Maracana%20Stadium.pdf.

[9] Spanne 2014a.

[10] See also Jacques Rogg's message cited on the front page of the report entitled *Olympic Games – Beyond 2024:* 'Creating sustainable legacies is a fundamental commitment of the Olympic Movement … Every city that hosts the Olympic Games becomes a temporary steward of the Olympic Movement. It is a great responsibility …'. Cf. http://www.olympic.org/content/the-ioc/olympic-games-candidature-process/bidding-for-the-games---beyond-2024/?tab=legacy.

[11] On the differences between the official green programme and the reality of damage to local ecosystems see for example http://www.olympic.org/news/sochi-2014-promoting-sustainable-development-in-russia/223083 and http://www.popsci.com/article/science/whats-environmental-cost-2014-winter-olympics.

Paradoxically, the more those involved seem to invest in thinking about environmental impact, the more they seem to pollute, as the scale of the events continues to expand. But what is worth emphasizing here is that in relation to modern mass sporting spectacles such as recent FIFA Soccer World Cups and Olympics Games, which all attempt and claim to respect the environment, whether by reducing their carbon footprint or by supporting environmentally sustainable projects, the calculations and methodologies used in order to assess overall environmental impact are still far from straightforward. Long-term sustainability often remains elusive, and the lack of uniformity in the types of criteria assessed and methods applied is quite striking.[12] Between 2006 and 2012, for different large-scale sporting events organised in Germany, Italy, Canada, and South Africa, calculations were based mainly on the event time, with the inclusion, but only in some cases and at different levels, of an evaluation of the emissions in connection with travel (sometimes of competitors only, sometimes of both competitors and spectators), waste management, accommodation, construction, and so on. For the London 2012 Summer Olympic Games, however, a new methodology was applied in order to try to calculate the 'bid to event' impact. The idea was to produce a reference footprint, put together as part of the initial bid to organize the games, and then to model calculations from the moment of the bid's acceptance right up to the closing ceremony of the Games, assuming development as set out in the original application dossier. On the basis of such calculations, some interesting observations are possible: the reference footprint, estimated at 3.4 million tons of CO_2e (carbon dioxide equivalent), corresponds to half of 1 % of annual UK emissions and, according to calculations, more than half of the reference footprint was probably produced by infrastructure and venue constructions in London's Olympic Park. The application of criteria such as the use of existing sporting venues (around 60 %), the promotion of renewable energies and the reduction of emissions, as well as low carbon technologies, were all intended to reduce the reference footprint by more than half (i. e. from 3.4 million tons CO_2e to 1.9, according to evaluations). The foot-printing served as a prospective impact assessment rather than as a simple post-event reporting tool, but the results did not match the original ambitious vision, thus demonstrating the fragility of both the calculations and of the model established to assess the ecological impact of a large-scale sporting spectacle in the contemporary world.

The fact that such great difficulties surround these matters even today means, of course, that it is not possible to estimate in a precise and scientific manner the environmental impact of mass spectacles in the ancient world. Even to dare ask the question in relation to a pre-industrial society may appear simplistic and anachronistic. Nevertheless, the methodology set up to assess the 'bid to event' ecological impact of a modern event may offer, in spite of the obvious differences in context, an interesting perspective from which to approach an enquiry aiming to take into consideration the possible environmental impact of Roman spectacle culture. In an attempt to apply a suitably modified 'bid to event' vision to the ancient context, I will deal in the rest of this paper with the following topics: 1) construction activities; 2) spectacles in context; 3) spectacle infrastructure in Rome; 4) buildings and spectacles in Italy and in the provincial world. Given the nature of the evidence, the city of Rome will be the main focus of attention, but it is important to take into account the implications of the Roman evidence for an empire-wide picture.

---

[12] These observations are based mainly on the case study of the 'forthcoming London 2012 Olympics' by HAYES and HORNE 2011.

I begin by providing two tables in which I present the results of a detailed but hardly exhaustive survey of the textual sources at our disposal. The first table concerns some of the most famous mass spectacles in which our sources record that large numbers of animals were used. The second concerns those that involved the participation of large numbers of human beings.[13] The main aim is to demonstrate the kinds of activities for which we have relatively rich sources of information, and to give an initial impression of the extent of those activities. Some of the evidence is well known and some anecdotes rightly famous, but some more obscure sources have also been included, and I know of no single place where all the evidence here accumulated can be easily found (See Tables 1 and 2, p. 223–9).

From all of these facts and figures, it will be useful to take a closer look at two examples. Doing so will enable us to have a sense of what certain types of Roman mass spectacle could involve, of the implications of some of the basic information provided by our sources, and of the gaps in our knowledge.

In his *Res Gestae*, Augustus emphasizes the splendour and the massive size of the games he sponsored. He records the hunting spectacles (*venationes*) involving wild beasts that he gave in Rome on 26 different occasions and during which around 3,500 beasts were killed.[14] The scale is indeed impressive, but it is noteworthy that this proud declaration does not remotely give a clear idea of the true scale of Augustus' mass spectacles. The figure 3,500 covers only *African* beasts and the spectacles in question are *only* those that Augustus gave under his own name and those of his sons and grandsons, and not all those he sponsored and promoted during his long reign.[15] Furthermore, he refers here only to hunting spectacles, and so does not cover, for example, all the other gladiatorial and circus games for which he will have been responsible. Elsewhere in his *Res Gestae*, for example, he records that 10,000 men fought in his gladiatorial shows and that 3,000 men took part in a staged sea-battle, not counting the actual rowers.[16]

A second example concerns Trajan's games given on the occasion of his triumph after his return to Rome from Dacia in 107 AD. In the space of 123 days, according to Cassius Dio, 11,000 animals were slain, both wild and tame, and 10,000 gladiators fought in front of a large and cosmopolitan audience, among which were present many embassadors 'from various barbarian countries', even from as far away as India.[17] These numbers can of course be questioned, but as surprising as it may appear, they are actually inferior to the 11,228 gladiators (or 5,614 *pairs* of gladiators) mentioned in the *Fasti Ostienses* in relation to Tra-

---

[13] Translations of the main literary sources used are included in two appendices, one for each table. I have mostly used Loeb translations throughout, sometimes slightly modified, without mentioning the names of individual translators. I have underlined some numbers or other remarkable features.

[14] Text 1.59 (= Table 1, text 59), *RGDA* 22.3: *[Ven]ation[es] best[ia]rum Africanarum meo nomine aut filio[ru]m meorum et nepotum in ci[r]co aut in foro aut in amphitheatris, popul[o d]edi sexiens et viciens, quibus confecta sunt bestiarum circiter tria m[ill]ia et quingentae.*

[15] Ville 1981: 127–8 rightly emphasizes that the number mentioned by Augustus concerns only the *bestiae Africanae* and not all the animals slain during the games he gave, i. e 'les bêtes d'Europe qu'il n'a pas comptabilisées dans ses *Res Gestae*' (p. 128). On the scale of arena spectacles under Augustus, see Coleman 2003: 73–7; see also Cooley 2009: 203 and 209. On the games given in 33 BC by Agrippa, see below n. 21.

[16] Texts 2.19 and 20.

[17] Text 1.84.

Table 1. Mass spectacles involving numerous animals – Selected textual testimonies
* The asterisk signals an explicit mention in the ancient text of the appearance of a beast or of a type of show for the first time

N°	Source	Date/editor, context	Location/game	Type of animal	Amount	Issue
1	*Sen., *brev.* 13.3	275/Curius Dentatus, trium.	Rome (triumph)	elephants		
2	Florus, *ep.*, 1.13.27–8	275/Curius Dentatus, trium.	Rome	elephants		
3	*Plin., *nat. hist.* 8.16	275 + 252 Dentatus + Metellus	Rome (trium. + vict.)	elephants		killed or chased
4	Plin., *nat. hist.* 8.16–7	252/L. Caec. Metellus, pont.	Rome, c. max.	elephants	142 or 140	
5	Plin., *nat. hist.* 18.17	250/L. Caec. Metellus triumph	Rome	elephants	plurimi	
6	*Liv. 39.22.1–2	186/M. Fulvius Nobilior	Rome (*venatio*)	lions + panthers		prob. killed
7	Liv. 44.18.8	169/Sc. Nasica+Lentulus	Rome, c. max.	Africanae+bears+eleph.	63/40/40	
8	Plaut., *persian*, 198–9		Rome	ostrich		
9	Plaut., *poen*, 1011–2		Rome (pompa aed.)	African (mice, joke)		
10	*Plin., *nat. hist.* 8.53	104/Q. M. Scaevola, aed. c.	Rome	lions	many	first fight
11	*Plin., *nat. hist.* 8.19–20	99/Cl. Pulcher, aed. c.	Rome, c. max.	elephant		fight
12 = 10	*Plin., *nat. hist.* 8.53	93/L. Sylla, praetor	Rome	lions	100	fight
13	*Sen., *brev.* 13.6	93/L. Sylla	Rome, c. max.	lions	loosed l.	exhibition
14	*Plin., *nat. hist.* 8.4	79/Pompey's Afr. trium.	Rome	chariot dr. by elephants		
15 = 11	*Plin., *nat. hist.* 8.20	79/two Luculli, aed. cur.	Rome, c.max.	elephant vs bulls		fight
16	Suet, *Iul.* 10.1	65/Julius Caesar, aedile	Rome	wild beasts		fight
17	Plin., *nat. hist.* 8.131	61/D. L. D. Ahenob., aed. cur.	Rome, c. max.	bears vs *venatores*	100 + 100	
18	App., *b.c.* 2.13	59/Caesar, consul	Rome	wild beasts		
19	*Plin., *nat. hist.* 8.96	58/M. Scaurus, aed.	Rome (temp. eur.)	hippo. + crocodiles	1 + 5	
20	*Plin., *nat. hist.* 8.64	58/M. Scaurus, aed.	Rome	*Africanae* (*variae*)	150	prob. killed
21	Amm. 15.24	58/M. Scaurus, aed.	Rome	hippopotames		

N°	Source	Date/editor, context	Location/game	Type of animal	Amount	Issue
22	Plin., *nat. hist.* 8.20–1	55/Pompey	Rome, circus	elephants *vs* Gaetulians	20 or 17	
23	Cass. Dio 39.38.1–3	55/Pompey	Rome, circus	beasts/lions/el. *vs* men	500 l./18el.	l. killed in 5 days
24	*Plin., *nat. hist.* 8.70	55/Pompey	Rome	lynx ? / cephic		
25	*Sen., *brev.* 13.6	55/Pompey	Rome, c. max.	elephants *vs* criminals	18 el.	killed
26	Cic., *ad fam.* 7.1.3	55/Pompey	Rome	*venationes* + elephants		10 ven. (5 days)
27 = 20	*Plin., *nat. hist.* 8.64	55/Pompey	Rome	*Africanae (variae)*	410	prob. killed
28	Plin., *nat. hist.* 8.71	55/Pompey	Rome	rhinoceros (1 horr.)		
29 = 10	Plin., *nat. hist.* 8.53	55/Pompey	Rome, c. max.	lions	600	fight
30	*Plin., *nat. hist.* 8.55	48/Mark Antony	Rome	lions/yoke		
31 = 10	Plin., *nat. hist.* 8.53	c. 49–44/Julius Caesar, dict.	Rome, c. max. ?	lions	40	fight
32	Suet., *Iul.* 37.2	46/Julius Caesar, triumphs	Rome	elephants with torches	40	
33	Suet., *Iul.* 39.3	46/Julius Caesar, triumphs	Rome	*venationes*	beasts	5 days
34	App., *b.c.* 2.102	46/Julius Caesar, triumphs	Rome	elephants	20 vs 20	fight
35	*Cass. Dio 43.22.1–4; 23.1–2	46/ J. Caesar, trium.+daughter	Rome	el., beasts, giraffe – gladiat.	40 el. + ?	fight men on el.
36	Plin., *nat. hist.* 8.22	46/5Julius Caesar, cons. III	Rome	elephants *vs* men	20 vs 500	fight
37	*Plin., *nat. hist.* 8.69	46/Julius Caesar, dictator	Rome (*circenses*)	camelopard/giraffe		
38	*Plin., *nat. hist.* 8.182	46/Julius Caesar	Rome	bulls *vs* Thess. on hors.		fight
39	Suet., *Iul.* 39.3	Julius Caesar	Rome, c. max.	elephants	20	
40	Cass. Dio 48.33.4	41 or 40/ knights, *ludi Apollin.*	Rome	knights *vs* beasts		killed
41	*Cass. Dio 51.22.5–9	29/Octavian-Augustus	Rome, ded. t. Caesar	var. beasts+rhino+hippo	many+1+1	killed
42	Cass. Dio 53.27.6	25/P. Servilius, praet.	Rome ?	bears + libyca	300+300	killed
43	Cass. Dio 54.8.5	20/Augustus' birthday, aediles	Rome	*circenses + beasts*		killed

N°	Source	Date/editor, context	Location/game	Type of animal	Amount	Issue
44	Suet., *Nero* 4.3	19+16/L. D. Ahenob. pr.+c.	Rome, c. max.	*venationes* +glad.		prob. killed
45 = 20	*Plin., *nat. hist.* 8.64	Augustus	Rome, c. max.	*Africanae* (*variae*)	420	killed
46	CIL 6.32323	17/Aug., *Ludi Saeculares*	Rome, c. max. ?	*venatio*		killed
47	Cass. Dio 54.26.1–2	13 or 11/Aug., ded. th. Marcel.	Rome, c. max.	libyca	600	killed
48 = 46	Cass. Dio 54.26.1–2	13 or 11/Ant. Iullus, praet.	Rome, c. max. ?	beasts for Aug.'s birthday	*circ.*+*venat.*	killed
49	*Plin., *nat. hist.* 8.25.65	13 or 11/Aug., ded. th. Marcel.	Rome	tiger in cage	1	killed
50	Cass. Dio 54.34.1–2	11/Drusus, praetor Aug.'s birth	Rome, c. m. + els.	beasts	*circ.*+*venat.*	killed
51	Suet., *Aug.* 43.4	Augustus	Rome, sev. loc.	rhino + tiger + snake	1 + 1+ 1	killed
52	Cass. Dio 55.10.6–7	2/Augustus, ded. Mars Ultor	Rome, c. max.	chariot races +lions	260	killed
53	Cass. Dio 55.10.8	2 BC/Aug., ded. Mars Ultor	Rome, c. Flaminius	crocodiles	36	exhib. + killed
54	Plin., *nat. hist.* 8.5	6/Germanicus	Rome	elephants (+ gladiators)		el. funambules
55	Aelian, *de nat. anim.*, 2.11	6?/Germanicus	Rome	elephants (born locally)		
56	Cass. Dio 56.27.4	12/Aug., *Lud. Mart.*	Rome, forum (flood)	horse-race +wild beasts		w. beasts killed
57	Cass. Dio 56.27.5	12/Aug.-Germanicus	Rome, c. max.	lions	200	killed
58 = 20	Plin., *nat. hist.* 8.64	Augustus	Rome, c. max. ?	*Africanae*/panthers	420	?
59	*RGDA* 22.3	Augustus	Rome (26 *venat.*)	(African) beasts	3,500 Af. b.	killed
60	Cass. Dio 59.7.1	37/Caligula, cons., ded. t. Aug.	Rome	*circ.* + *venat.* bears+libyca	400+400	killed
61	Cass. Dio 59.13.8	39/Caligula/Drusilla's birthday	Rome, c. max.	chariot with elephants		killed
62	Cass. Dio 59.13.9	39/Caligula/Drusilla's birthday	Rome, c. max.	horse-races+bears + libyca	500 b+500 l	killed
63	Cass. Dio 59.20.1–2	39/Caligula's birthday	Rome	horse-race + wild beasts		killed
64	Suet., *Cal.* 18.1–5	Caligula	Rome, sev. loc.	Afric., lights, *missilia*, colors		
65	Suet., *Cal* 20	Caligula	Syracuse, Lyon			
66	Cass. Dio 60.7.3	41/Claudius	Rome, circ.	camels+hors../bears+lybica	300 + 300	bears+ly. killed

N°	Source	Date/editor, context	Location/game	Type of animal	Amount	Issue
67	Cass. Dio 60.13.4	Claudius	Rome	lion trained to eat men		
68	Cass. Dio 60.23.4–6	44/Claudius, triumph on Brit.	Rome, var. loc.	horse-races+bears/men		bears killed
69	Cass. Dio 60.27.2	Claudius, c. max.	24 horse-races, c. m.	increase of number of races		
70	Tac., *ann.*12.3.2	48/Claud., Octavia's engag.	gladiatorial contest			
71	Suet., *Claud.* 21.1–4	Claudius	Rome, sev. loc.	h.races+Afr/w. bulls vs men	div. shows	animals + men
72	Plin., *nat. hist.* 8.65	Claudius	Rome	elephants vs men		
73 = 49	Plin., *nat. hist.* 8.22	Claudius	Rome	tigers in cage	4	
74 = 72	Plin., *nat. hist.* 8.22	Claudius + Nero	Rome	elephants vs men		
75	Suet., *Nero* 7.9	54/Nero's wed. Octav.+sal.Cla	Rome, c. m.	*circenses+venatio*		
76	Suet., *Nero* 11.1–2	Nero	Rome, c. m. + sev. l.	beasts ; 4 camels, *missilia*	1000 birds	some given
77	Suet., *Nero* 12.1	Nero	Rome, wood. amphi.	men vs beasts, el., *missilia*	700 men	s. animals given
78	Calp. Siculus 7.23–73	57?/Nero?	Rome	bears+seals, bulls, boars etc.	various	
79	Jos., *b.j.* 7.132	71/Titus (and Vesp.)	Rome (triumph)	wild beasts, var. species		
80	Cass. Dio 66.25.1–4	80/Titus	Rome, sev. locat.	an., cranes, el., horses, bulls	c. 9,000	(100 d.)
81	Suet., *Tit.* 7.3	80/Titus	Rome	various beasts	5,000	(1 day)
82	Mart., *spect.* 20 +*passim*	80–5/?Titus?/ Dom.	Rome	eleph.rhino,etc.+naumach.		
83	Suet., *Dom.* 4	Domitian	Rome	beasts, h.-races, battles		
84	Cass. Dio 68.15.1	post 107/Trajan	Rome	animals + 10,000 gladiators	11,000	(123 days)
85	HA, *Hadr.* 7.12	Hadrian/ … birthday	Rome	gladiators /wild beasts	1000	(6 days)
86	HA, *Hadr.* 19.2–8	Hadrian	Athens, stadium Rome, c. m.	*venatio*, wild beasts beasts, lions	1000 often 100	killed
87	HA, *Ant. Pius* 10.9	149/Ant. Pius	Rome	elep.+ cor.+tig.+ rhino+lions	100 lions+t.	1 performance

N°	Source	Date/editor, context	Location/game	Type of animal	Amount	Issue
88	Hdn. 1.15.5–6	Commodus	Rome	various beasts fr. all world		
89	*Cass. Dio 76.1.3–5	202/Septimius Severus	Rome, c. m.	boars+w.b+el.+coro.	700 total	killed
90	HA, *Gord.* 3.5–8	a. 213/Gordian I aedileship	Rome	lions/bears/gladiat. / etc.	100/1000/	12 s/year; killed
91	HA, *Gord.* 4.5	213/ Gordian I praet. +consuls.	Rome	horses (given to factions)	200	
92	HA, *Gord.* 33.1–2	248/Philip Arab, *Lud. Saec.*	Rome, c.m. + amphi.	m. var. anim.; glad.	22 el+60 l.	
93	HA, *Aurel.* 33–4	274/Aurelian, triumph	Rome, c.m.	variousanim.+gladiat.+pris.	200 + 800gl	partly given
94	HA, *Prob.* 19.1–8	276–82/Probus, triumph	Rome, c.m.+ amphi.	m. var. anim., forest	thousands	killed
95	Claud., *Theod.* 291–332	399/Theodorus, consulhip	Milan	bears,lions,leopard,mecha.		naumachy
96	Claud., *Stil.* 3.262–32	400/Stilicho, consulship III	Rome	beasts, lions; eleph.		pompa w. world
97	Luxorius	5th–6th cent.	Africa, amphi.	wild-beasts vs agriculture		

Table 2. Mass spectacles involving large numbers of people (soldiers, prisoners, gladiators, and *noxii*) – Selected textual testimonies

N°	Source	Date/editor-resp.	Location	Amount/type	Issue
1	Dion. Hal., *ant.* 6.17.2	496/499?/A. Postumius	Rome	5,500/prisoners	?
2	Eutropius 2.2	356/C. Marius Tuscus	Rome	8,000/Galli prisoners	?
3	Liv. 7.27.8–9	346/M. Valer. Corvus	Rome	4,000/prisoners	sold (or enslaved)
4	Val. Max. 2.7.14	168/L. Aemilius Paullus		deserters + elephants	crushed by eleph.
5	Val. Max. 2.7.13	146/Scipio Aemilianus		deserters	*ad bestias*
6	Diod. Sic. 36.10.2	100/Aquilius, proc. Sic.	Rome	1,000 prisoners vs beasts	suicide of rebels
7	Sen., *brev.* 13.6–7	55/Pompey	Rome	criminals; 18 elephants	
8	Cic., *ad fam.* 7.1.3	55/Pompey	Rome	600 mules; caval.+infantry	
9	Plut., *Caes.* 5	65/Jul. Caesar, aedile	Rome	320 pairs of gladiators	
10	Suet, *Iul.* 10.2	65/Jul. Caesar, aedile	Rome	gladiators	limit. of gladiat.
11	Plin., *nat. hist.*, 8.22	46/Jul. Caesar, cons. III	Rome	500 (and more) vs elephants	
12	Suet, *Iul.* 39.1–4	Jul. Caesar	Rome, v.loc.	various spectacles	
13	Suet, *Iul.* 39.3	Jul. Caesar	Rome, c.m.	530m/20 el./beasts races/Troy	
14	Suet, *Iul.* 39.4	Jul. Caesar	Rome.Trast.	naumachy Tyrians /Aegypt.	many deaths
15	Cass. Dio 43.23.4–6	46/Jul. Caesar, triumphs	Rome	naumachy, Troy, venat.	pris., *noxii*, sold.
16	App., *b. c.* 2.102	46/Jul. Caesar, triumphs	Rome	c. 8400 men battles + naum.	
17	Cass. Dio 51.22.4	29/Aug. ded. t. Caes.	Rome	Troy (patricians)	
18	Cass. Dio 51.22.6–9	29/Aug. ded. t. Caes.	Rome	prison. Dacians vs Suebians	
19	RGDA 22	29 BC–AD 6/Augustus	Rome	c. 10,000 gladiators	
20	RGDA 23	2 BC/Augustus	Rome	*navalis proeli spectaculum*	
21	Cass. Dio 55.10.7–8	2 BC/ Augustus	Rome	naval battle Persians-Ahenians	
22	Cass. Dio 59.10.3–5	28/Caligula, Saepta etc.	Rome, Saepta	spectators thrown to beasts	

N°	Source	Date/editor-resp.	Location	Amount/type	Issue
23	Cass. Dio 60.30.3	47/Claudius, ov. Plaut,	glad. contests	foreigner freemen+prisoners	many
24	Suet., *Claud.* 21.6	Claudius, C. Martius	Rome	repres/ storm+sack of town	
25	Suet., *Claud.* 21.6	52/Claudius, naumachy	Lake Fucine	12 vs 12 triremes	
26	Tac., *ann.* 12.56.1–3	52/Claudius, naumachy	Lake Fucine	19,000 men convicts +praet.	many wounded
27	Cass. Dio 61.33.3–4	52/Claudius, naumachy	Lake Fucine	100 ships, criminals	forced to destruct.
28	Tac., *ann.* 15.44	65/Nero	Rome, N.'s.	num. christians ; dogs	killed
29	Suet., *Nero* 12.1–2	58/Nero	Rome, amp.	4 camels, *missilia*	
30	Suet., *Nero* 12.1	58/Nero	Rome	naumachy+fatal charades	
31	Jos., *b.j.* 7.118	71/Titus (+Vesp.) tr.	Rome	700 Jews prisoners	
32	Jos., *b.j.* 7.139–48	71/Titus (+ Vesp.) tr.	Rome	prisoners / repr. war +naum.	killed
33	Cass. Dio 66.25.2	80/Titus, Aug's stag.+Col.	Rome	i. battle +horse race+ naum.	c. 3,000 men
34	Suet., *Dom.* 4	Domitian	Rome, amp.	2 battles + naum.	
35	HA, *Gord.* 3.5–8	213/Gordian	Rome	500 pairs gladiators	1,000 gladiators
36	HA, *Aurel.* 33–4	274/Aurelian, triumph	Rome	800 pairs glad.+ pris.; naum.	1,600 gladiators+ ?

jan's third triumph.[18] By any standards, we are dealing with spectacles on a remarkable scale, and even as we focus on these two cases we must not forget that between Augustus and Trajan, during sixty 'extraordinary years', emperors like Nero, Titus, and Domitian also played an active role in the development of Roman spectacle culture.[19]

Obviously, one can always worry about the precision of the figures relayed by our sources. Cassius Dio certainly did so.[20] But for our purposes in this paper the important thing is to take into account that, even if many of the recorded figures may be inexact and even exaggerated to some extent, the sense of the scale of the extraordinary events being described is more important. And crucially, behind the exceptional size of events organized by emperors like Augustus and Trajan on special occasions, we must also take into account the sheer number of ordinary events taking place across the empire over a very long period of time, but going unrecorded in the surviving sources. That is, even if we add up all the figures given in the two tables for different spectacles, they may in fact represent only the tip of an iceberg. As Roman spectacle culture develops throughout the Republican period from essentially local roots to become what can be considered a truly global phenomenon under the Empire, we should not expect the regular public shows, given each year by magistrates in charge of organizing games, to be systematically recorded in our textual sources, especially if they contained no particular novelty and were not of remarkable scale.[21] The same is true of the spectacles held in the provinces, obviously less well documented in the literature and for knowledge of which we depend mostly on epigraphical documentation. The relative scarcity of epigraphical testimonies commemorating spectacles in the Western provinces

---

[18]   COLEMAN and NELIS-CLÉMENT 2012: xxi-ii; the total amount of gladiators involved in the various contests in Rome between 107 AD and 109, according to the *fasti*, corresponds to 11,520; see also MIGLIORATI 2003: 111–3. On Dio questioning exaggerations in numbers: text 1.35; on this question and in general on Dio and the games: NEWBOLD 1975: 591. I owe these references to G. Urso.

[19]   See texts 1.75–83 and 2.28–34; '60 years…': PURCELL 2013: 456–8. On Titus and Domitian's spectacles: COLEMAN 2006 (see also BUTTREY 2007) distinguishes (p. xlix table 4) nine different types of events on the basis of all the textual evidence, revealing the diversity and the quantities of natural and human resources required for these spectacles: (1) *venationes* (in which 9,000 animals were killed, 5,000 in one day); (2) beast-fights involving many exotic animals of all sorts (rhinoceros vs bull, bear, bullocks, aurochs, bison, lion; bull vs elephant; hounds vs doe; cranes; fights between cranes and between 4 elephants); (3) performances of trained animals (tame lion and tiger; elephant kneeling; 'flying' bull, tracted with stage-devices; trained horses and bulls, various animals performing in water and on land); (4) horse-racing (at Augustus' *stagnum*); (5) gladiatorial combats (among them one *munus apparatissimum largissimumque*) and one infantry battle (at Augustus' *stagnum*); (6) *naumachiae* (among them, one between Corcyra and Corinth, in the Flavian amphitheatre, and one between Athens and Syracuse, at Augustus' *stagnum*); (7) mythological enactments (Pasiphae, Daedalus, Orpheus, Leander, Nereids); (8) parade of *delatores*; (9) lottery (wooden balls thrown in the arena as tokens to be exchanges for clothes, vessels, horses, pack-animals, cattle or slaves).

[20]   Text 1.35 and above n. 18.

[21]   For a discussion of the different attitudes of Roman magistrates in relation to aedileship and the *cura ludorum sollemnium* at the end of the republican period see DAGUET-GAGEY 2015: 235–318, and also p. 319–33 for a list of the *venationes* and *munera* given by the aediles between 169 BC and 33 BC. Note that in 33 BC, during Agrippa's aedileship, 59 days of games were given, which Agrippa wanted to be recorded in his memoirs: Plin., *nat. hist.* 36.121–2; see also Cass. Dio 49.43 with RODDAZ 1984: 152–4, and WISEMAN 2015: 122–3.

– only c. 500 Latin inscriptions mention games and their benefactors over more than 5 centuries – does not mean that spectacles were not mounted in the cities of these provinces. On the contrary, the presence of many entertainment buildings is clearly attested and the signs of ongoing maintenance suggest that they were in use for centuries. The relatively limited amount of epigraphical evidence can be explained by the fact that only the shows given on a voluntary basis (*ob liberalitates*) were worthy of being commemorated in an inscribed monument, as demonstrated recently by G. CHAMBERLAND. The statutory games (solemn or public games) which had to be given every year by the provincial magistrates, as required, for example, by the municipal constitutions such as the *Lex Ursonensis* (*colonia Genetiva Iulia*) in Baetica, go unrecorded.[22]

As well as probing what may lie behind statistics, it is also necessary to take into consideration what is involved in staging a spectacle and what lies behind a statement such as Augustus' that he organized *venationes*. First and foremost, of course, obvious though it may seem, the actual venue had to be in existence and had to be capable of accommodating the event in question. Temporary enclosures could be thrown up rapidly and then dismantled, but if we are dealing with a major stone-built circus or amphitheatre in Rome or elsewhere we must have in mind all the requisite infrastructure both for the organisation of the event in question and for the spectators assembling to see it. For any given day all the necessary animals and human participants had to be in place at the required time, thus implying all the hunting, the transporting, the feeding, and so on. And after many events, there will have been large numbers of dead animals and dead humans to be disposed of. Overall, it is a long and complex set of activities that make it possible for any single event to take place, and only by taking into account the whole process can we in any way hope to apply a suitably modified 'bid to event' approach to the spectacle culture of ancient Rome and to the assessment of its possible environmental implications. Limitations on space mean that only a few examples will be discussed in any detail in what follows, but I hope at least to be able to convey some idea of the sheer amount of the ancient evidence and of the extraordinary scale of the activities we are dealing with. It will of course in the end not be possible to measure accurately any environmental impact, but it seems at least worth looking at the ancient evidence from this modern point of view.

1. Construction activities

A very large variety of resources, in enormous quantities, went into constructing both temporary structures and permanent buildings for staging spectacles. Among the most common materials one thinks immediately of wood, stone, iron, marble, bricks, and concrete, in some cases found or made locally, in others imported from far away. When we read that

---

[22] *Lex coloniae Genetivae*: *CIL* 2²/5.1022=*ILS* 6087=*EAOR* 7.1 r. LXX and LXXI; CRAWFORD 1996.1: 393–454, esp. 423–4; CHAMBERLAND 2012: 267: 'les inscriptions des *duumviri*, édiles et *quattuorviri* qui ne mentionnent pas les spectacles (réglementaires ou libres) se comptent par milliers'; see also p. 263 and p. 295–303 for the distinction between *ludi publici*, games 'réglementaires' or 'publics', and *privati* and a discussion on the commemoration and reglementation of games; see also recently CARTER and EDMONSON 2014: 544–7; on games given in Italy and in the provincial world, see below p. 242–54.

among the constructions that amazed the emperor Constantius II during his visit to Rome in 357 AD was the Colosseum, '… the great solid mass of the amphitheater, built of travertine and so tall that human sight can scarcely reach the top', it seems worth adding that R. Lanciani calculated that 100,000 m³ of travertine marble were required to build it.[23] Lanciani even went so far as to estimate that over four centuries the Romans quarried 5.5 million m³ of travertine. More recent archaeological research attempts to measure the human effort and the natural resources exploited in the production, the transformation, and the transportation of all the materials needed for major construction sites in ancient Rome.[24] It would be interesting to try to apply these modern approaches systematically to the various kinds of entertainment buildings as a discrete group, taking into account their specificities as sites for mass entertainment.[25] One case where such work has already been done, at least partially enables us to turn our attention from the Colosseum to a provincial circus in the Roman colony of Arles, in Gaul.

On the basis of recent discoveries, it has been estimated that around 25,000 trees (and among them 16,000 oaks) were needed for the construction of the wooden substructure of Arle's medium-sized circus (c. 440 × 101 m).[26] Located close to the banks of the Rhône, it needed to have firmly reinforced foundations. Dendrochronological analyses have shown that almost all the trees were cut during the winter of 148–9 AD. The choice and use of the different species shows that the project was well planned and involved many people, employed for example in choosing the trees before cutting them, trimming and transporting them, all in a limited period of time. Weaker woods (mostly *pinus halepensis* and *pinus pinea*, with some *pinus silvestris*) were used to support the lighter parts of the structure, while stronger oaks were placed to support heavier loads. According to the local experts, the oak grove must have occupied a surface of c. 200 hectares.[27] Taking into account that trees need in general from 20 to 50 or 60 years to grow, the oaks growing rather more slowly than the pines, the cutting of a single winter must have had some impact on the landscape of the region of Arles. One could think that this impact was perhaps very limited, if we take it as a unique or exceptional case of deforestation in the region, and if we formulate the hypoth-

---

[23] Amm. 16.10.14–5. On the estimations, see DeLaine 2000: 135–6; Hopkins and Beard 2005: 127.

[24] For studies of raw energy involved in ancient constructions, see DeLaine 2000; Graham 2013; on the use, development, and dating of Roman concrete: Mogetta 2015.

[25] For a recent attempt to do exactly this kind of thing, see Hopkins and Beard 2005: 127–48 (according to their estimations, 250,000 m³ of material were required merely to fill the hole excavated for building the Colosseum), with the important review of Lancaster 2007 for rectifications concerning in particular the ways in which the arena could be flooded and then drained for particular occasions, and also with additional bibliography. According to recent estimations, c. 3,000 tons of iron were necessary to fix the blocks of travertine together: Bomgardner 2013: 30. The same kind of estimations could be made for the wood used for the arena, the machinery or lifts to bring the animals up from the basements, the poles supporting the *velum*, as well as the tons of linen for the *velum*, the ropes etc., or the bronze for the statues placed in the arcades of the second and third storeys.

[26] Sintès 2008; for comparisons of sizes, see Crummy 2008: 229–30 and below p. 243.

[27] This corresponds to c. 3.7 % of an annual surface of deforestation (i. e. 5,420 hectares) considered as 'very alarming' by Meiggs 1982: 380, who also notes that most parts of Gaul remained thickly forested during the Roman imperial period. In general on deforestation, see Meiggs 1982: 371–403; see also Harris 2013: 174; Hughes 2014 chap. 5, and in this volume p. 203–16.

esis, possible though rather uncertain, that new trees would have been quickly replanted in the following years and in the same space, instead of leaving the soil free for agriculture. But even in doing so, one would ignore the potential indirect effects of the cutting of such a surface on the ecosystem of the region, particularly on the fauna living in its forests and on the flora.[28] It is worth noting also that Arles had, in addition to its circus, a theatre and an amphitheatre, thus implying even greater local efforts in terms of the numbers of trees and other materials selected, prepared, transported, and used in construction.[29]

We have traces of how Romans could react when they felt that the construction of an entertainment venue had been taken to extremes. Pliny the Elder harshly criticizes M. Aemilius Scaurus, aedile in 58 BC, in relation to the construction of his theatre, which he considered as totally excessive and even immoral in the choice of various precious materials, many transported from far away and used in massive quantities (360 marble columns; 3,000 bronze statues adorning the *frons scaenae*). His criticisms also concern the use of massive marble instead of veneer, the over-capacity of the building (seating for 80,000 spectators, twice that of Pompey's theatre), and its cost (30 million sesterces). Pliny also points out that it was '… used barely for a month', and even laments the absence of laws against such excess, '… and the laws were silent' (36.4).[30] From a related perspective, Tacitus sees in the building of permanent edifices an evident economical avantage preventing the spending of money year after year on temporary structures.[31]

2. Spectacles in context

The venues for the different types of spectacles included a theatre, mainly used (but not exclusively) for dramatic performances, theatre and pantomimes (*ludi scaenici*), an odeon, in which were presented poetic recitals and declamations (*recitationes*), or a stadium, for the agônes or *certamina Graeca*, athletic competitions, boxing-matches, or foot races (*athletarum certamina*), and also for various types of musical or poetic contests. But the biggest spectacles were the games and other events usually presented in a circus or in an amphitheatre. In the city of Rome, the capacities of the Circus Maximus and the Flavian Col-

---

[28] On the impact of local ecosystems of wood production see Horden and Purcell 2000: 335–41, esp. p. 256 and 335 (relating to the area of Roman Arles and Narbonne); see also Graham 2013 (with bibliography), who underlines the need for more studies related to the timber industry (p. 290–1); on the Mediterranean environment: Wilson 2013.

[29] For epigraphic evidence attesting the presence of a *venatio* in the local amphitheatres: *CIL* 12.697 + *ILGN* 109=*AE* 1965, 270=*EAOR* 5.7. Since Arles was also a port on the Rhône near the Mediterranean, wood was also required for the construction of the ships for its *navicularii*, famous for their trade activities and their extended network around the Mediterranean: see for example *CIL* 12.6720=*ILS* 1432, and *CIL* 3.14165.8=*ILS* 6987, in Beirut (Lebanon).

[30] On Scaurus' theatre, Plin., *nat. hist.* 34.36 and 36.5; 36.50; 36.113–5. 36.113: 'As aedile he constructed the greatest of all the works ever made by man, a work that surpassed not merely those erected for a limited period but even those intended to last for ever'. On Pliny's exaggerations see Sear 2006: 96–7. Note also Pliny's complaint that 'no law was ever passed forbidding us to import marble and to traverse the seas for its sake' (36.5).

[31] Tac., *ann.* 14.21; on the notion of luxury in the architecture of private villas on the Bay of Naples see Zarmakoupi 2014: 111–339.

osseum represented respectively c. 15% and 5% of the population of the city (estimated at one million in the first century AD). The shows displayed in a circus, the *ludi circenses*, consisted mainly of chariot-races, the *maxima spectacula* as Ovid calls them, for which the monument was especially designed.[32] The oldest and biggest of this kind of entertainment building in Rome was the Circus Maximus. Livy and Dionysius of Halicarnassus associate its origins with Tarquinius Priscus, who held games in the space between the Aventine and Palatine hills, and this place was destined to become in time one of 'the most beautiful and most admirable structures in Rome', with a capacity of 150,000 spectators, a figure in line with modern estimations.[33] Due to its size, shape and huge capacity, the circus was also the perfect venue for other grandiose performances, such as staged beast hunts (*venationes*) and triumphal celebrations. These could take place on both sides of the central *euripus*, all along the two immense tracks, the total space measuring c. 81,200 m² (c. 580 × 140 m), the equiv-alent of around 11 contemporary soccer pitches.[34] Staged hunts more usually took place in an amphitheatre, often with the *venatio* in the morning, gladiators in the afternoon, with the *meridiani* as an interlude in the middle of the day for public executions such as fights be-tween men (*bestiarii*) and various types of animals, *damnationes ad bestias* or other similar spectacles, like parades of *delatores* or 'fatal charades' in which prisoners and condemned criminals were killed in the course of elaborate spectacles staged as re-enactments of myth-ological scenes.[35] Of course some *venationes* were also given independantly during a whole day in a circus or in an amphitheatre, and a *munus* could be held during several days, the best parts being kept for the last day, the *summus dies*.[36]

In addition to these types of buildings other structures *ad hoc* must be mentioned. Their size and therefore their impact could vary a lot, from simple wooden structures installed for the period of a show in a public place or forum, to a circus made of almost nothing except a flat field (like that of Dougga in Africa known by an inscription and a graffito, later monu-mentalized), to the threatre of Scaurus already mentioned, built with great magnificence but used for no more than one month, or to the constructions made especially to hold imperial mock sea-battles (*naumachiae*), for which remarkable skills in engineering were required,

---

[32]   Ov., *Am.* 3.2.65–6: *maxima iam vacuo praetor spectacula circo/quadriiugos aequo carcere misit equos.* For an up to date survey about circus spectacle, see BELL 2014.

[33]   Liv. 1.35.8; see also 1.56.2; Dion. Hal., *ant.* 3.68.1–4: the capacity of 150,000 spectators he mentions is supported by many scholars: HUMPHREY 1986, 126; COLEMAN 2000: 213; CIANCIO ROSSETTO 2008: 22; but ZANKER 2001: 152 does not exclude the figure of 250,000 given by Plin., *nat. hist.* 36.102, and M. BUONFIGLIO now agrees (personal communication, March 2017). For different views: DODGE 2014b: 564 considers (on the basis of *LTUR* 1: 272–7) the capacity of 350,000 as 'quite credible for the Trajanic building'; 385,000 and 485,000 *loca* figure in the Regionary Cata-logues of the Chronographer of 354 AD; VERGNIEUX 2008: 240, on the basis of the computerized model of the Institut Ausonius (CNRS-University of Bordeaux Montaigne), proposes a low estima-tion of c. 80,000.

[34]   For spectacles displayed explicitly in one of the various circuses of Rome, mostly in the Circus Maximus, see for example texts 1.4, 7, 11 (=15), 13, 17, 22–3, 25, 29 (=10), 37, 39, 44–8, 50, 52–3, 57, 61–2, 66, 69, 75–6, 86, 89, 92–4 and 2.13.

[35]   On mythological enactments or 'fatal charades': COLEMAN 1990 and 2006: esp. lxxiv (Icarus), 62–5 (Pasiphae), and 174–5 (Orpheus); on parades of *delatores* COLEMAN 1990: 72.

[36]   See VILLE 1981:129 on the difficulty of counting separately gladiatorial fights and *venationes*; but gladiatoral games could still be given without a *venatio*, or a *venatio* without gladiators; on the tri-partite programme, see p. 392–5.

particularly in order to assure the provision of huge quantities of water, all for temporary or even one-off performances on an extraordinary scale.[37] For example, the lake excavated for Julius Caesar's naval displays in the Campus Martius was filled in after his death for health reasons, since the stagnation of water was later seen as a suspected cause of plague.[38] The *stagnum* of Augustus was erected in Trastevere for a *naumachia*, in which 30 ships and 3,000 men were involved in battles inside a massive pool of 536 m × 357 m with a depth of 1.7 m, according to K. COLEMAN's estimations; her calculations also suggest that it took 17 days to top it up from the *Aqua Alsietina* (i. e. 676,2 m^3 per hour). We also know of a sea battle staged by Claudius in the natural environment of the Lake Fucine, which was 'like a theatre' (*in modum theatri*), according to Tacitus, providing ideal viewing conditions for the numerous spectators who came from the local area, but also from Rome to see a spectacle involving 19,000 men and 100 ships. Such a show, not possible without impressive technological expertise, took place in the context of the huge project of draining Lake Fucine, implying digging through a mountain, which was carried out in order to create land for agricultural purposes. This massive task took 11 years and 30,000 men, for a cost of 350 million sesterces. It was recorded as a remarkable achievement by Pliny, who also criticized the absence of maintenance by Nero.[39]

3. Infrastructure for spectacles in the city of Rome and its environs

In the city of Rome and its environs, entertainment buildings were numerous, ubiquitous, and imposing.[40] In this survey, before turning our attention to the provincial world, a rapid enumeration will be enough to give an idea of the amount of resources required for the construction of the entertainment buildings in Rome and of their maintenance throughout the centuries.[41] Leaving aside some of the temporary structures and wooden buildings, which

---

[37] On the circus of Dougga MAURIN 2008: 101–4 fig. 15-7.

[38] Caesar's sea-battle: texts 2.14–5 with Cass. Dio 45.17.7–8 (plague).

[39] Augustus' *stagnum*: texts 1.80 and 2.20–1 and Aur. Vict., *Caes.* 28.1; Claudius' sea-battles: texts 2.25–7 and Plin., *nat. hist.* 124; on the venues of the different imperial *naumachiae* and on the question of the problems related to the provision of water see COLEMAN 1993 (with fig. 2 p. 54, Augustus' *stagnum*) and TAYLOR 1997; see also HAMMER 2010 (on spectacles and technology), LEVEAU 1993 (on Claudius' project and the environment), and BRANDON ET ALII 2014 (for the use of concrete in construction sites involving water, i. e. harbours etc.).

[40] For an insightful survey of the topography of entertainment in Rome see COLEMAN 2000 (with bibliography), on which this summary is mainly based. Among various studies published since, see also on the Campus Martius as 'theatric district' and on the odeon and stadium of Domitian: CALDELLI 2012 and 2014; on the Colosseum and earlier amphitheatres in Rome: COARELLI 2001a; HOPKINS and BEARD 2005; WELCH 2007; GOLVIN 2007; on the *Amphitheatrum Castrense:* Barbera 2001; on the Circus Maximus: CIANCIO ROSSETTO 2008; BUONFIGLIO 2008; GOLVIN 2008a; MARCATTILI 2009; on the Circus of Maxentius: PISANI SARTORIO 2008; on Domitian's stadium and odeon in Piazza Navona: CALDELLI 2014; other specific studies are cited below.

[41] On the resources imported to Rome from the whole world, note for example the mention by Pliny of the largest tree ever seen in Rome, exhibited under Tiberius and used for Nero's amphitheatre: Plin., *nat. hist.* 16.200.

have not survived, we will focus mainly on a selection of the major building projects (mostly from the 1st century BC on), among which we can count:

- 6 circuses: the *Circus Maximus*; the *Circus Flaminius*, which is more an open multi-functional space than a monumental circus; the circus built by Caligula and transformed by Nero, known as *Circus Vaticanus*; the *Circus Varianus*, built by Caracalla and used by Elagabalus for his own chariot-racing performances; the *Circus of Maxentius*, along the via Appia (with a capacity of 10,000); the *Arval Circus* in the grove of the *Fratres Arvales*, along the via Campana, c. 7.5 km outside Rome;
- 4 theatres: the already mentioned theatre of Scaurus (58 BC), the theatres of Pompey (55 BC), of Balbus (13 BC) and of Marcellus (dedicated 13 BC, or perhaps 11 BC, but already in use in 17 BC for the *Ludi Saeculares*), with capacities estimated of 11,000, 7,700 and 13,000 or 14,000;
- 5 amphitheatres: three in the Campus Martius, one built in stone by Statilius Taurus (29 BC), burned down in 64 AD in Nero's fire, and two in wood, the first, whose construction was begun by Caligula and already abandoned by Claudius, and then Nero's wooden amphitheatre, completed within a single year (57 AD), which was also damaged or destroyed in the fire of 64 AD.[42] Two other buildings were the *Amphitheatrum Castrense* built by Elagabalus in the area between Porta Maggiore and the Laterano, near the *Circus Varianus* and the *horti Variani*, and the famous Colosseum, whose construction on the site occupied by Nero's *domus aurea* was planned by Vespasian and financed by the booty brought to Rome in 70 AD at the end of the campaign in Judaea. This enormous building with a capacity of 50,000 was dedicated ten years later by Titus with lavish spectacles, also financed by the gold and the money taken from Judaea, as the spectators on their way to the arena and passing through the Arch of Titus and looking at the iconographical representations were reminded.
- 4 artificial pools used for the *naumachiae* given by Caesar, Augustus, Domitian, and Trajan (109 AD).[43]
- 2 *stadia,* one in wood, erected by Augustus in the Campus Martius in 28 BC, and the stadium built by Domitian for hosting the athletic contests of the Agôn Capitolinus he created, located in what is today Piazza Navona.[44]
- 1 *odeum* also built by Domitian and also located on the Campus Martius, with a capacity estimated of 7,000, which was probably partially roofed and offered the acoustics required by the *recitationes* pronounced there.[45]

In addition to the resources required for all these constructions, those used for support buildings must also be considered. We hear of four gladiator barracks or training schools

---

[42] On this building and the amphitheatre of *Leptis Magna*: GOLVIN 2011.

[43] Suet., *Domit.* 5: 'He [Domitian] restored many splendid buildings which had been destroyed by fire … Furthermore, he built … a stadium, an odeon, and a pool for sea-fights. From the stone used in this last the Circus Maximus was afterwards rebuilt, when both sides of it had been destroyed by fire.' On these *naumachiae*, presumably all different, and on their localisation: see above with COLEMAN 1993 and 2000: 240–1.

[44] Augustus' wooden *stadium*: Cass. Dio 53.1.5 with HASELBERGER 2002: 234 and CALDELLI 2014: 43; Domitian's *stadium* and odeon: CALDELLI 2014, who does not support COLINI's hypothesis of a *stadium* built by Nero on the site where Domitian's *stadium* would later be built (p. 44).

[45] CALDELLI 2014.

in Rome, the biggest, the *Ludus Magnus*, connected to the *Colosseum*, in which up to 1,000 gladiators could be accommodated, three or four to a cell. It had its own central training-arena of c. 2,000 m², and three smaller barracks known as *Ludus Matutinus, Ludus Dacicus* and *Ludus Gallicus*.[46] The *Summum Choragium* must also be mentioned, in which were stored the scenery and equipment for the games held in the Colosseum, as well as the different imperial *vivaria* in which were kept the animals between their arrival in Rome and their transport to the amphitheatre or to the circus, where *venationes* continued to be held, even after the construction of the Colosseum. In spite of the scarcity of conclusive archaeological evidence, one of these *vivaria* was presumably located at *Laurentum* or *Vicus Laurentium Augustanus,* not far away from the harbour of Ostia, where the ships would disembark the animals, some 20–25 km south of Rome. The epigraphic evidence and mosaics with representations of beasts (e. g. a boar, a leopard with a *venator*) found in *Laurentum* suggest the presence of various species of animals, some of them kept together in different enclosures, according to the care they required (for example camels with elephants) under the supervision of an imperial procurator or of military officers. One of these enclosures in which were kept the lions and other animals is thought to have been located beside the Porta Praenestina, between the *Via Labicana* and the Aurelian wall, c. 2 km from the Colosseum, whither the animals could be transported along the *Via Praenestina* the night before the games, in order to avoid encountering too many people on their way. Some other enclosures were possibly scattered in several places around the city of Rome, as well as in different *vivaria,* elsewhere in Italy.[47]

Among the support-buildings related to other types of spectacles, some must have been very impressive, such as the headquarters or *stabula* for each of the four *factiones* in charge or supplying the horses, charioteers and other circus performers, as well as all the personnel involved in training and taking care of the horses and in the organization of the races. In addition to storage space for all the necessary materials, these buildings must have included the stables for the horses, as well as the *Trigarium* reserved for equestrian training (and in somes cases also used for chariot races), even if this was more an open space rather than a proper building. Archaeological evidence, though rather limited, suggests that these infrastructures (or at least part of them) may have been in the north-western part of the *Campus Martius*.[48]

As well as all the resources used for the construction of these different buildings, we must also take into account all the additional resources needed for the props and scenery and all the machinery required: lifts for raising the animals in the Colosseum; awnings (e. g. in the theatre of Pompey or in the Colosseum) requiring tons of rope and cordage (and in some cases pigments to taint them); all sorts of other machines, used occasionaly, for example to diffuse perfumes (*sparsiones* of saffron-scented water) or tokens (*missilia*) among the

---

46　Coleman 2005; see also Coleman 2000: 235–6 and Coarelli 2001b; *EAOR* 6.35 p. 539 with *CIL* 6.41388 (possible connection between the *Ludus* Magnus and the Circus Maximus at the end of the fifth century AD when the spectacles of the circus gradually replaced those of the amphitheatre).

47　Epplet 2003: 78–84 and 2014a: 511–4 (with bibliography); Guasti 2007: 148–9; Scobie 1988: 202–3; camels and elephants in *Laurentum*: *CIL* 6.8583=*ILS* 1578 (*procurator Laurento ad elephantos*), *AE* 1955, 181 (*praepositus camellorum*); *vivarium* and imperial horses under military supervision: *AE* 1973, 39; Procop., *de bel.* 5.22–3, mentions a *vivarium,* where were kept 'lions and other wild animals' near Porta Praenestina.

48　Humphrey 1986: 558–60 with fig. 268; Coarelli 1997; Haselberger 2002: 234 (*stabula factionum)* and Pentiricci 2009; see also Pisani Sartorio 2008: 73.

spectators, to imitate the rain, or to cover the terrible odours of blood, transpiration, urine, faeces and corpses; hydraulic systems by which the arena could be successively flooded and drained for nautical displays; quantities of oil and lamps for lightening the spectacles given by night.[49] In terms of impact, human involvement must be tracked in a similar all-embracing manner: food provision and sanitation requirements for, at least on some occasions, vast numbers of spectators as well as all those involved from start to finish of the whole process, including builders and engineers, hunters and trainers, arena personnel, performers of all kinds, charioteers and their teams, gladiators, condemned prisoners and so on.[50] What kinds of smells, what noise levels, what blockages to the rhythms of normal daily life affected those who were not involved in the spectacles?[51] What was the impact of a full day of mayhem and slaughter attended by many thousands of people in the Colosseum or in the Circus on the drainage systems of Rome and on the Tiber, which was used as an evacuation system?[52] Note that the relative scarcity of bones found near the amphitheatres and entertainment buildings, in Rome and elsewhere, should be used with caution when it comes to thinking about how many animals were actually slaughtered. On the one hand, it was necessary to take away quickly the corpses and carcasses, so that the show could go on, and on the other hand, the meat was transported to be consumed by the population and by the livestock in the *vivaria*, and the bones were either discarded outside the limits of the cities or reused for the fabrication of objects and artefacts. There was in fact a kind of recycling process in place.[53]

---

[49] Lifts: BEST 2001; HUFSCHMID 2009: 488 fig. 268 (Pozzuoli) and GOLVIN 2015; awnings: MADELEINE 2014 and 2015; *sparsiones*: texts 1.78 and 86 (saffron) and FLEURY 2008; *missilia*: texts 1.64 and 94, and BRIAND-PONSARD 2007; see also ROBERT 1940: 176 (saffron and roses); in general on machine-technology: HAMMER 2010; fires in shows: texts 1.76 and 94; 2.32; floods in shows: texts 1.95 and 96; on importation of saffron and plants: MARZANO 2014: 242–4.

[50] On the staff of a big amphitheatre with lifts for animals (over 100 according to estimations): GOLVIN 2015: 41, and on jobs related to the circus (staff and performers): NELIS-CLÉMENT 2002; THUILLIER 2012 and 2015.

[51] See SCOBIE 1986 (sanitation) and 1988. On the substructures and hydraulic system (drainage, drawing of water, latrines etc.) of the Circus Maximus and of the Colosseum: BUONFIGLIO 2008 and LOMBARDI 2001; despite archaeological evidence of latrines in the circus, see Tert., *Spect.* 21: 'So it comes about that a man who will scarcely lift his tunic in public for the necessities of nature'; compare with VEASH 2000, who notes that in c. 50 years, the Maracanã stadium was 'being eaten away by corrosion by football fans relieving themselves on the terraces. Not wanting to miss even a second of a match by using the lavatories, the Brazilians have long held to a tradition of urinating in the stadium's access ramps. The problem came to light after officials noticed that the entrances' concrete and steel structures were eroding badly'. On the soundscape of the circus: NELIS-CLÉMENT 2008; on the behaviour of the spectators: FAGAN 2011 (mostly in the amphitheatre) and FORICHON forthcoming (in the circus).

[52] KYLE 1998: 184–212; BODEL 2000; PANCIERA 2000; DUPRÉ RAVENTÓS and REMOLÀ 2000; POTTER 2001; DAVIES 2012; TUCK 2013; BALLET ET ALII 2003; see also GOWERS 1995. On the geoarchaeology of the Tiber and Portus: GOIRAN ET ALII 2011; SALOMON 2012; see also DELILE 2014 (Rome and Ephesos). It would be interesting to know more about possible detection of pollution in connection with games or aquatic displays, e. g. exotic parasitoses possibly brought with the animals; in general on germs and diseases: SCHEIDEL 2003–13; see also above n. 38 (plague).

[53] On the disposal of corpses from the arena and on the consumption of animals killed in games: KYLE 1994: 155–83 and 184–202 with KYLE 2014. On bones of dromedary (or camel?), horse, bull, and

To these various human activities must be added those required for the maintenance of the buildings themselves. Much textual and archaeological evidence shows that they were in constant need of repair, whether simply because of normal wear and tear or as a result of natural catastrophes, such as earthquakes. Problems could also be caused by human negligence or by personal ambition or pride, provoking transformation of buildings to adapt them to the specific needs of spectacles as sponsors grew ever ambitious and tried to offer enticing novelties.[54] All these activities of maintenance required the production and transport of materials, even if evidence shows that some were reused or recycled from old structures that had been damaged or destroyed.[55]

To give an idea of what this maintenance might mean through the centuries, we can take the example of the Circus Maximus. After several interventions made during the Republican period, the Circus was much transformed and adorned by first Julius Caesar and then Augustus, with Agrippa's help: construction of the lap-counters with eggs and dolphins, of new seating and the *pulvinar*, of the *euripus* with its towers, statues and pools, and in the middle of which was erected the huge obelisk of some hundreds of tons of granite, transported from Heliopolis to Rome, as a symbol of the conquest of Egypt by Augustus. Next, Tiberius spent 100 million sesterces for the victims of the damage caused by a fire 'in the area of the Circus and in the Aventine', but it does not appear that the Circus itself was much affected. Caligula's interventions were limited to colouring the whole surface of the circus in red and in green (the colours of two of the factions) for some special races, using presumably enormous quantities of *minium* (red lead) and *chrysocolla* (malachite or carbonate of copper). As for Claudius, we hear that he covered the *carceres* with gold. Damaged and partly destroyed in the fire of 64 AD, the Circus was restored during the reigns of Nero, Vespasian, Titus, and Domitian, which means during a period running from 64 AD to 96. Trajan undertook large structural work on the *cavea* and probably also on the *pulvinar*, increasing seating by 5,000 places, making of the Circus 'a setting worthy of the people victorious over nations', to cite Pliny the Younger's *Panegyric* (fig. 1).[56] But the circus was again seriously damaged under Antonius Pius (*circi ruina*),

---

bear found in the excavations of the Colosseum (or nearby?), and bones of lions, panthers and bears, from a less clear context: Rea 2001a: 239–41; see also MacKinnon 2006: esp. 154–5, who states that 'Available animal bone data (…) attest to much less grandiose versions of Roman *venationes* than those presented through the media of ancient texts or art.' (p. 155). But this is not a necessary conclusion since carcasses were disposed of, as he is well aware, and since bones could be reused to make objects and artefacts: see for example Deschler-Erb and Gostenčnik 2008; Landes 2008. On bones of animals found in proximity to entertainment buildings in the provinces, see n. 107; on bones discarded outside the limits of cities, as in Ostia: MacKinnon 2014.

[54] Fires: text 1.71 and Tac., *ann.* 15.38; floods: Cass. Dio 58.26.5 and Aldrete 2007; earthquakes (circus of Antioch): Humphrey 1986: 456; human negligence: Plin., *nat. hist.* 36.117–20 (on Curio's amphitheatre, more dangerous than the gladiatorial shows); fights and riots: Tac., *ann.* 1.77; 14.17.

[55] On the reuse of building material by Domitian, see Suet., *Domit.* 5 (cited above n. 43) and in general Bradley and Williams 1998.

[56] In general on the work done by all the emperors mentioned, Humphrey 1986: 73–84, 117; Golvin 2008a: 81–4; see also Cass. Dio 58.26.5 (a fire during Tiberius' reign); on the colours used on the surface of the circus by Caligula: text 1.64; in general on the use of *minium* and carbonate of copper: Plin., *nat. hist.* 33.4–5, with his remark: 'man has learnt to challenge nature'; since the *minium* is highly toxic when inhaled, the effect on the charioteers and those placed near the track must have been quite strong. Plin., *pan.* 51; Plin., *nat. hist.* 36.70–5 (on the difficulties of transporting obelisks to Puteoli and Ostia and then on another ship to Rome).

Fig. 1: Sestertius of Trajan, 103–111 AD. Obv.: Head of Trajan, laureate, right. Rev.: View of the Circus Maximus from the Palatine showing a colonnaded front, with arch of Titus surmounted by a *quadriga* on left and the *carceres* on the right; tetrastyle temple of Sol in the back colonnades; in the interior set on the euripus, a large obelisk and the *metae*. BMC 1872,0709.529. (Photo courtesy of the Trustees of the British Museum).

according to the *Historia Augusta*, a disaster that caused numerous casualties (numbered at 1,112).[57] Septimius Severus made only some minor adaptations in order to set up the lavish seven-day display he gave in 202 AD, during which a total of 700 animals were slain.[58] As for Caracalla, he increased and expanded the gates and doors, probably enlarging the openings of the *carceres*. We then do not hear of much work done until the erection of a second obelisk in 357 AD by Constantius II.[59] However, it is worth mentioning the magnificent shows given in the Circus by Gordian III and by Probus, since they will have required a huge amount of human and natural resources: considerable numbers of trees with their foliage were apparently brought in to transform the entire surface into a forest and to create a 'staged nature' in which was held a *venatio* with thousands of beasts, killed by hundreds of prisoners brought to Rome for Probus' triumph. Note that, despite the enormous efforts made to present a spectacle challenging the limits between reality and artifice, its reception did not match its expectations, at least according to the author of the *Historia Augusta*: the beasts did not show much interest in leaving their cages, perhaps because of poor health, like most of Symmachus' crocodiles, who did not want to eat and died, except the two in the games given in 401 AD.[60]

---

[57] HA, *Ant. Pius* 9.1, with CHASTAGNOL 1994: 101 n. 10 (citing MOMMSEN, *Chron. Min.*, I, 146 for the number of victims).

[58] Text 1.89; despite the use of ἐν τῷ θεάτρῳ (instead of ἐν τῷ ἱπποδρόμῳ as in 76.2), this show was certainly given in the Circus Maximus, whose euripus was transformed into an immense boat which fell apart, as the re-enactment of a real *naufragium* or shipwreck; on this show: BAJARD 2008; in general on aquatic displays: COLEMAN 1993, TAYLOR 1997, and BELAN-BAJARD 2006 with the review of COLEMAN 2008b.

[59] Caracalla: HA, *Alex.* 24.3; see HUMPHREY 1986: 99 and 117; see also ROYO 2008; on the transport and erection of Constantius II's obelisk, which required much wood and human resources: Amm. 17.4.12–5 and IVERSEN 1968.1: 55–64.

[60] Text 1.94; see also text 1.90; crocodiles: Symm., *Ep.* 6.43 (and 9.41; 9.151) with JONES 2012: 305–6; on natural landscape used as decor in entertainment buildings: CARTER 2015; for archaeological

The evolutions of the Circus Maximus thus give an idea of the scale of certain aspects of the spectacle culture in Rome and of its history.[61] Enriched over time by contacts with different traditions, the development of Roman spectacular culture was also strongly influenced by its associations with the celebrations of the imperial cult.[62] From the Augustan age on, the number of spectacles increased considerably over the years by the successive addition of new celebrations for the emperor and members of the imperial family (e.g. birthdays and other anniversaries, as well as commemorations of imperial victories, consecrations of new temples, and so on): from the 77 days inscribed in the calendar at the beginning of the first century AD, we pass to the 177 days of celebrations inscribed in the calendar of Filocalus of 354 AD.[63] Games were still regularly given in the Circus at this period; in 400 AD, the spectacle given for Stilicho's third consulship seems to have been very impressive according to Claudian's poetic description of what figures as one of the last pieces of evidence concerning mammoth games in this building. It was almost a kind of grand finale, in which appeared creatures from all over a world pacified by Stilicho, whose 'laws have given peace to the cities so let his shows give peace to the mountains'; even the lions, coming now from as far away as the Atlas Mountains and Ethiopia, are happy to join the immense *pompa* parading towards Rome, blocking the roads, to celebrate Stilicho's triumph, as well as the elephants, whose tusks have been used for the ivory plaques engraved with Stilicho's name. Later evidence shows, nevertheless, that reparations were still made by the Vandal kings on several other entertainment buildings during the 6th century in Rome, and that chariot-races were held in Ravenna, capital of the Vandal kingdom, in the newly built circus, until the middle of the sixth century, whereas in Constantinople, they persisted through the Byzantine period.[64]

---

evidence of commercial production of trees in pots in Aegypt, see KENAWI ET ALII 2012.

[61] On special occasions, various types of spectacles and celebrations could be held simultaneously at different locations in Rome (on the streets, in a forum, in entertainment buildings), for several days, by day or by night, as stated in the protocol of the *Ludi Saeculares* given in 17 BC. On these *Acta* engraved in bronze and marble: SCHNEGG-KÖHLER 2002; BEARD, NORTH and PRICE 1998.1: 203 and 2: 139–44, 5.7b; on shows given by night; texts 1.60 and 83; 2.28; Tac., *ann.* 15.37.7 with COLEMAN 1993: 51, on Nero's aquatic display, with music and lights.

[62] Influences: COLEMAN 1996; cultural dialogue between Greece and Rome: BERGMANN 1999, EDMONSON 1999, FERRARY 2014²: 517–26; FERRARY 2011: 19 (*agones*); on the *Capitolia* founded by Domitian: CALDELLI 1993, 1997 and 2014; see also DUNBABIN 2015; on Hadrian's letters in relation to the organization of the circuits of *agones*: COLEMAN and NELIS-CLÉMENT 2012: xi–xiv (with bibliography).

[63] By the mid-first century BC, they were seven festivals or *Ludi* spread out during the year, besides other festivities: the *Ludi Megalenses* (4–9 April), the *Ludi Cereales* (12–18 April), the *Ludi Florales* (28 April–2 May), the *Ludi Apollinares* (6–12 July), the *Ludi Romani* (4–12 September), the *Ludi Victoriae* (26–31 October), and the *Ludi Plebei* (4–12 November). On the 177 days of celebrations: SALZMAN 1990; 119–49; see also BEARD, NORTH and PRICE 1998, 1: 262–3 and 2: 67–9; the majority of these days were marked as days with *ludi, c(ircenses) m(issus)*, or *munus,* among which 10 days celebrated with a *munus* and 63 with chariot-races (58 days with 24 races, 2 days with 12, 1 with 30, 1 with 36 et 1 with 48 races). The number of races leads on to consideration of the numerous charioteers, horses, and chariots involved for each of the 63 days indicating *circenses*; spectacles for imperial birthdays or members of the imperial family: texts 1.43, 47, 50, 60, 61, 63, 85; Cass. Dio 60.5 and Cass. Dio 78.19.3–4 (a gladiatorial contest given by Caracalla in *Nicomedia* in 215 AD for his birthday).

[64] Text 1.96; on ivory used for diptychs commemorating the consuls and their games: DELBRÜCK 1929; Procop., *b. goth.* 3.37.4 (549 AD); according to JONES 2012: 316, the 'coup de grâce in the

4. Entertainment buildings and spectacles in Italy and in the provincial world

In the framework of this investigation, it is of course not possible to go into great detail about all aspects of the infrastructure associated with spectacles across the Empire. Nevertheless, it is worth considering the amount of material and resources necessary for the construction of the numerous theatres and odeons, amphitheatres, stadia, and circuses, which existed in the entire Roman world, and for their long term maintenance. Considering that the Roman world included several thousand of cities for the period from Augustus until the 5th century, and that many of them had at least one entertainment building (and often more, as we have seen before at Arles), their number must have been extemely high overall.[65] Support-buildings must also be considered, as illustrated by the discovery in 2011 of the barrack or *ludus* of gladiators located near to the municipal amphitheatre of *Carnuntum* (Austria), provincial capital of Pannonia Superior and seat of a legionary fortress, where a military amphitheatre is also attested near the camp.[66]

The fact that the several hundreds of theatres and odeons, amphitheatres, stadia, and circuses we know today represent only a relatively small part of the buildings constructed, transformed or restored, means we must tread carefully in our analyses.[67] Several aspects come into play, such as types, sizes, locations, choice of materials, durability and maintenance, when we try to gauge the impact of all these buildings on the urban environment. Just to give a rough idea of numbers and distribution: according to recent surveys, over 230 amphitheatres are known today in the entire Roman world, among which c. 210 are located in Italy, Gaul and Germany, North Africa, Britain and in the Danubian provinces; in Italy, Roman theatres are counted by hundreds, whereas in the Western provinces, as regards the number of theatres and amphitheatres, Gaul is at the top with its 115 theatres and 50 amphitheatres.[68]

West' was given by Honorius; on games and maintenance of buildings in Late Antiquity (and under the Vandals) in the Roman world: JIMÉNEZ SÁNCHEZ 2008: 140; HUGONIOT 2008 and GHADDHAB 2008 (Africa); DUMASY 2008 (Gauls); FAUVINET-RANSON 2008 and 2006: 202, 221–3 (chariot-races in *Catane*, in the seventh-eighth century), and p. 386–94; the last *venationes* are attested in Rome in 523 AD and in Constantinople in 536 AD, BOMGARDNER 2013: 220. On the circuses of Ravenna and Constantinople, see below p. 244.

[65] Both FRIER 2000: 813 and DUNCAN-JONES 1974: 245–6 speak of 'several thousand cities'; according to INGLEBERT 2005: 67, there were c. 400 cities in Italy, 1,000 cities in the Eastern part of the Empire, 500 in Spain and 500 in Africa, 100 in Gaul and Germany, and c. 20 only in Britain. Even a *vicus* could have its theatre as suggests the dedication of a *proscaenium* by an *aedilis vici Petu[ar(iensis)]*: *RIB* 707 (140–4 AD), with MOURITSEN 2015: 245.

[66] NEUBAUER ET ALII 2014: esp. 181; according to estimations (cf. MEIJER 2007 cit. p. 188, *non vidi*), over 100 *ludi* must have been built in the Roman world; in Pompei, the *ludus* could hold c. 280 according to COLEMAN 2005: 6–7; on the presence of a *ludus* in *Alexandria*: ROBERT 1940: 242 and 125 n. 70 (*CIL* 10.1685=*ILS* 1397: *procur(ator) ludi famil(iae) glad(iatoriae) Caes(aris) Alexandriae ad Aegyptum*); *Carnuntum* and *Aquincum*: GOLVIN 1988: 135–8; on gladiators and the military life: LE ROUX 1990 [2011].

[67] DODGE 2014a (amphitheatres) and DODGE 2014b (circuses, *stadia*, and other venues). Italy: TOSI 2003. For specific studies on theatres: SEAR 2006 (with odeons), CIANCIO ROSSETTO and PISANI SARTORIO 1994; DI NAPOLI 2013; on amphitheatres: GOLVIN 1988; WELCH 2007; BOMGARDNER 2013.

[68] DUMASY 2011.

In the colony of Augst (*Colonia Augusta Rauricorum*), for example, near Basel in Switzerland, archaeological evidence has confirmed the presence of two successive amphitheatres built in c. 110 AD and in c. 170 AD, the construction of the second having begun before the abandonment of the first. Their cost is estimated at 400,000–500,000 sesterces, which gives an idea of the huge efforts deployed locally, mainly by local benefactors, for the infrastructure of the games, which were intended to be closely associated to the monumental complex dedicated to the celebrations of the imperial cult. Interestingly, the use of painting imitating polychromic and exotic marble, archaeologically attested on the podium of the building, represents a local alternative, cheaper than imported marble that could be seen, from a modern perspective, as more ecological.[69] In any attempt to explore systematically the potential environmental impact of spectacles, every one of the numerous excavated buildings would have to be considered both in relation to its local environment and ecosystem, but also in comparison with the thousands of other cases, in all their diversity, but each one forming part of an empire-wide network.

Of all the venues for Roman spectacles, the circuses were the largest but also the least numerous. Their length could vary from over 500 m, such as in El Djem (*Thysdrus*) and in Antioch, to 269 m for the circus of *Gerasa*, the smallest known. The majority of the circuses were in the middle category between 420 and 485 m in length, including those in Toledo, *Antinoupolis*, Bostra, Merida, Alexandria, *Caesarea Maritima*, Colchester, Trier, Vienne in Gaul, Aquileia, *Leptis Magna*, Carthage, Cherchel, Sétif, Tyr and Milan.[70] One of them, the circus of Colchester (c. 447 m), built in the 2nd century and discovered in 2004, is the only circus known in Britain. The building material was local and rather modest: its *cavea* was made of earth, a practice characteristic in Britain where it is observed for many of its entertainment buildings, and a gravelled area was set around its perimeter. Eventually, it appears that the outer wall of the cavea was dismantled and the materials removed and reused elsewhere.[71] Circuses are particularly well attested in Italy as well as in North Africa and in the Iberian Peninsula, two regions famous for their horse farms, from where many of the best horses were sent to Rome for the chariot races.[72] In North Africa, 16 circuses at least are known (perhaps even 19), and 8 are confirmed archaeologically, among which 7 are monumental circuses; they are located in Carthage, *Thysdrus*, *Hadrumetum,* and *Utica* (in Tunisia), *Leptis Magna* (in Libya) and *Caesarea* (in Algeria), as well as in *Sitifis* (also in Algeria). In the Iberian Peninsula, there are 19 circuses known, 12 archaeologically confirmed; the best preserved are in *Augusta Emerita* (Merida, in Lusitania) and in Tarragona, two provincial capitals and each also equipped with an amphitheatre. Interestingly, epigraphic evidence shows that the circus of Merida, greatly damaged by the effect of time (*vetustate conlapsus*) at the end of the reign of Constantine, was restored between 337 AD and 340. The mention of new columns, ornaments, and the provision of water (*aquis inundari disposuit*) gives an idea of the scale of the work and indicates that aquatic displays were still intended to be

---

[69] Hufschmid 2009: 171–95, 283–5, 543–8; the presence of animal displays or *venationes* is confirmed by analyses of the sandy arena; on importation of marble, see Plin., *nat. hist.* 36.8.50 and n. 30 (critics on Scaurus' theatre).

[70] Crummy 2008: 229.

[71] Crummy 2008.

[72] Pisani Sartorio 2008 (Italy); Nogales Bassarate 2008 (Spain), with map p. 163 fig. 2; Maurin 2008 (Africa), with map fig. 1 p. 91.

held in the circus in the middle of the fourth century.[73] At a later period, circuses were often integrated with an imperial palace, on the model of the complex Circus Maximus/imperial house on the Palatine, like the Circus of Maxentius in Rome for example, or Diocletian's circus at *Nicomedia*, which was dedicated in 303 AD or 304. Such complexes are attested in several imperial capitals of the fourth century, at Trier, *Sirmium*, Milan, Aquileia, Thessalonika, Antioch, Ravenna, and at Constantinople.[74]

Amphitheatres are rather well attested in Italy and in the Western provinces and we know of a reasonable number of cities equipped with both a circus and an amphitheatre. In North Africa for example, all the above mentioned cities with a monumental circus (with the exception of *Sitifis*) also had their amphitheatre. The most notable are in *Leptis Magna* and in *Thysdrus,* the former built in 56 AD, whereas the latter, which is the largest amphitheatre known in North Africa, was constructed only in the first half of the third century. Both amphitheatres, like the Colosseum and the amphitheatres of Pozzuoli and Augst, for example, were provided with a system of cages for moving animals and also personnel, capable of delivering them into the arena and of removing them.[75] In Britain, for example, where most of the amphitheatres known today survived only as earthworks (except the recently discovered amphitheatre of London, rebuilt as a stone construction in the second century), the ratio of amphitheatres to cities is very well balanced (c. 19 vs 20).[76]

A very different situation is observed in the Eastern part of the Empire, where estimations suggest a ratio of c. 20 to c. 1,000. The scarcity of amphitheatres in the Eastern cities contrasts with epigraphic evidence, which seems to indicate that gladiatorial contests, *bestiarii*, and *venationes* were held locally on a regular basis. This is particularly striking in Ephesos, for example, the provincial capital of Asia, where no amphitheatre is archaeologically attested, whereas gladiatorial contests must have been rather frequently held, as suggested by epigraphic evidence and confirmed archaeologically by the discovery in 1993 of around 67 corpses identified as those of young gladiators, buried together in a special cemetery. Interestingly, several inscriptions attest the presence of games involving beasts imported from Africa to Ephesos, one of them indicating that 25 ζῷα Λιβυκά or African beats were killed during 5 days of games given by an *asiarcheus*.[77] The same observations can be made for several cities like *Cibyra* (Gölhisar, South-East Turkey), for example, where the discovery of many reliefs and inscriptions (27 blocks of limestone) shows the variety of games held locally, such as gladiatorial contests, *venatores* fighting with various types of animals (bears,

---

[73] Chastagnol 1976; *AE* 1975, 472; on the entertainment buildings in Spain during the second and fourth centuries, see Brassous 2015, who shows that the amphitheatres were generally abandoned after the theatres, around the third/fourth century, whereas many circuses were still debout in the fourth/fifth centuries, sign of the interested manifested in that region for the *ludi circenses*.

[74] Humphrey 1986: 578–633; circus at Constantinople: Golvin 2008; chariot-races: Dagron 2000; 2011.

[75] Crummy 2008 (Colchester, Britain); see also Dodge 2008 (Eastern part of the Empire); Golvin 2011 (amphitheatre of *Leptis Magna*), and Golvin 2015 (amphitheatre of *Thysdrus*).

[76] Dodge 2014a: 554.

[77] Aldrete 2014: 447–8; Robert 1940: 313–4 and n. 198 (25 ζῷα Λιβυκά in Ephesos, with n. 199 and n. 200 (c. 180–250 AD) p. 195–6; among others, games involving local and foreign beasts (διὰ παντοίων ζῴων ἐντοπίων καὶ ξενικῶν καὶ μονομαχίας) are attested at *Beroia* (n. 15), and bears, panthers and lions at *Sagalassos* (n. 108). In Sicily: *Panormos: CIL* 10.7295=*ILS* 5055: *[omni] genere herbariarum et numerosas orientales (bestias),* as well as chariot-races (*bigae*).

bulls, dogs, ostriches), and perhaps also some type of mythological re-enactment or fatal charade, in which an animal attacks a condemned man in costume.[78]

The limited number of amphitheatres attested in the Eastern part of the Roman world finds its main explanation in the multifunctional character of many theatres, odeons or *stadia*, usually used for dramatic performances or athletics contest, but also for other types of shows like animal displays, *venationes*, or gladiatorial contests.[79] In Aphrodisias (modern Turkey), for example, one the extremities of the stadium was adapted and transformed into a sort of small amphitheatre, still perfectly visible today, where the gladiatorial contests and other types of games, well attested in the local epigraphic evidence, were held.[80] Among several of these multifunctional buildings, one may cite for example the Herodian amphitheatre/hippodrome of Jerusalem, the temporary wooden structure theatre/gymnasium of Jericho, used as stadium and also as hippodrome for chariot-races, despite the absence of the *carceres,* or the circus/stadium of *Caesarea Maritima* built by Herod.[81] It was in this building, attached to the palace complex and located near the sea, that he celebrated the end of the ten years spent on building the city of *Caesarea* in 10 AD. According to Josephus, the constructions were made 'not of ordinary material but of white stone', of huge blocks 'brought from outside at great expense', which were possibly blocks of *pulvis puteolanus* imported from Pozzuoli, as has been suggested recently. On this occasion, Herod gave in honour of Augustus some 'lavish shows that are to be seen at Rome and in various other places', including musical contests and athletic exercises, gladiators in great numbers, wild beasts and horse races, for which he received from Livia a contribution of 500 talents (i. e. 3 million drachmas or 9 million sesterces). He also arranged for games to be celebrated every fifth year.[82]

The development of hybrid types combining different entertainment buildings, for various reasons (cultural, practical, or economical) was not restricted to the Eastern part of the Empire. A comparable phenomenon is noticeable with the construction, in many cities in Gaul, Germany and Britain, of various types of multifunctional buildings designated as '*édifice mixte, semi-amphithéâtre, théâtre-amphithéâtre,* and *théâtre mixte*', or as 'Gallo-Roman'.[83] From a modern ecological perspective, these types of polyvalent buildings, particu-

---

[78] BERNS and EKINCI 2015: esp. 149 (animals), 169–73 with fig. 26–32 (representations of the three types of cages in which were transported the animals; elimination of dead animals taken away from an arena), and 173–4 C7 (fatal charade, rarely attested in the provinces); on the local theatre and odeon: SEAR 2006: 331–2.

[79] This extended use of many entertainment buildings explains some of the ambiguities as regards the terminology chosen to describe them.

[80] ROUECHÉ 1993; DODGE 2008: 141–3; COLEMAN 2008; KONTOKOSTA 2008.

[81] DODGE 2008: 137–41.

[82] Jos., *j. a.* 15.331–5 (stone imported) and 16.136–8 (games); on Herod's constructions: RODDAZ and GOLVIN 2014: esp. 130–3 and 144–57, suggesting the importation of material and of construction teams from Italy, esp. Pozzuoli, p. 124 and 154 (blocks of over 90 tons of *pulvis puteolanus*, wood from Cyprus); see also TAYLOR 2014: 150–6, 178, for a discussion of Herod's 'Italian inspiration' in his programme of construction and of the Circus Maximus as model. On Roman games, and on distribution of entertainment buildings in ancient Palestine and Arabia : WEISS 2014: 59 fig. 2.1; on stone trade: RUSSEL 2014; on Roman ships as potential source of lead pollution: LE ROUX ET ALII 2005; ROSEN and GALILI 2007: 303–6; see also BRESSON in this volume p. 196.

[83] See DODGE 2014a: 554–5, referring to GOLVIN 1988.1: 226–36 (French terminology) and SEAR 2006: 98–101; for the distribution of amphitheatres in the Roman World: HUFSCHMID 2009: map B. 48.

larly when there were built with local and cheap material, could be seen as a 'green' alternative to the multiplication of separate venues.

Imperial visits or stays, though occasional, must have had an impact on the environment and on the finances of the cities visited, since they had to ensure that the local entertainment buildings would be suitable for such circumstances and that animals would be supplied for the games. During Hadrian's visit to Athens, for example, 1,000 beasts seem to have been killed in the stadium, at least according to the *Historia Augusta*.[84] Interestingly, we hear of criticisms and accusations addressed approximately one century later by several Eastern cities towards Caracalla, after he cancelled his planned visit, in prevision of which they had been forced to finance the construction of circuses and amphitheatres, and to provide numerous animals for the games.[85] The fact that these new buildings were apparently quickly demolished invites the question why such a radical decision could have been taken, and leaves open other questions concerning the types of construction and their material form. Were they perhaps conceived originally as temporary structures, and then quickly demolished without even having been used? In any case, it is worth noting that the criticisms were not directed against the waste of natural resources (at least not specifically), but rather against the waste of money imposed by Caracalla himself, who was accused of ruining on purpose the finances of the cities in question.[86]

Obviously, the spectacles held in a middle-sized city in Italy or in the provincial world could not have competed with the magnificence of the shows given by emperors in the capital of the Empire, or those given by king Herod, for example, thanks to Livia's millions of sesterces. Local cultural traditions as well as varying levels of facilities for provision of animals could explain some of the differences observed between different regions and provinces. In Africa and in Spain, for example, provinces well known, along with Sicily and Cilicia, for the breeding of horses for spectacles in Rome, circus games were preferred to athletics.[87]

The sums spent on games are good indicators of their scale, but we must also take into account the types of games in question and their location. The financial resources of local élites, the status and the traditions of different cities varied widely, even within a region or province. In a Severan colony like *Auzia*, for example, in Mauretania Caesariensis, a donation assuring a small annual revenue of 270 denarii (i.e. 1,080 sesterces) was sufficient to

---

[84] Text 1.86; see also text 1.65 for Caligula's games given in Sicily and in Lyon.

[85] Cass. Dio 77.9.7–10: 'Moreover, we constructed amphitheatres and race-courses (θέατρα κυνηγετικὰ καὶ ἱπποδρόμους πανταχοῦ) wherever he [Caracalla] spent the winter or expected to spend it, all without receiving any contribution from him; and they were all promptly demolished, the sole reason for their being built in the first place being apparently, that we might become impoverished. The emperor himself kept spending the money upon the soldiers, as we have said, and upon wild beasts and horses; for he was for ever killing vast numbers of animals, both wild and domesticated, forcing us to furnish most of them, though he did buy a few.' On games given in *Nicomedia*, see n. 63. Compare these criticisms with Cass. Dio 59.10.6 (Caligula's destructions of buildings and disdain for Scaurus' theatre).

[86] Cass. Dio 77.6.2, 10.4, 19.3–4 (on Caracalla's performances and its relationship with games and performers). Like Nero, Caligula, or Commodus (see Cass. Dio 72.10.2–3, 72.19.1 and Toner 2014), Caracalla was perceived by the senators as one of the 'bad emperors', often involved in cruel and degrading games, killing men and animals in the arena, or socializing with despised performers.

[87] On the provision of racehorses: Rossiter 1992 (stables in Africa; on cross-breeding p. 47), Matter 2012 and Wilson 2014 (Sicily); see also Toynbee 1973: 177–83 and n. 103.

provide six chariot races twice a year for the birthday of a veteran, a member of the local council, and for that of his wife. This rather modest sum suggests that the races involved the participation of local charioteers, trainers and locally bread horses, invited to perform on special occasions in the local circus.[88] We do not know the cost of the spectacles given in Hr ech Schorr (near Siliana, centre of Tunisia) by a local magistrate, in which were presented games with a mixed programme including a boxing bout, some chariot races, and some dramatic performances, and also a distribution of gifts and a banquet for the entire population.[89] In the same province of Africa Proconsularis, in Hr Bou Cha (or Hr Fraxine), it cost 8,000 sesterces to provide circus games including a banquet and a distribution of baskets of food and money (*ludos circenses itemq(ue) epulum et sportulas*). In both cases it is worth noting the absence of beasts and of gladiators.[90] In the mid-3rd century AD, the same amount of money should have been spent by Magerius for the four leopards displayed in the games he sponsored as a *munerarius*, possibly in the small amphitheatre of *Leptiminus* (Lemta), c. 20–25 km south-southwest of Smirat, between between Sousse (*Hadrumetum*) and El Djem (*Thysdrus*), where a famous mosaic has been found, and where he may have lived (fig. 2). But Magerius seems to have agreed, as a mark of his generosity and also of his influence, to pay for them twice the original asking price, in cash, perhaps at an auction or in pre-arranged, public display. Instead of paying 500 denarii or 2,000 sesterces for each beast, he gave to the *Telegeni* four sacks of coins, each containing 1,000 denarii (as indicated with the symbol ∞ visible on each bag of money), which corresponds to a total sum of 16,000 sesterces.[91] This is made explicit by the text inscribed in the mosaic with the acclamations he received from the audience for his act of generosity. Interestingly, this text mentions the name of the *Telegenii*, one of the African *sodalitates* or trading societies which were involved in the import and export trade between North Africa, Rome, and the Mediterranean coast, selling all sorts of African products, among which lions, panthers and leopards (*Africanae*), captured in the province to be sold locally, but also in Rome, in Italy, and in other cities of the Roman world.[92]

---

[88] The presence of a circus in *Auzia* (still not localized) is attested by epigraphic evidence: *CIL* 8.9052 (*circenses edere*); *CIL* 8.9065=*ILS* 5661 (227 AD): *... perfectis metis et ovaris itemque tribunali iudicum ... (anno) prov. CLXXXVIII*; on the circus: Humphrey 1986: 329–31; Maurin 2008: 104.

[89] *CIL* 8.11998=*ILS* 5072: *sportulas decurionib(us) eisdemque et universo populo epulum et gymnasium dedit itemque spectaculum pugilorum et aurigarum et ludorum scaenicorum edidit.*

[90] *ILTun* 746: *[---] (sestertium) octo mil(ibus) n(ummum) posuit, ob cuius dedication(em) ludos circenses itemq(ue) epulum et sportulas condecurionibus suis dedit l(oco) d(ato) d(ecreto) d(ecurionum)*; the context could be related to the dedication of a race track, rather than a monumental circus; see Humphrey 1986: 321 and 665 n. 38; Maurin 2008: 104.

[91] This mosaic has given rise to various interpretations: see particularly Bomgardner 2009 suggesting that the setting of the game was the amphitheatre of *Leptiminus* (with a capacity of 9,100 spectators, p. 172); see also Dunbabin 1978: 67–9 Pl. 52–3; Beschaouch 2006; Carter 2004; Coleman 2006: 230–2, and recently Adams 2016: 354–66 (for a linguistic commentary). Note that in 301 AD, according to Diocletian's Edict of Maximum Prices, a leopard of top quality cost in Rome 400,000 sesterces and a top quality lion 600,000 sesterces, which is equivalent to six times the price of a bear of the same category: see Giacchero 1974: 211–3 n. 32; in comparison (Giacchero 1974: 208–9 n. 30) an *equus curulis* cost 400,000 sesterces, a *camelus optimus Bactrianus* 100,000, a *dromadarius optimus* 80,000, and a bull of the best category 20,000 sesterces.

[92] Much attention has been devoted to the question of the wild beast trade and providing and preparing animals for the games, for example Aymard 1951; Bertrandy 1987; Deniaux 2000; Epplet

Fig. 2: Smirat (Tunisia), 'Magerius' mosaic, showing a contest of *bestiarii* and four leopards, with a herald displaying four money-bags on a *lanx*. Probably mid-third century AD. Sousse Museum. (Photo by Hazel Dodge).

The types, sizes, and cost of games given in a provincial city also contrasted with what could be expected in a provincial capital like Carthage, where we hear of a spectacle of gladiators with a display of African beasts given for four days in the big amphitheatre (*spectaculum in amphi[theatro] gladiatorum et Africanaru[m] quadriduo*), for which was donated the huge sum of 200,000 sesterces (without the *ampliatio*).[93] This was perhaps not so common, and it seems to be one of the highest sums known to have been spent by a magistrate for games in a city in the Western provincial world. As a comparison, the sum represents the equivalent of five times the amount of money spent by a freedman in Praeneste for providing five days of

2001a; 2003 (*vivaria*); 2004; 2014a; GUASTI 2007; on the role of the army: VELKOV and ALEXAN-
DROV 1988; KNOEPFLER 1999; REA 2001b; EPPLET 2001b; HAMDOUNE 2012; FAURE 2013: 126–7;
PALME 2006; for very stimulating studies combining various types of data: BOMGARDNER 1992 and
MACKINNON 2006. On competition between African trading societies: *AE* 1977, 847 with BES-
CHAOUCH 2011: esp. 318; see also BOMGARDNER 1992 and 2009. For examples of African beasts
involved in provincial games, see in Ephesos (cited n. 77) and in Alife (Italy): *CIL* 9.2350=*ILS*
5059=*EAOR* 3.26 and *CIL* 9.2351=*EAOR* 3.27.

93 *ILS* 9406=*ILAfr* 390=*AE* 1910, 78: ... *qui ob honorem cum HS CC mil(ia) promisisset inla[tis] aera-
r(io) HS XXXVIII mil(ibus) leg(avit) am[pliata] pec(unia) spectaculum in amphi[theatro] gladiatorum et Africanaru[m] quadriduo dedit d(ecreto) d(ecurionum) p(ecunia) p(ublica)*, with CHAMBER-
LAND 2012: 283; for other examples of donations for funding games and additional acts of generosity in Africa, see *ILTun* 1066=*AE* 1977, 85 (Carthage) and *AE* 1999, 1781 (*Ammaedara*? mid-second
century AD).

ludi (40,000, i. e. 8,000 sesterces per day, like in the examples seen in Africa).[94] Gladiatorial combats and *venationes* with African beasts were expensive, even in Africa, and this could explain the huge difference in price between the spectacle held in Carthage under Hadrian and the *ludi circenses* held in Praeneste a century earlier. Strikingly, the sum of 200,000 sesterces paid in Carthage corresponds to a hundred times the personal contribution of 2,000 sesterces legally required, a century earlier, from each duumvir and aedile of the *colonia Genetiva Iulia* in Baetica in charge of organizing each year the spectacles for the gods. Note that, even with the additional sum of respectively 2,000 and 1,000 sesterces received from public funds, a magistrate wishing to offer lavish spectacles in Urso with many gladiators and beasts, like those of Carthage, would have needed to pay a lot more than the minimum required.[95] The obligation to provide games seems to have weighed on some magistrates and priests of the imperial cult who perceived it as a burden. Is this a sign of an audience more and more difficult to satisfy, expecting to see always more exotic animals and experienced gladiators? Is it possible to perceive behind these difficulties the effects of economic tensions, of prices increasing due to general inflation, or perhaps a connection with a shortage of gladiators and animals caused by an excessive demand of resources, in order to provide imperial games in Rome?

Facing difficulties in sponsoring gladiatorial games, some magistrates tried to escape their obligations, whereas others looked for alternatives solutions. Among them, the example of a candidate to a priesthood of Aphrodisias is particularly interesting for our perspective: instead of organizing expensive games, which he found difficult to afford, he was allowed to choose to contribute to the collective financing of the local aqueduct, a choice which required the emperor's approval, as we learn from the letter he received from Hadrian, engraved on a stone later re-used as a paving-stone.[96] Comparable alternatives received the support of several emperors, and legal measures were taken to support similar acts of munificence.[97] Although chosen primarily for its financial advantages, this alternative pre-

---

[94] *CIL* 14.3015=*ILS* 6256=*EAOR* 4.19: *L. Vruineio L. l. Philomuso, mag(istro) conl(egii) libert(inorum), publice sepulturae et statuae in foro locus datus est quod is testamento suo lavationem populo gratis per triennium gladiatorumque paria X et Fortunae Primig(eniae) coronam auream p(ondo) I dari, idemque ludos ex HS XL(milia) per dies V fieri iussit. Philippus l(ibertus) monumentum de suo fecit;* see Chamberland 2012: 284. The sum of 40,000 must be related to the *ludi* only, which means that the funding of a *munus* of ten pairs of gladiators is not included; see Chamberland 2012: 283–5; Duncan-Jones 1974: 245–6; see also *CIL* 11.6377=*EAOR* 2.9, an endowment whose interest should provide a *munus* with gladiators every 5 years in Pesaro (Umbria): *... decies centena / mil{l}ia num(-mum) dedit ita ut per sing(ulos) / annos ex sestertiorum CCCC(milium) / usuris populo epulum die / natali Titi Maximi filii(i) eius / divideretur et ex sestertiorum / DC(milium) usuris quinto quoque an/no munus gladiatorium ederetur / plebs urbana.*

[95] On the law of Urso, see the references n. 22: the *duumviri* were required to organize a gladiatorial show or some dramatic performances (*munus ludosve scaenicos*) for four days; the *praetor* had to provide three days of games and one day of spectacles given in the circus or in the forum (*... et unum diem in circo aut in foro*); see also the *lex Irnitana*: *AE* 1986, 333 lxxvii and lxxxi=*EAOR* 7.2 (but with no mention of cost); in general on games and municipal laws: Carter and Edmonson 2014; note also that a complete *munus* (*iustum atque legitimum*) implies gladiatorial games and wild beasts fights (*venationes*): Bomgardner 2009: 57.

[96] *SEG* 50.1095 with Coleman 2008.

[97] Kokkinia 2012: 117–24, for her discussion of 'buildings vs. *munera*'. It is worth wondering if it was perhaps in a comparable context that the aedile Sex. Iulius Ianuarius decided to finance 500 *loca* in

Fig. 3.1: Villa Romana del Casale (Piazza Armerina, Sicily), mosaic of the 'Great Hunt' Corridor. South Apse, figure of Ethiopia (?) with elephants tusks and animals. 4th century AD. (Photo courtesy of the Museo Regionale della Villa Romana del Casale di Piazza Armerina).

sented nevertheless what we would qualify today as positive effects for the environment, helping to reduce, even if in a very modest way, the waste of resources generally used for games. Considering the example of Aphrodisias, one may wonder to what extent the difficulties felt in 125 AD in Asia, or elsewhere, could be connected with the effects of the massive importation to Rome, only a few years before, of the thousands of animals and gladiators which were displayed in Trajan's grandiose games, discussed above. Rome was the centre of a connected world, as is particularly well emphasized in the hunt-mosaic of Piazza Armerina (fourth century AD), which represents all sorts of animals captured and taken from various regions of the empire, Africa, Egypt, Ethiopia or perhaps even from India, to be transported to Rome (fig. 3.1 and 3.2).[98] A similar impression of power appears also in

---

the circus of Lyon (*CIL* 13.1919), instead of offering games. Interestingly, the perennity of his act was assured when a corporation of *centonarii* (variously taken to be makers of patchwork, dealers in rags or makers of mats) decided to restore the same *loca*, in the third century, and mentioned his name on their inscription (*CIL* 13.1805). On these aspects of munificence, see also NG 2015: 111–9 (longevity; buildings vs. spectacles).

[98] See most recently WILSON 2004 (with bibliography and map fig. 3 p. 162), who identifies the female figure in the South apse not as Africa, Libya, Arabia, the Earth, Egypt, the Orient, Asia or India, as proposed before in various interpretations, but as Ethiopia (modern Ethiopia, Sudan, Erythrea and Somalia), an area which produced a vaste range of animals for Rome, including (p. 161–2) elephants, hippopotami, giraffes, Nubian ibexes, leopards and lions, as well as many spices, particularly frankincense and myrrh. See also his discussion related to the presence of the phœnix which figures on the mosaic, in which he sees a symbol of Ethiopia, rather than of Arabia or India (p. 163–4); CARANDINI ET ALII 1982. On animals, see also EPPLET 2001a: 233–344; LETZNER 2009: 104 (list

Fig. 3.2: Villa Romana del Casale (Piazza Armerina, Sicily), mosaic of the 'Great Hunt' Corridor, transportation of exotic animals for the games. 4th century AD. (Photo courtesy of the Museo Regionale della Villa Romana del Casale di Piazza Armerina).

Claudian's description of Stilicho's almost world-wide *pompa* which portrays animals coming from the edges of the empire towards Rome.[99] Since regions were very much interconnected and trading societies active all around the Mediterranean, it seems not impossible that at least in some periods, when the demand was extreme, as it surely was in the case of Trajan's great celebrations, the effects could be felt in faraway places.

In 177 AD, the emperors Marcus Aurelius and Lucius Commodus took legal measures by means of the *senatus consultum de pretiis gladiatorum minuendis*, in order to limit the expenses of games by regulating the prices of gladiators.[100] That decision must have had an impact on the choice of the displays. Besides, the use in games of condemned criminals, some of them specially trained to fight beasts, could offer a cheap alternative to gladiators, while still ensuring tax returns.[101] Among various legal measures taken to protect the imperial games given in Rome, the law passed by Honorius and Theodosius (414 AD) restricting the hunting of lions in Africa is particularly interesting from our perspective, since it may suggest that these animals, always associated with the élite, were becoming scarcer in this region, and therefore needed to be preserved as an imperial privilege.[102] Noteworthy also are the legal measures aimed at ensuring the provision of numerous race horses for imperial games held in Rome, where chariot races were still frequent in the mid-fourth century, as we have already seen, according to the calendar of Philocalus, as well as in the other imperial capitals equipped with a circus, and particularly in Constantinople.[103] On the other hand, evidence suggests that the number of exotic wild beasts, like lions or leopards, seems to decrease progressively, whereas trained and local animals, such as bears for example, seem to be appreciated by audiences. We learn from two inscribed dedications to Diana, dating to the second c. AD, that a centurion captured 50 bears in six months, while another centurion was involved in the construction of the enclosure of a *vivarium*. It is not known whether the animals were kept in it to be displayed in games mounted in Cologne, capital of the province of Germania Inferior, or if the *vivarium* was used as a temporary place where they were kept before being transported to Rome for imperial games.[104] The use and re-use of local and trained animals could be seen as the result of a shortage of African beasts and as a sign of the impact of games in some regions. In the animal display figured in a mosaic of Radès

---

of animals for the games and map whit their provenances); on the griffin: Manganaro 1959: 245; Settis Frugoni 1975; Witts 1994; Blásquez 1997; see also Cordovana p. 11-8 in this volume.

[99] Text 1.96; see also texts 1.88 and 89.

[100] *CIL* 2.6278=*ILS* 5163; Oliver and Palmer 1955; Carter 2003; 2014.

[101] Ebner 2012: esp. 267–8, for a general discussion (with a particular emphasis on legal aspects) of the different types of *damnati* (*ad ludum*, *ad gladium* and *ad bestias*) displayed in spectacles. All of them were not necessary killed, and some of them, e. g. the *damnati ad ludum,* were even trained as gladiators and as *venatores* to fight beasts (p. 258 and 265); I owe this reference to Professor Wacke. See also Teyssier 2009: 21; Gueye 2013: 173–6.

[102] *CTh.* 5.11.1; see also *CTh.* 6.4.19 (372 AD); for the price of lions in Rome in 301 AD, see above n. 91. For different legal measures concerning the beasts involved in games, see for example Plin., *nat. hist.* 8.64 (a ban on African beasts, 118 BC), Cass. Dio 54.2.4 (Augustus), Tac., *ann.* 13.31 (Nero, 57 AD).

[103] *CTh.* 15.7.6 (381 AD); Soler 2008: 60–1.

[104] *CIL* 13.12048=*ILS* 9241=*AE* 1910, 61=*EAOR* 5.48; *CIL* 13.8174=*ILS* 3265=*EAOR* 5.49, for other examples of *ursarii*: *CIL* 12.8639=*EAOR* 5.50 (third century AD, Birten near Xanten-*Vetera*), *CIL* 13.5703=*EAOR* 5.51 (second/third century AD, Langres), *CIL* 13.5243=*ILS* 3267=*RIS* 2 192=*EAOR* 5.52 (third century AD, Zürich).

Fig. 4: Radès (*Maxula*, Tunisia), mosaic representing a display of animals, among them trained bears. Probably mid-third century AD. Bardo National Museum, Tunis. (D-DAI ROM 63.350).

(*Maxula*), in Tunisia, the bulls could have been the most numerous among several species of animals, if we consider that the figure XVI inscribed on the body of the only bull which survived indicates their number, rather than a stable-mark (fig. 4).[105] The presence of names beside some of the bears figured, such as Braciatus, Gloriosus, Simplicius, Alecsandria, Fedra, seen climing up a pole, [---]itus (and perhaps also near a leopard, if the name Nilus is related to him?) could suggest that the audience was familiar with some of this troupe of animals exhibited locally. Note that this 'animal-catalogue' pavement, to use Dunbabin's term, shows no *bestiarius*, no blood, no sign of killing: the pleasure of the spectacle seems to have come from the number and diversity of animals exhibited in the show. The discovery of bear bones in late-antique levels in a site near Carthage seems to indicate that bears were not extinct from

---

[105] DUNBABIN 1978: 72–3 pl. XXIV fig. 58; BALL 1984: 131–2 fig. 5; VISMARA 2001: 206 fig. 5 and most recently ROSSITER 2016: 255–7 fig. 13 (published while this chapter was at proof stage); see also COLEMAN 2006: 87–90 on trained bears in Martial; on the price of bears, see above n. 91.

North Africa, despite what Pliny says.[106] They could have been local (no testimony indicates importation of bears to Africa), and possibly trained to take part in animal shows presented to the local population, like the bears of the mosaic.[107] Interestingly, a study by a specialist on zoology, who has analyzed the fauna figured on mosaics from Syria and Africa, though well aware of the existence of the circulation of models to be used by the artists, has noted that the depictions are generally quite accurate and mostly very exact in representing anatomy, often even with details concerning the sub-species and the sex of animals. It has also been noted that in many cases the artists seem to have represented the local fauna.[108] In one case a scholar claims to be able to identify the precise location, in the region of *Bulla Regia* and *Utica*, represented in a hunt mosaic.[109] The relative modesty of the events presented in some localities, even in a temporary structure, could be compensated by the variety put on show. In Egypt, for example, several programmes of shows preserved on papyrus reveal that in a city like *Oxyrhynchus* (or *Antinoupolis*), in the fifth or sixth century AD, mixed spectacles were presented with successive performances, such as chariot races, athletic, acrobats, dramatic, and singing performances, and in some case animal displays. Interestingly the only animals mentioned are gazelles and dogs (the presence of horses being implicit), which is striking in a province known for having provided so many animals for Roman games, and where the tradition of animal displays was very ancient, dating back to Ptolemaic times.[110]

Clearly, spectacle culture was changing, perhaps because of the results of excessive exploitation of natural resources, but probably also because of the transformation of Roman society more generally. The question which must be asked is the following: were the Romans aware of the demands they made on nature, and of the impact of the sustained exploitation of natural resources used in order to provide lavish spectacles? Awareness of the over-exploitation of natural resources by Rome appears in some sources, and the demands imposed on nature in relation to the provision of mass spectacles are clearly identified as one of the main sources of worrying trends and problems.[111] In the fourth century AD, Ammianus

---

[106] See text 1.17.

[107] MacKinnon 2006: esp. 152, noting: 'Presumably this bear was involved in a *venatio*, given the close proximity of the site to the Carthage amphitheatre'; I would be perhaps a little more cautious about the nature of the show and the cause of death. However, this type of investigation is extremely important and opens new perspectives. MacKinnon mentions also the discovery in the area of Carthage of bones of various species: elephant, deer, gazelle, ostrich, wild boar and wild sheep. Note also that bones of camels of the species *camelus dromedarius* were found in or near the site of an amphitheatre both in *Vindonissa* (Switzerland) and in Carthagene (Spain), two sites associated with a military presence; in both cases, it is quite possible that these animals were connected with games displayed locally, rather than used as pack animals; see Schmid 1952/53 and Sorge 2006: 250–1 map fig. 2; on the amphitheatre of *Vindonissa*: Frei-Stolba et alii 2011; see also Deschler-Erb and Akeret 2010; on that of Carthage: Bomgardner 2013, and now Rossiter 2016.

[108] Matthews 1988–1989.

[109] Hanoune 2009: 68.

[110] P. Oxy. 34.2707; P. Oxy. 79.5215–5218; P. Bingen 128; P. Harrauer 56, with Remijsen 2014; see also Decker 2008; Nelis-Clément 2002: 295–7. On hunts in Egypt for providing animals for games (not exclusively for the circus) by Roman soldiers: Palme 2006: 307–8. For the details relating to Ptolemy Philadelphus' procession: Rice 1983; Coleman 1996. On zoological magnificence in Ptolemaic Egypt: Jennison 2005²: 30–5.

[111] Animals were also of course hunted in great numbers simply because they were dangerous predators and could cause considerable damages to crops and agricultural production generally and

Marcellinus refers to the fact that hippopotami are moving towards the south of Egypt, as far south as the second cataract, driven out because of the continuous and massive process of hunting. The local population was clearly aware of the connection between the supply of animals for games in Rome and the effect on their local environment. Ammianus' mention of Scaurus' name as the aedile who initiated the hunts of hippopotami dates the beginning of this activity to 58 BC, about a quarter of a century before the conquest of Egypt by Octavian. According to Pliny, Scaurus displayed indeed hippopotami, with five crocodiles kept in a temporary basin, as well as 150 other beasts.[112] Hippopotami are also mentioned by Ammianus' contemporary Themistius, among other species of animals, such as elephants and lions, in relation to their disappearance respectively from Egypt, Africa, and Greece.[113]

As stated at the beginning of this chapter, games form a kind of mirror in which a society can, if it so desires, inspect itself. In its imperialist culture the image that Rome wants to give of itself through its mass spectacles is one that knows no limits: everything must be put to use in order to illustrate and emphasize Roman power and grandeur. As a result, its spectacles become an essential defining element of Roman culture, helping to display the extent of Rome's control over the known world. For all their violence and cruelty the games underpin the *pax romana* and a sense of world domination, order, and control. Part of what it is to be Roman involves attending the games, going to watch in the arena elements of the dominated world, enslaved enemies and convicted criminals, beasts hunted from all over the empire.[114] But there is an essential paradox inherent in the dynamics underpinning Rome's triumphalist urge to put on display in the arena its control of the whole world. As each successive event must be seen to surpass in extravagance its predecessors, so the demands put on the natural world become ever greater, until eventually they spiral almost out of control.

Giving ourselves the opportunity, however anachronistically, to think about the environmental impact of Roman spectacle culture allows us to appreciate how the games, that are such an important part of the expression of imperial domination, come over time to play a role in the over-exploitation of the world's natural resources. That we cannot measure in an accurate scientific manner the overall negative impact of this phenomenon should hardly come as a surprise, given that experts today cannot agree on how to measure the ecological footprint of a Soccer World Cup or the Olympic Games. Nor should we underestimate the frightful paradoxes at the heart of our own contemporary spectacle culture, given that we live in a world in which almost everyone agrees on the importance of protecting our planet, and yet in 2022 the World Cup of Soccer will take place in Qatar, where hundreds of lives have already been lost in the construction of ecologically advanced stadia.

---

more particularly in North Africa with its lions, leopards, and other wild beasts; on this question, see BOMGARDNER 1992; 2013: 216–7.

[112] Text 1.21; on Scaurus, see also texts 1.19 and 20. Ammianus' statement is concordant with Pliny's, but seems to contradict Dio's observation on the first appearance of hippopotami in Rome under Augustus: text 1.41.

[113] Themistius, *Or.,* 10.140: 'And so we spare for ourselves the fiercest animals … so that by procreation they may be preserved and remain, and we are angry at the elephants removed from Libya, the lions from Thessaly, the hippopotami from the marshes around the Nile'. On the interpretation of this text and in general on the question of animal extinction, EPPLET 2001a: esp. 228 and 225–33.

[114] See for example on animals in triumphs: ÖSTENBERG 2014; on spectacles of animals abuse: SHELTON 2014.

Appendix 1

Mass spectacles involving numerous animals.
Texts mentioned in Table 1

1. *Sen., brev.* 13.3: Curius Dentatus was the first who had <u>elephants</u> led in his triumph.

2. Florus, *ep.*, 1.13.27–28: if you looked upon the procession, you saw gold, purple statues, pictures and all the luxury of Tarentum. But upon nothing did the Roman people look with greater pleasure than upon those <u>huge beasts [elephants]</u>, which they had feared so much, with towers upon their backs, now following the horses which had vanquished them, with heads bowed low not wholly unconscious that they were prisoners.
On these elephants, also Florus 1.13.6: elephants, till then unknown in Italy, threatening Rome by land and sea, with men, horses and arms and the added terror of wild beasts.

3. *Plin., nat. hist.* 8.6.16: Italy saw <u>elephants</u> for the first time in the war with King Pyrrhus, and called them Lucan oxen because they were seen in Lucania, in the year 474 of Rome [280 BC]; but Rome first saw them at a date 5 years later, in a triumph [i. e. Curius Dentatus, 275 BC], and also a very large number that were captured from the Carthaginians in Sicily by the victory of the pontiff Lucius Metellus.

4. Plin., *nat. hist.* 8.4.16–17: There were <u>142 of them [elephants], or by some accounts 140</u>, and they had been brought over on rafts that Metellus constructed by laying decks on rows of casks lashed together. Verrius records that they fought in the Circus and were killed with javelins, because it was not known what use to make of them, as it had been decided not to keep them nor to present them to native kings; Lucius Piso says that they were merely led into the Circus, and in order to increase the contempt felt for them were driven all round it by attendants carrying spears with a button on the point. The authorities who do not think that they were killed do not explain what was done with them afterwards.

5. Plin., *nat. hist.*18.4.17: Marcus Varro states that at the date when Lucius Metellus led <u>a very large number of elephants</u> in his triumph.
On these African elephants taken from the Carthaginians after Metellus' victory in Palermo, Jennison 2005[2]: 44 and Toynbee 1973: 34–8 and 348–9.

6. *Livy 39.22.1–2: About the time that these reports were brought from Spain, the Taurian Games (*ludi Taurii*) were performed for 2 days for religious reasons. Then for 10 days, with great magnificence, Marcus Fulvius gave the games which he had vowed during the Aetolian war. Many actors too came from Greece to do him honour. Also a contest of athletes was then for the first time made a spectacle for the Romans (*tum primo Romanis spectaculo fuit*) and <u>a hunt of lions and panthers</u> was given (*venatio data leonum et pantherarum*), and the games, in number and variety, were celebrated in a manner almost like that of the present time.
According to Aymard 1951: 74–6, this passage does not mention the first *venatio* ever organized in Rome (*primo* is linked only to the *spectaculum* of athletes, and not to *venatio*); besides, despite the presence of the plural *leonum*, A. expresses some doubts about the number of the lions, but without more precise comment than (p. 74): 'Notons que les lions furent sans doute en petit nombre, peut-être même s'agissait-il d'un seul animal'. The *pantherae* 'probably came from the East, since Livy does not mention Africa', according to Toynbee 1973: 82.

7. Livy 44.18.8: Display being now on the increase, it is recorded that at the games in the circus games (*ludis circensibus*) by the curule aediles Publius Cornelius Scipio Nasica and Publius Lentulus, <u>63 leopards and 40 bears and elephants</u> participated. (trans. Loeb modified).

8. Plaut., *per.* 198–9: Fly running. That's what an ostrich does in the circus.

9. Plaut., *poen.* 1011–2: What does he say? Can't you hear? He states that he wants to present <u>African mice</u> (*mures Africanos*) for the parade at the games of the aediles.

10. *Plin., nat. hist.* 8.20.53: A fight with <u>several lions</u> at once was first (*leonum simul plurium … princeps*) bestowed on Rome by Quintus Scaevola, son of Publius, when consular aedile, but the first of all (*primus omnium*) who exhibited a combat of <u>100 maned lions</u> (*centum iubatorum*) was Lucius Sulla, later dictator, in his praetorship. After Sulla Pompey the Great showed in the Circus <u>600, including 315 with manes, and Caesar when dictator 400.</u> (On this show: VILLE 1981: 88).

11. *Plin., *nat. hist.* 8.7.19–20: Fenestella states that the first <u>elephant</u> fought in the circus at Rome in the curule aedileship of Claudius Pulcher and the consulship of Marcus Antonius and Aulus Postumius [99 BC] and also that <u>the first fight of an elephant against bulls was 20 years later</u> in the curule aedileship of the Luculli. (On Pulcher's *venatio*, see VILLE 1981: 89).

12. *Plin., *nat. hist.* 8.20.53: see text 10.

13. *Sen., *brev.* 13.6: Perhaps you will permit someone to be interested also in this – the fact that Lucius Sulla was the first to <u>exhibit loosed lions</u> in the Circus, though at other times they were exhibited in chains, and that javelin-throwers were sent by King Bocchus to despatch them?

14. *Plin., *nat. hist.* 8.2.4: At Rome they <u>[elephants] were first used in harness</u> to draw the chariot of Pompey the Great in his African triumph, as they are recorded to have been used before when Father Liber went in triumph after his conquest of India. Procilius states that at Pompey's triumph the team of elephants were unable to pass out through the gate.

The date of this African triumph is 79 BC according to ITGENSHORST 2005: CD: file 3 (p. 345) and 80–1 BC according to BEARD 2007: 14 and 16.

15. *Plin., *nat. hist.* 8.7.20: see text 11.

16. Suet., *Iul.* 10.1: When aedile, Caesar … exhibited <u>combats with wild beasts</u> (*venationes*) and stage-plays (*ludos*) too, both with his colleague and independently.

Note the absence of *venationes* in Plut., *Caes.* 6.5 and in Plin., *nat. hist.* 33.16 in relation to these games, as underlined by cf. VILLE 1981: 89–90.

17. Plin., *nat. hist.* 8.131: It is noted in the Annals that on 18 September in the consulship of Marcus Piso and Marcus Messala, Domitius Ahenobarbus as curule aedile provided in the circus <u>100 Numidian bears and the same number of Ethiopian huntsmen</u> (*venatores Aethiopas*). I am surprised at the description of the bears as Numidian, since it is known that the bear does not occur in Africa. (trans. Loeb modified). The mention of *venatores* suggests that a large number of these performers (bears and men) were killed. On the erroneous comment of Pliny in relation to the absence of bears in Africa, see JENNISON 2005[2]: 49–50.

18. App., *b. c.* 2.13: He [Caesar] gave <u>spectacles and combats of wild beasts</u> (κυνηγέσια θηρίων) beyond his means, borrowing money on all sides, and surpassing all former exhibitions in lavish display and splendid gifts, in consequence of which he was appointed governor of both Cisalpine and Transalpine Gaul for five years, with the command of four legions.

19. *Plin., *nat. hist.* 8.96: A <u>hippopotamus was exhibited</u> at Rome for the first time, together with <u>5 crocodiles</u>, by Marcus Scaurus at the games which he gave when aedile; a temporary channel was made to hold them (*temporario euripo ostendit*).

20. *Plin., *nat. hist.* 8.64: But <u>Scaurus in</u> his aedileship first sent <u>150 African beasts</u> [i. e. *Africanae*] (*varias CL universas misit*) in one flock, then <u>Pompey the Great 410</u>, and the late lamented <u>Augustus 420</u>. (trans. Loeb modified).

21. Amm. 15.24: This monstrous and once rare kind of beast [i. e. <u>hippopotami</u>] the Roman people first saw when Scaurus was aedile, … and for many ages after that more hippopotami were often brought to Rome. <u>But now they can nowhere be found</u>, since, as the inhabitants of those regions conjecture, <u>they were forced from weariness of the multitude that hunted them to take refuge in the land of the Blemmyae.</u>

22. Plin., *nat. hist.* 8.20–21: Also in Pompey's second consulship, at the dedication of the Temple of Venus Victrix, <u>20 [i. e. elephants], or, as some record, 17,</u> fought in the Circus, their opponents being Gaetulians armed with javelins, one of the animals putting up a marvellous <u>fight</u> – its feet being disabled by wounds it crawled against the hordes of the enemy on its knees, snatching their shields from them and throwing them into the air, and these as they fell delighted the spectators by the curves they described, as if they were being thrown by a skilled juggler and not by an infuriated wild animal. There was also a marvellous occurrence in the case of another, which was killed by a single blow, as the javelin striking it under the eye had reached the vital parts of the head. The whole band attempted to burst through the iron palisading by which they were enclosed and caused considerable trouble among the public.

23. Cass. Dio 39.38.1–3: During these same days Pompey dedicated the theatre in which we take pride even at the present time. In it he provided an entertainment consisting of music and gymnastic contests, and in the Circus a <u>horse-race and the slaughter of many wild beasts of all kinds</u>. Indeed, <u>500 lions were used up in 5 days</u>, and <u>18 elephants</u> fought against men in heavy armour. <u>Some of these beasts were killed at the time and others a little later.</u> For some of them, contrary to Pompey's wish, were pitied by the people when, after being wounded and ceasing to fight, they walked about with their trunks raised toward heaven, lamenting so bitterly as to give rise to the report that they did so not by mere chance, but were crying out against the oaths in which they had trusted when they crossed over from Lybia (ἐκ τῆς Λιβύης), and were calling upon Heaven to avenge them. (trans. Loeb slighthy modified). This *venatio* was mounted in the circus (κἀν τῷ ἱπποδρόμῳ), probably the Circus Maximus, and not 'nel teatro di Pompeo', as states COARELLI 1997: 358, n. 65. On Dio's terminology, see below text 35. According to Cassius Dio, the elephants (and probably all the others beasts) were from Africa, where Pompey, before coming back to Rome and giving staged hunts, had hunted for some days in the South of Numidia: Plut., *Pomp.* 12.

24. *Plin., *nat. hist.* 8.70: The games of Pompey the Great first displayed the <u>*chama*</u>, which the Gauls used to call the <u>lynx</u>, with the shape of a wolf and leopard's spots; the same show exhibited what they call <u>*cephos*</u> <u>from Ethiopia</u>, which have hind feet resembling the feet of a man and legs and fore feet like hands. Rome has not seen this animal subsequently. (trans. Loeb slightly modified). On the *cephoi*, see JENNISON 2005[2]: 55.

25. *Sen., *brev.* 13.6: … Pompey was <u>the first to exhibit the slaughter of 18 elephants</u> in the Circus, pitting criminals against them in a mimic battle.

26. Cic., *ad fam.* 7.1.3: There remain the <u>wild-beats hunts, 2 a day for 5 days</u> – magnificent; nobody says otherwise. But what pleasure can it possibly be to a man of culture, when either a puny human is torn to pieces by a powerful animal, or a splendid beasts is transfixed by a hunting spear? Anyhow, if these sights are worth seeing, you have seen them often; and I, who was a spectator, saw nothing new in it. The last day was for the <u>elephants</u>, and on that day, the mob and crowd were greatly impressed, but manifested no pleasure. There was even an impulse of compassion, a kind of feeling that the monsters had something human about them. (trans. Loeb modified). See also text 8 Table 2.

27. *Plin., *nat. hist.* 8.64: see text 20.

28. Plin., *nat. hist.* 8.71: At the same games [i. e. Pompey's] there was also <u>a rhinoceros with one horn on the nose</u> such as has often been seen. Another bred here to fight matches with an elephant gets ready for battle by filing its horns on rocks, and in the encounter goes specially for the belly, which it knows to be softer. It equals an elephant in length, but its legs are much shorter, and it is the colour of box-wood. The rhinoceros with one horn is an Indian animal, the one with two horns being African; see JENNISON 2005[2]: 34, 54–5, 74; on the rhinoceros in Flavian Games: COLEMAN 2006: lvi, 102–3, 101, 110; see also BUTTREY 2007:107–11.

29. Plin., *nat. hist.* 8.53: see text 10.

30. *Plin., *nat. hist.* 8.55: Mark Antony broke <u>lions to the yoke</u> and was the first person at Rome to harness them to a chariot, and this in fact during the civil war, after the decisive battle in the plains of Pharsalia, not without some intention of exhibiting the position of affairs, the portentous feat signifying that generous spirits can bow to a yoke.

31. Plin., *nat. hist.* 8.53: see text 10.

32. Suet., *Iul.* 37.2: As he [Caesar] rode through the Velabrum on the day of his Gallic triumph, the axle of his chariot broke, and he was all but thrown out; and he mounted the Capitol by torchlight, with <u>40 elephants bearing lamps</u> on his right and his left.

33. Suet., *Iul.* 39.3: <u>Combats with wild beasts</u> were presented on 5 successive days. (On Caesar's triumphal spectacles, see Table 2).

34. App., *b. c.* 2.102: There was a <u>combat of elephants, 20 against 20</u>.

35. *Cass. Dio 43.22.1–4, 43.23.1–2: Thence he [Caesar] proceeded homeward with practically the entire populace escorting him, while <u>many elephants carried torches</u> … He built a kind of hunting-theatre of wood, which was called an amphitheatre (θέατρόν τι κυνηγετικὸν ἰκριώσας, ὃ καὶ ἀμφιθέατρον) from the fact that it had seats all around without any stage. In honour of this and of his daughter he <u>exhibited</u>

combats of wild beasts and gladiators; but anyone who cared to record their number would find his task a burden without being able, in all probability, to present the truth; for all such matters are regularly exaggerated in a spirit of boastfulness. I shall accordingly pass over this and other like events that took place later, except, of course, where it may seem to me quite essential to mention some particular point, but I will give an account of the so-called camelopard [giraffe], because it was then introduced into Rome by Caesar for the first time and exhibited to all. This animal is like a camel in all respects except that its legs are not all of the same length, the hind legs being the shorter.

For a discussion on giraffes and on their African provenance: Toynbee 1973: 141–2. See also texts 1.37, 92 and 93.

36. Plin., *nat. hist.* 8.22: Elephants also fought for the dictator Caesar in his third consulship, 20 being matched against 500 foot soldiers, and on a second occasion an equal number carrying castles each with a garrison of 60 men, who fought a pitched battle against the same number of infantry as on the former occasion and an equal number of cavalry.

37. *Plin., *nat. hist.* 8.69: The Ethiopians give the name of *nabun* to one that has a neck like a horse, feet and legs like an ox, and a head like a camel, and is of a ruddy colour picked out with white spots, owing to which it is called a camelopard [i. e. giraffe]; it was first seen at Rome at the games in the Circus given by Caesar when dictator. From this it has subsequently been recognized to be more remarkable for appearance than for ferocity, and consequently it has also got the name of 'wild sheep'.

38. *Plin., *nat. hist.* 8.182: We have seen bulls, when fighting a duel under orders and on show for the purpose, being whirled round and caught on the horns as they fall and afterwards rise again, and then when lying down be lifted off the ground, and even stand in a car like charioteers with a pair of horses racing at full speed. It is a device of the Thessalian race to kill bulls by galloping a horse beside them and twisting back the neck by the horn; the dictator Caesar first gave this show at Rome.

On games with bulls and Thessalians, see also below text 71 and on epigraphic evidence: Robert 1940: 319.

39. Suet., *Iul.* 39.3: Combats with wild beasts were presented on 5 successive days, … there was a battle between two opposing armies, in which 500 foot-soldiers, 20 elephants, and 30 horsemen engaged on each side. (See text 13 Table 2).

40. Cass. Dio 48.33.4: In the year preceding this, men belonging to the order of knights had slaughtered wild beasts at the games in the Circus on the occasion of the *Ludi Apollinares*.

41. *Cass. Dio 51.22.5–9: Wild beasts and tame animals were slain in vast numbers, among them a rhinoceros and a hippopotamus, beasts then seen for the first time in Rome. As regards the nature of the hippopotamus, it has been described by many and far more have seen it. […] These beasts, accordingly, were brought in, and moreover Dacians and Suebi fought in crowds with one another. […] The whole spectacle lasted many days, as one would expect, and there was no interruption, even though Caesar fell ill, but it was carried on in his absence under the direction of others.

This *venatio* for the dedication of the *aedes Caesaris* was the first of the *venationes* given by Augustus in his own name (see below text 59) cf. Ville 1981: 108.

42. Cass. Dio 53.27.6: Publius Servilius, too, made a name for himself because while praetor he caused to be slain at a festival 300 bears and other African wild beasts equal in number.

43. Cass. Dio 54.8.5: The aediles gave games in the Circus and a slaughter of wild beasts on Augustus' birthday.

44. Suet., *Nero* 4.3: He [L. Domitius Ahenobarbus] gave beast-baitings both in the Circus and in all the regions of the City; also a gladiatorial show, but with such inhuman cruelty that Augustus, after his private warning was disregarded, was forced to restrain him by an edict. (On these shows, Ville 1981:101).

45. *Plin., *nat. hist.* 8.64: see text 20.

46. *CIL* 6.32323: l. 162–163: Three days before the Ides of June [i. e. 11 June] an edict was posted in the following words: the *quindecimviri sacris faciundis* proclaimed 'on the day before the Ides of June [i. e. 12 June] we shall give a hunting display [in … and we shall commence circus games].' *(a(nte) d(iem) III Eid(us) Iun(ias) edictum propo[situm est in haec verba XVvir(i) s(acris) f(aciundis) dic(unt)] / pr(idie) Eid(us) Iun(ias) venationem dabim[us ---])*. On this document: Schnegg-Köhler 2002; see also Beard, North & Price 1998: vol. 2:139–44 n. 5.7b.

47. Cass. Dio 54.26.1–2: He [Augustus] next dedicated the theatre named after Marcellus. In the course of the festival held for this purpose the patrician boys, including his grandson Caius, performed the eques-trian exercise called 'Troy', and 600 wild beasts from Africa were slain. And to celebrate the birthday of Augustus, Iullus, the son of Antony, who was praetor, gave games in the Circus and a slaughter of wild beasts.
48. Cass. Dio 54.26.1–2: see text 47.
49. *Plin., *nat. hist.* 8.65: Augustus also in the consulship of Marcus Tubero and Paullus Fabius, at the dedication of the Theatre of Marcellus, on May 7, was the first of all persons at Rome who exhibited a tamed tiger in a cage (*tigrim primus omnium Romae ostendit in cavea mansuefactam),* although his late Majesty Claudius exhibited four at one time.
On this show: VILLE 1981: 110–11. According to Cass. Dio 54.9, tigers (among other presents) were given to Augustus by Indian ambassadors, during his stay in Samos, and 'this was the first time these animals had been seen by the Romans and also I think by the Greeks'. See also below text 51 (Suet., *Aug.* 43); on this passage and in general on contacts and exchanges between Rome and the Far East: McLAUGHLIN 2010, 115 and 164–8 (on Roman revenues).
50. Cass. Dio 54.34.1–2: While Drusus was thus occupied, the festival belonging to his praetorship was celebrated in the most costly manner; and the birthday of Augustus was honoured by the slaughter of wild beasts (θηρίων σφαγαῖς ἐτιμήθη) both in the Circus and in many other parts of the city. This was done almost every year by one of the praetors then in office, even if not authorised by a decree; but the *Augustalia,* which are still observed, were then for the first time celebrated in pursuance of a decree.
51. Suet., *Aug.* 43.4: Furthermore, if anything rare and worth seeing was ever brought to the city, it was his habit to make a special exhibit of it in any convenient place on days when no shows were appointed. For example a rhinoceros in the Saepta, a tiger on the stage and a snake of 50 cubits in front of the Comitium.
52. Cass. Dio 55.10.6–7: And they [Gaius and Lucius] did, in fact, have the management of the Circensian games on this occasion, while their brother Agrippa took part along with the boys of the first families in the equestrian exercise called 'Troy'. 260 lions were slaughtered in the Circus. There was a gladiatorial combat in the Saepta, and a naval battle between the 'Persians' and the 'Athenians' was given on the spot where even to-day some relics of it are still pointed out.
53. Cass. Dio 55.10.8: Afterwards water was let into the Circus Flaminius and 36 crocodiles were there slaughtered. See also Strab. 17.44: When the crocodiles were brought to Rome for exhibition, they were attended by the Tentyritae; and when a reservoir and a kind of stage above one of the sides had been made for them, so that they could go out of the water and have a basking-place in the sun, these men at one time, stepping into the water all together, would drag them in a net to the basking-place, so that they could be seen by the spectators, and at another would pull them down again into the reservoir.
It seems that the inhabitants of Tentyra in Egypt came to Rome with the crocodiles; they were choosen because, according to Strabo, there was a mutual antipathy between them and the crocodiles and so the beasts would not eat them. It was probably in 2 BC during one of the *venationes* given by Augustus for the dedication of the temple of Mars Ultor that the show described by Strabo was mounted in the Circus Flaminius. See VILLE 1981: 112.
54. Plin., *nat. hist.* 8.5: At the gladiatorial show given by Germanicus Caesar some [elephants] even per-formed clumsy movements in figures, like dancers. It was a common display for them to hurl weapons through the air without the wind making them swerve, and to perform gladiatorial matches with one another or to play together in a sportive war-dance. Subsequently they even walked on tight-ropes, four at a time actually carrying in a litter one that pretended to be a lady lying-in; and walked among the couches in dining-rooms full of people to take their places among the guests, planting their steps carefully so as not to touch any of the drinking party.
55. Ael., *nat. anim* 2.11: Germanicus Caesar was about to give some shows for the Romans. (He would be the nephew of Tiberius.) There were in Rome several full-grown male and female elephants, and there were calves born of them in the country; and when their limbs began to grow firm, a man who was clever at dealing with such beasts trained them and instructed them with uncanny and astounding dexterity.
56. Cass. Dio 56.27.4: A special festival was also held by the actors (i. e. pantomimes) and the horse-breed-ers. The *Ludi Martiales,* owing to the fact that the Tiber had overflowed the Circus, were held on this oc-

casion in the Forum of Augustus and were celebrated in a fashion by a <u>horse-race and the slaying of wild beasts</u>.

57. Cass. Dio 56.27.5: They were also given a second time, as custom decreed, and Germanicus this time caused <u>200 lions to be slain in the Circus</u> (ἐν τῷ ἱπποδρόμῳ).

58. Plin., *nat. hist.* 8.64: see text 20.

59. *RGDA* 22.3: In my own name, or that of my sons or grandsons, on <u>26 occasions</u> I gave to the people, in the <u>circus</u>, in the <u>forum</u>, or in the <u>amphitheatre</u>, <u>hunting shows of African wild beasts</u> (*[vena]tion[es] best[ia]rum Africanarum*), in which about 3,500 beasts (*bestiae*) <u>were slain</u>.

60. Cass. Dio 59.7.1: There were <u>spectacles of all sorts</u>. For not only all kinds of <u>musical entertainments</u> were given, but also <u>horseraces </u>took place on 2 days, 20 heats the first day and 40 the second [or 24 ? cf. Ville 1981: 130], because the latter was the emperor's birthday, being the last day of August. And he [Caligula] exhibited the same number of events on many other occasions, as often as it suited him; previously to this, it should be explained, not more than ten events had been usual. He also caused <u>400 bears to be slain </u>on the present occasion together with an <u>equal number of wild beasts from Libya</u>. The boys of noble birth performed the <u>equestrian game of 'Troy'</u>, and 6 horses drew the triumphal car on which he rode, something that had never been done before.

61. Cass. Dio 59.13.8: Later he [Caligula] returned to celebrate the birthday of Drusilla, brought her statue into the Circus on a <u>car drawn by elephants</u>, and gave the people a free exhibition for two days. On the first day, besides <u>the horse-races, 500 bears</u> were slain, and on the second day <u>as many Libyan beasts</u> were accounted for; also athletes competed in the pancratium in many different places at the same time.

62. Cass. Dio 59.13.9: On the first day, besides <u>the horse-races, 500 bears were slain</u>, and on the second day <u>as many Libyan beasts </u>were accounted for.

On the festivities associated with Drusilla's birthday, see also 59.11.3: and on her birthday a festival equal to the *Ludi Megalenses* should be celebrated; and 59.24.7–8: Among other votes passed was one providing that the birthdays of Tiberius and Drusilla should be celebrated in the same manner as that of Augustus. The people connected with the stage also exhibited a festival, furnished a spectacle, and set up and dedicated images of Gaius and Drusilla. All this was done, of course, in response to a message from Gaius.

63. Cass. Dio 59.20.1–2: He [Caligula] immediately appointed Domitius consul, after removing those who were then in office because they had failed to proclaim a thanksgiving on his birthday (the praetors, it is true, had held a horse-race and had slaughtered some wild beasts, but this happened every year) and because they had celebrated a festival to commemorate the victories of Augustus over Antony, as was customary.

On the absence of proclamation of his birthday see also Suet., *Cal.* 26: When the consuls forgot to make proclamation of his birthday, he deposed them, and left the state for three days without its highest magistrates.

64. Suet., *Cal.* 18.1–5: He [Caligula] gave several <u>gladiatorial shows</u>, some in the amphitheatre of Taurus and some in the Saepta, in which he introduced pairs of <u>African and Campanian boxers</u>, the pick of both regions … He exhibited stage-plays (*scaenicos ludos*) continually, of various kinds and in many different places, sometimes even by night, lighting up the whole city (*nocturnos accensis tota urbe luminibus*). He also threw about gifts of various kinds (*sparsit et missilia variarum rerum*), and gave each man a basket of prepared food … He also gave many games in the <u>Circus</u>, lasting from early morning until evening, <u>introducing between the races now a baiting of panthers</u> (*modo Africanarum venatione*) and now the manœuvres of the <u>game called Troy</u>. In some special games, the arena was scattered with red and green and all the chariots were driven by men of the senatorial order. (trans. Loeb modified in light of Edwards). See also Suet., *Cal.* 23.1: and not content with this slur on the memory of Augustus, he [Caligula] forbade the celebration of his victories at Actium and off Sicily by annual festivals (*vetuit sollemnibus feriis celebrari*), on the ground that they were disastrous and ruinous to the Roman people.

65. Suet., *Cal.* 20: He [Caligula] also gave shows in foreign lands, a city festival at <u>Syracuse in Sicily</u>, and mixed games at <u>Lugdunum in Gaul</u>.

66. Cass. Dio 60.7.3: In the Circus (ἐν δὲ τῷ ἱπποδρόμῳ) there was one contest (ἠγωνίσαντο) with <u>camels</u> and <u>12 with horses</u>, and <u>300 bears </u>and the same number of <u>Libyan beasts</u> were slain (ἄρκτοι τε τριακόσιαι καὶ Λιβυκὰ θηρία ἴσα αὐταῖς ἐσφάγη).

67. Cass. Dio 60.13.4: He [Claudius] used to delight especially in watching those who were cut down during the intermission in the spectacle at lunch time; and yet he had put to death a lion that had been trained to eat men and therefore greatly pleased the crowd, claiming that it was not fitting for Romans to gaze on such a sight.

68. Cass. Dio 60.23.4–6: The festival was celebrated in both theatres at the same time (ἐν τοῖς δύο ἅμα θεάτροις). [...] He [Claudius] had announced as many horse-races as could take place in a day, yet there were not more than ten of them. For between the different races bears were slain, athletes contested, and boys summoned from Asia performed the Pyrrhic dance. Another festival, likewise in honour of his victory, was given by the artists of the stage with the consent of the Senate.

69. Cass. Dio 60.27.2: for the expenses connected with the Circensian games (τὰ γὰρ ἀναλώματα τὰ ἐν ταῖς ἱπποδρομίαις γιγνόμενα) had greatly increased, since there were usually 24 races.

70. Tac., *ann.*12.3.2: Caesar [Claudius] had betrothed Octavia to L. Silanus and by means of triumphal insignia and the magnificence of a gladiatorial show had brought the young man ... to the enthusiastic attention of the public. (trans. WOODMAN 2004).

71. Suet., *Claud.* 21.1–4: He also gave several splendid shows (*complura et magnifica edidit*), not merely the usual ones in the customary places, but some of a new kind and some revived from ancient times, and in places where no one had ever given them before. He opened the games at the dedication of Pompey's theatre, which he had restored when it was damaged by a fire, from a raised seat in the orchestra, after first offering sacrifice at the temples in the upper part of the auditorium and coming down through the tiers of seats while all sat in silence. He also celebrated secular games ... He often gave games in the Vatican Circus also, at times with a beast-baiting (*venatio*) after every five races. But the Great Circus he adorned with barriers of marble and gilded goals ... In addition to the chariot races he exhibited the game called Troy (*super quadrigarum certamina Troiae lusum*) and also African beasts (*Africanae*), which were hunted down by a squadron of the praetorian cavalry under the lead of the tribunes and the prefect himself; likewise Thessalian horseman, who drive wild bulls all over the arena, leaping upon them when they are tired out and throwing them to the ground by the horns. He gave many gladiatorial shows and in many places: one in yearly celebration of his accession, in the Praetorian Camp without wild beasts (*sine venatione*) and fine equipment, and one in the Saepta of the regular and usual kind; another in the same place not in the regular list, short and lasting but a few days. (On fights between Thessalian and bulls, see above text 38).

72. Plin., *nat. hist.* 8.2: ... and subsequently for the emperors Claudius and Nero elephants versus men single-handed, as the crowning exploit of the gladiators' careers.

73. Plin., *nat. hist.* 8.65: see text 49.

74. Plin., *nat. hist.* 8.22: see text 72.

75. Suet., *Nero* 7.9. Shortly afterwards he took Octavia to wife and gave games in the Circus and a beast-baiting that health might be vouchsafed Claudius. (trans. Loeb modified).

76. Suet., *Nero* 11.1–2: He [Nero] gave many entertainments of different kinds: the Juvenales, chariot races in the Circus, stage-plays and a gladiatorial show ... For the games in the Circus he ... even matched chariots drawn by 4 camels. At the plays which he gave for the 'Eternity of the Empire', which by his order were called the *Ludi Maximi*, parts were taken by several men and women of both the orders; a well known Roman knight mounted an elephant and rode down a rope; a Roman play of Afranius, too, was staged, entitled 'The Fire', and the actors were allowed to carry off the furniture of the burning house and keep it. Every day all kinds of presents were thrown to the people; these included 1,000 birds of every kind each day, various kinds of food, tickets for grain, clothing, gold, silver, precious stones, pearls, paintings, slaves, beasts of burden, and even trained wild animals; finally, ships, blocks of houses, and farms.

77. Suet., *Nero* 12.1: At the gladiatorial show, which he gave in a wooden amphitheatre, erected in the district of the Campus Martius within the space of a single year, he had no one put to death, not even criminals. But he compelled 400 senators and 600 Roman knights, some of whom were well to do and of unblemished reputation, to fight in the arena. Even those who fought with the wild beasts and performed the various services in the arena were of the same orders. He also exhibited a naval battle in salt water with sea monsters swimming about in it; besides pyrrhic dances (...).

78. Calp. Sic. 7.23–73: I saw a theatre that rose skyward on interwoven beams and almost looked down on the summit of the Capitoline. Passing up the steps and slopes of gentle incline, we came to the seats, where in dingy garments the baser sort viewed the show close to the women's benches. For the uncovered parts, exposed beneath the open sky, were thronged by knights or white-robed tribunes. Just as the valley here expands into a wide circuit, and, winding at the side, with <u>sloping forest background all around</u>, stretches its concave curve amid the unbroken chain of hills, so there the sweep of the amphitheatre encircles the level ground, and the oval in the middle is bound by twin piles of building. Why should I now relate to you things which I myself could scarcely see in their several details? So dazzling was the glitter everywhere. Rooted to the spot, I stood with mouth agape and marvelled at all, nor yet had I grasped every single attraction, when a man advanced in years, next me as it chanced on my left, said to me: 'Why wonder, country-cousin, that you are spellbound in face of such magnificence? you are a stranger to gold and only know the cottages and huts which are your humble homes. Look, even I, now palsied with age, now hoary-headed, grown old in the city there, nevertheless am amazed at it all. Certes, we rate all cheap we saw in former years, and shabby every show we one day watched'. Look, the partition-belt begemmed and the gilded arcade vie in brilliancy; and withal just where the end of the arena presents the seats closest to the marble wall, wondrous ivory is inlaid on connected beams and unites into a cylinder which, gliding smoothly on well-shaped axle, could by a sudden turn balk any claws set upon it and shake off the beast. Bright too is the gleam from the nets of gold wire which project into the arena hung on solid tusks, tusks of equal size; and (believe me, Lycotas, if you have any trust in me) <u>every tusk was longer than our plough</u>. Why narrate each sight in order? <u>Beasts of every kind I saw;</u> here I saw <u>snow-white hares</u> and <u>horned boars</u>, here I saw <u>the elk, rare even in the forests which produce it.</u> <u>Bulls</u> too I saw, either those of heightened nape, with an unsightly hump rising from the shoulder-blades, or those with shaggy mane tossed across the neck, with rugged beard covering the chin, and quivering bristles upon their stiff dewlaps. Nor was it my lot only to see <u>monsters of the forest: sea calves</u> also I beheld with <u>bears</u> pitted against them and the unshapely herd called by the name of horses, bred in that river whose waters, with spring-like renewal, irrigate the crops upon its banks. Oh, how we quaked, whenever we saw the arena part asunder and its soil upturned and <u>beasts plunge out from the chasm cleft in the earth</u>; yet often from those same rifts the golden arbutes sprang amid a sudden fountain <u>spray (of saffron?)</u>.

The Neronian dating of Calpurnius (see L. Watson in OCD[4] 2012, with bibliography, and Karakasis 2016) is debated; some prefer a third century dating (see Coleman 2006: 85 n. 5).

79. Jos., *b.j.* 7.132: It is impossible adequately to describe the <u>multitude of those spectacles</u> and their magnificence under every conceivable aspect, whether in works of art or <u>diversity of riches or natural rarities</u> … <u>Beasts of many species</u> were led along all caparisoned with appropriate trappings. <u>The numerous attendants conducting each group of animals</u> were decked in garments of true purple dye, interwoven with gold … Moreover, even among the <u>mob of captives</u>, none was to be seen not unadorned, the variety and beauty of their dresses concealing from view any unsightliness arising from bodily disfigurement.

Followed many tableaux and figured <u>representations of landscapes and cities devastated by the war</u>, of 'rivers flowing not on a cultivated land, nor supplying drink to man and beast, but across a country still on every side in flames.' And then a number of <u>ships</u>, with the spoils, and at the end of the *pompa,* a copy of the Jewish Law, the images of Victory, in ivory and gold, and at the end the emperors, Vespasian followed by Titus, and beside them Domitian 'in magnificent apparel and mounted on a steed that was itself a sight'.

80. Cass. Dio 66.25.1–5: in dedicating the hunting-theatre and the baths that bear his name (τὸ δὲ δὴ θέατρον τὸ κυνηγετικὸν τό τε βαλανεῖον τὸ ἐπώνυμον αὐτοῦ) he [Titus] produced many remarkable spectacles. <u>Cranes</u> fought against each other, and also <u>4 elephants; all together, 9,000 animals, both tame and wild</u> (καὶ βοτὰ καὶ θηρία), were despatched, and women (not those of any prominence, however) took part in despatching them. (2). As for the men, many fought in single combat, and several groups competed against each other in groups in <u>infantry and naval battles.</u> Titus suddenly <u>filled this same theatre with water</u> and brought in <u>horses and bulls and other domesticated animals</u> that had been taught to do in water just as on land. He also brought in people on ships; (3) they engaged in a <u>naval battle</u> there, impersonating Corcyreans versus Corinthians. Others gave a <u>similar display outside the city</u> in the grove of Gaius and Lucius, a place which Augustus had once excavated for this very purpose. There, too, on the first day there was a <u>gladiatorial display and a slaughter of wild-beasts</u>, the lake in front of the images having first been covered over with a platform of planks and wooden stands erected around it. (4) On the second day there was a <u>horse-race</u>, and on the third day <u>a naval battle between 3,000 men</u>, followed by an infantry battle. The 'Athenians' conquered the 'Syracusans' (these were the names the combatants used) landed on the island, and assaulted and captured a wall that had been erected around the monument. These were the spectacles that were offered, and they continued for <u>100 days</u>; but Titus also provided some things that were of practical use to the people. (5) He would throw down into the theatre (ἐς τὸ θέατρον) from above little <u>wooden balls variously inscribed,</u> one designating some article of food, another clothing, another a silver vessel or perhaps a gold one, or again horses, pack-animals, cattle, or slaves. Those who seized them were to take them to the dispensers of the bounty, from whom they would receive the article named. (trans. Loeb modified in light of COLEMAN 2006: xlvii-xlviii); on these battles, see also Table 2.

81. Suet., *Tit.* 7.3: At the dedication of the amphitheatre and of the baths, which were hastily built near it, he put on a most sumptuous <u>gladiatorial</u> show on a very lavish scale; he also gave a <u>naval battle in the old bassin</u> (*navale proelium in veteri naumachia*), and in the same place he also displayed and, <u>in a single day, 5,000 wild beasts of every kind</u>.

82. Mart., *spect.* 20 (17): Respectful and suppliant, the <u>elephant</u> that was recently so formidable to a bull worships you, Caesar. It does not do this on command, nor on instruction from any trainer: believe me, it too feels the presence of our god. (trans. COLEMAN).

On the variety of animals in Martial's *Book of Spectacles*, see 22 (19): <u>elephant</u>; 11 (9) and 26 (22+23): <u>rhinoceros</u>; 8 (6b), 12 (10), 17 (15), 21 (18), 26 (22+23): <u>lion</u>; 10 (8), 13 (11), 15 (15), 24 (21): <u>bear</u>; 14–16 (12–14): <u>sow</u> (pregant, mother); 17 (15): <u>boar</u>; 11 (9), 17–20 (15–17), 26 (22+23): bull; 26 (22+23): <u>auroch</u>; 26 (22+23): <u>bison</u>; 21 (18): <u>tigress</u>; various beasts, <u>wild and domesticated</u>, with Orpheus: 24 (21). On fights between animals: COLEMAN 2006: 109; for all the details related to these spectacles, with a discussion concerning the identity of the anonymous 'Caesar' of Martial's *Book of Spectacles* and its attribution to Titus' inaugural games of 80 AD and to Domitian's games: COLEMAN 2006: xlv-lxiv; for the attribution to Domitian's games only see BUTTREY 2007: 111–2.

83. Suet., *Dom.* 4: He [Domitian] constantly gave grand and costly entertainments, both in the amphitheatre and in the Circus, where in addition to the <u>usual races</u> between two-horse and four-horse chariots, he also exhibited <u>two battles</u>, one between forces of infantry and the other by horsemen; and he even gave a <u>naval battle in the amphitheatre.</u> Besides he gave <u>hunts of wild beasts, gladiatorial shows at night</u> by the light of torches, and not only combats between men but between women as well … He often gave <u>sea-fights almost with regular fleets, having dug a pool near the Tiber and surrounded it with seats</u>; and he continued to witness the contests amid heavy rains. He also celebrated Secular games … In the course of these, to make it possible to finish <u>100 races on the day</u> of the contests in the Circus, he diminished the number of laps from 7 to 5. He also established a quinquennial contest in honour of Jupiter Capitolinus of a threefold character, comprising music, riding, and gymnastics, and with considerably more prizes than are awarded nowadays.

84. Cass. Dio 68.15.1: Upon Trajan's return to Rome ever so many embassies came to him from various barbarians, including the Indi. And he gave spectacles on 123 days, in the course of which some 11,000 animals, both wild and tame, were slain, and 10,000 gladiators fought.

85. HA, *Hadr.* 7.12: He gave gladiatorial combats (*gladiatorium munus*) for 6 days in succession, and on his birthday, a 1,000 wild beasts. (trans. Loeb modified).

86. HA, *Hadr.* 19.2–8: In <u>almost every city</u> he did some building and gave public games (*ludos*). At <u>Athens</u> he [Hadrian] exhibited in the stadium a hunt involving <u>1,000 wild beasts,</u> but he never called away from Rome a single hunter or actor. In Rome, in addition to popular entertainments of unbounded extravagance, he gave spices to the people in honour of his mother-in-law, and in honour of Trajan he had essences of balsam and saffron poured over the seats of the theatre … In the Circus he had <u>many wild beasts killed</u> and <u>often 100 lions</u>. He frequently gave the people exhibitions of military Pyrrhic dances, and he frequently attended <u>gladiatorial shows</u> (*gladiatores spectavit*). (trans. Loeb modified).

87. HA, *Ant. Pius* 10.9: He held games at which he displayed <u>elephants</u> and the animals called <u>*corocottae*</u> and <u>tigers</u> and <u>rhinoceroses</u>, even <u>crocodiles</u> and <u>hippopotami</u>, in short, <u>all the animals of the whole earth</u>; and he presented at a single performance as many as <u>100 lions together with tigers</u>.
If among all these animals there were bisons, they may have been those seen by Pausanias (Paus. 9.21.3) in Rome at the *decennalia* of 148 AD. On the capture of bisons and bears by Roman soldiers in 147 AD near *Montana* in *Moesia Inferior* (Bulgaria) for a *venatio Caesariana*, see AE 1987, 867=AE 1999, 1327: *Dianae / Ti(berius) Claudius Ulpianu(s) / trib(unus) coh(ortis) I Cili(cum) cum vexilla/tionib(us) leg(ionum) I Ital(icae) XI Cl(audiae) class(is) / Fl(aviae) Mo(esicae) ob venationem / Caesarianam iniunc/ tam a Cl(audio) Saturnino leg(ato) / Aug(usti) pr(o) pr(aetore) ursis et vison/tibus(!) prospere captis / aram consecra/vit Largo et Mes/sallino co(n)s(ulibus)*, with Knoepfler 1999: 503–4.

88. Hdn. 1.15.5–6: Commodus now gave orders for the celebration of public shows, at which he promised he would <u>kill all the wild animals with his own hand</u> and engage in gladiatorial combat with the stoutest of the young men. As the news spread, people flocked to Rome from all over Italy and the neighbouring provinces to be spectators at something they had never seen or heard of before. The topic of conversation was about Commodus' marksmanship and how he made sure he never missed with his spear or arrow. He had the finest <u>Parthian archers and Mauretanian spearsmen</u> with him as his teachers, but he was more skilful than any of them. At last the day of the show came and the <u>amphitheatre was packed</u>. A special raised enclosure was put up for Commodus' benefit so that he could spear the animals safely from above without endangering himself from close quarters, a demonstration of his skill but not of his courage. He shot down <u>deer</u> and <u>gazelles</u> and other <u>horned animals (apart from bulls)</u> by pursuing them as they ran and stopping them dead in their tracks with a well-timed shot just as they made a dash for it. The <u>lions</u> and <u>leopards</u> and other fine animals of this kind he speared from above by running around the enclosure … <u>Wild beasts were brought from all over the world for him to kill, species which we had admired in pictures but saw for the first time on that occasion; from India and Ethiopia, from the North and South they came. All of them, if any were previously unknown, were now on show for the Romans to see as they were killed by Commodus</u>. His marksmanship was generally agreed to be astonishing. For instance, on one occasion he used some arrows with crescent-shaped heads to shoot at <u>Mauretanian ostriches, birds</u> that can move tremendously quickly because of the speed at which they run and because of their folded back wings. Commodus decapitated the birds at the top of their necks with his arrows, so that they went on running around as though they had not been touched, even when their heads had been cut off by the sweep of the arrow. Once when a <u>leopard</u> had dashed out and pounced on a victim summoned into the arena, Commodus pierced it with a javelin just before it savaged the man. The animal was killed, the man was saved; Commodus forestalled the point of the teeth with the point of his spear. On yet another occasion <u>100 lions were released simultaneously from the subterranean traps</u> and Commodus killed them all with exactly 100 spears. The bodies lay in a long line where they fell for everyone to count at leisure and see that not a single extra spear had been used. So far Commodus was still quite popular with the mob even if his conduct, apart from his courage and marks-manship, was unfitting for an emperor. But when he ran into the amphitheatre stripped and carrying his weapons for a gladiatorial fight, the people were ashamed to see a Roman emperor of noble lineage, whose father and forebears had all celebrated great triumphs, now disgracing his office with a thoroughly <u>degrad-ing exhibition</u>, instead of using his weapons to fight the barbarians and prove himself worthy of the Roman empire. He had no difficulty in overcoming his opponents in gladiatorial fights by merely wounding them, since they all looked upon him as the emperor rather than as a gladiator and let him win. But his madness reached such a stage that he even refused to stay in the palace any longer and was intending to go and live with the gladiators in their barracks. Orders were issued that he was no longer to be addressed as Heracles but by the name of a famous gladiator, now dead. He removed the head from the enormous statue of the Co-lossus which represents the sun and is revered by the Romans. On the base he inscribed the usual, imperial family names but, instead of the title 'Germanicus', he put 'Victor of a Thousand Gladiators'; see also Cass. Dio 72.10.2–3 and 72.19.1. On this famous episode: Epplett 2014: 510 and Toner 2014: 7–18.

89. * Cass. Dio 76.1.3–5: At this time there occurred, too, all sorts of spectacles in honour of Severus' return, the completion of his first 10 years of power, and his victories. At these spectacles <u>60 wild boars</u> of Plautianus fought together at a signal, and <u>among many other wild beasts that were slain were an elephant</u>

and a *corocotta*. This last animal is an Indian species, and was then introduced into Rome for the first time, so far as I am aware. It has the colour of a lioness and tiger combined, and the general appearance of those animals, as also of a dog and a fox, curiously blended. The entire receptacle in the entertainment building (ἐν τῷ θεάτρῳ) had been constructed so as to resemble a boat in shape, and was capable of receiving or discharging 400 beasts at once; and then, as it suddenly fell apart, there came rushing forth bears, lionesses, panthers, lions, ostriches, wild asses, bisons (this is a kind of cattle foreign in species and appearance), so that 700 beasts in all, both wild and domesticated, at one and the same time were seen running about and were slaughtered. For to correspond with the duration of the festival, which lasted 7 days, the number of the animals was also 7 times 100.

This show with so many animals was presumably exhibited in the Circus Maximus despite the use of ἐν τῷ θεάτρῳ (instead of ἐν τῷ ἱπποδρόμῳ like in 76.2); on this spectacle, its location and its datation: Bajard 2008, and in general on aquatic spectacles: Coleman 1993, Taylor 1997, and Belan-Bajard 2006 with Coleman 2008.

90. HA, *Gord.* 3.5–8: When he was an aedile he gave the Roman people 12 exhibitions (*munera*), that is one for each month, at his own expense; at times, indeed, he provided 500 pairs of gladiators, and never less than 150. He produced a 100 wild beasts of Libya (*feras libycas*) at once, and likewise at one time 1,000 bears. There exists also a remarkable wild beasts hunt of his staged in a forest, pictured in Gnaeus Pompey's 'House of the Beaks' (*domus rostrata*). Follows the description and the list on animals, including 200 stags with antlers and stags of Britain, 30 wild horses, 100 wild sheep, 10 elks, 100 Cyprian bulls, 300 red Moorish ostriches, 30 wild asses, 150 wild boars, 200 ibises (*ibices*), 200 fallow deer. And all this he handed over to the people to be killed on the day of the sixth exhibition he gave. (trans. Loeb modified).

91. HA, *Gord.* 4.5: With the emperors' permission he distributed 100 Sicilian and 100 Cappadocian horses among the factions.

92. HA, *Gord.* 33.1–2: There were 32 elephants at Rome in the time of Gordian (of which he himself had sent 12 and Alexander 10), 10 elk, 10 tigers, 60 tame lions, 30 tame leopards, 10 *belbi* or hyenas, 1,000 pairs of imperial gladiators, 6 hippopotami, 1 rhinoceros, 10 wild lions, 10 giraffes, 20 wild asses, 40 wild horses, and various other animals of this nature without number. All of these Philip presented or slew at the Secular Games. All these animals, wild, tame, and savage, Gordian intended for a Persian triumph; but his official vow proved of no avail, for Philip presented all of them at the Secular Games, consisting of both gladiatorial spectacles and races in the Circus, that were celebrated on the 1,000th anniversary of the founding of the City, when he and his son were consuls.

See also Aur. Vict., *Caesars* 28: They [Philipp and his son Philipp] constructed a reservoir on the other side of the Tiber because that region used to be plagued by a shortage of water, and they celebrated the 1,000th anniversary of the city with games of all kinds (transl. Bird). These games were presumably given on the site where was held 200 years before Augustus' naumachy. According to Chastagnol 1994: 692, the lifes of the three Gordians of *HA* are totally conform to Herodian's text.

93. HA, *Aurel.* 33–34: It is not without advantage to know what manner of triumph Aurelian had, for it was a most brilliant spectacle. There were three royal chariots, of which the first … had belonged to Odaenathus, the second … had been given to Aurelian by the king of the Persians, and the third Zenobia had made for herself, hoping in it to visit the city of Rome. And this hope was not unfulfilled; for she did, indeed, enter the city in it, but vanquished and led in triumph. There was also another chariot, drawn by four stags and said to have once belonged to the king of the Goths. In this … Aurelian rode up to the Capitol, purposing there to slay the stags, which he had captured along with this chariot and then vowed, it was said, to Jupiter Best and Greatest. There advanced, moreover, 20 elephants, and 200 tamed beasts of divers kinds from Libya and Palestine, which Aurelian at once presented to private citizens, that the privy-purse might not be burdened with the cost of their food; furthermore, there were led along in order 4 tigers and also giraffes and elks and other such animals, also 800 pairs of gladiators besides the captives from the barbarian tribes. […] Then came the Roman people itself, the flags of the guilds and the camps, the mailed cuirassiers, the wealth of the kings, the entire army, and, lastly, the senate (albeit somewhat sadly, since they saw senators, too, being led in triumph) – all adding much to the splendour of the procession. Scarce did they reach the

Capitol by the 9th hour of the day, and when they arrived at the Palace it was late indeed. On the following days amusements were given to the populace, <u>plays in the theatres, races in the Circus, wild-beast hunts, gladiatorial fights and also a naval battle.</u>

On the funding of the games for its consulship in 271 AD, see HA, *Aurel.* 12.1: To Aurelian, whom we have named for the consulship, because of his poverty … you will supply for the performance of the races in the Circus 300 aurei of Antoninus, 3,000 silver minutuli of Philip, 5,000,000 bronze sesterces, 10 finely-woven tunics of the kind used by men, 20 tunics of Egyptian linen, 2 pairs of Cyprian table-covers, 10 African carpets, 10 Moorish couch-covers, 100 swine, and 100 sheep.

On the unrealistic character of that description, see CHASTAGNOL 1994: 1004 n. 2 (citing E. MERTEN, *Zwei Herrscherfeste in der H. A.*, Bonn 1968: 101–40, *non vidi*).

94. HA, *Prob.* 19.1–8: He [Probus] celebrated a triumph over the Germans and the Blemmyae, and caused companies from all nations, <u>each of them containing up to 50 men,</u> to be led before his triumphal procession. He gave <u>in the Circus a most magnificent wild-beast hunt,</u> at which all things were to be the spoils of the people. Now the manner of this spectacle was as follows: <u>great trees,</u> torn up with the roots by the soldiers, were set up on a platform of beams of wide extent, on which earth was then thrown, and in this way <u>the whole Circus, planted to look like a forest,</u> seemed, thanks to this new verdure, to be putting forth leaves. Then through all the entrances were brought in <u>1,000 ostriches,</u> <u>1,000 stags</u> and <u>1,000 wild-boars,</u> then <u>deer, ibexes, wild sheep</u> and <u>other grass-eating beasts,</u> as many as could be reared or captures. The populace was then let in, and each man seized when he wished. Another day he brought out in the Amphitheatre, at a single performance <u>100 maned lions, which woke the thunder with their roaring.</u> All of these were <u>slaughtered</u> as they came out of the doors of their dens, and being killed in this way they afforded no great spectacle. For there was none of that rush on the part of the beasts which takes place whe they are let loose from cages. Besides, many, unwilling to charge, were despatched with arrows. Then he brought out <u>100 leopards from Libya,</u> then <u>100 from Syria,</u> then <u>100 lionesses</u> and at <u>the same time 300 bears;</u> all of which beasts, it is clear, made a spectacle more vast than enjoyable. He presented, besides, <u>300 pairs of gladiators,</u> among whom fought many of the Blemmyae, who had been led in his triumph, besides many Germans and Sarmatians also even some Isaurian brigands.

These lifes are mostly based on Zosimus and Aurelius Victor, but the description of the spectacles, which shows much fantasy, does not appear elsewhere: CHASTAGNOL 1994:1067–8.

95. Claud., *Theod.*, 291–332: Fly, Clio, to Taygetus' heights and leafy Maenalus and beg Diana not to spurn thy petition but help the amphitheatre's pomp. Let the goddess herself choose out brave hunters cunningly to <u>lasso the necks of wild animals</u> and to drive home the hunting-spear with unfailing stroke. With her own hand let her lead forth from their caverns <u>fierce beasts and captive monsters,</u> laying aside her bloodthirsty bow. Let <u>bears</u> be gathered together … Let smitten <u>lions</u> roar till the people turn pale, lions such as Cybele would be fain to harness to her Mygdonian chariot or Hercules strangle in his mighty arms. May <u>leopards,</u> lightning-swift, hasten to meet the spear's wound, beasts that are born of an adulterous union what time the spotted sire did violence to the nobler lion's mate: of such beasts their markings recall the sire, their courage the dam. Whatsoever is nourished by the fields of Gaetulia rich in monsters, whatsoever lurks beneath Alpine snows or in Gallic woods, let it fall before the spear. <u>Let large streams of blood enrich the arena and the spectacle leave whole mountains desolate</u> … Let the counterweights be removed and the mobile crane descend, lowering on to the lofty stage men who, wheeling chorus-wise, scatter flames; let Vulcan forge balls of fire to roll innocuously across the boards, let the flames appear to play about the sham beams of the scenery and a tame conflagration, never allowed to rest, wander among the untouched towers. <u>Let ships meet in mimic warfare on an improvised ocean and the flooded waters be lashed to foam by singing oarsmen.</u>

96. Claud., *Stil.* 3.262–355: From us too let Stilicho receive the favour we justly owe him; the task needs no javelin; let our arrows remain bloodless in our unopened quivers. Let every bow refrain from its wonted hunting and the blood of our prey be spilled but in the arena. Not for now their death; close the glades with net and cages and <u>lead the beasts captive;</u> withhold your impatient arrows; spare the monsters of the forest whose death shall win applause for our consul […] My purpose is to traverse the unfruitful desert; <u>Mauretania has given ere now her animals</u> to other consuls as a gift, to this consul alone she owes them as a conquered land owes tribute. While we track out the dread progeny of Libya do you hunt the glades and rocks of Europe. Let joy banish fear from the shepherd's breast and his pipe hymn Stilicho in the <u>dreadless</u> <u>forests</u>. <u>As his laws have given peace to the cities so let his shows give peace to the mountains</u> […] Thero with her dogs explores the caves of Spain and from their recesses ousts the horrid bears of whose bloody jaws full off Tagus' flood has failed to quench the thirst, and whose bodies, numbed with cold, the holm-oak of the Pyrenees o'ershadows with its leaves. The manlike maiden Nebrophone <u>hunts the mountains of Corsica and</u> <u>Sicily</u> and captures <u>deer</u> and other harmless <u>beasts, beasts that are the joy of the rich amphitheatre and the</u> <u>glory of the woods</u>. Whatsoever inspires fear with its teeth, wonder with its mane, awe with its horns and bristling coat – all the beauty, all the terror of the forest is taken. Guile protects them not; neither strength nor weight avails them; their speed saves not the fleet of foot. Some roar enmeshed in snares; some are thrust into wooden cages and carried off. There are not carpenters enough to fashion the wood; leafy prisons are constructed of unhewn beech and ash. <u>Boats laden with some of the animals traverse seas and rivers;</u> bloodless from terror the rower's hand is stayed, for <u>the sailor fears the merchandise he carries. Others are</u> <u>transported over land in wagons that block the roads with the long procession, bearing the spoils of the</u> <u>mountains. The wild beast is borne a captive</u> by those troubled cattle on whom in times past he sated his hunger, and each time that the oxen turned and looked at their burden they pull away in terror from the pole. By now Phoebus' sister had wandered <u>over the torrid plains of Libya and chosen out superb lions who</u> <u>had often put the Hesperides to flight, filled Atlas with alarm at their wind-tossed manes, and plundered far</u> <u>and wide the flocks of Ethiopia</u>, lions whose terrible cries had never struck upon the herdsmen's ears but as heralding their destruction. To catch them had been used no blazing torches, no twigs strewn over turf undermined; the voice of a tethered kid had not allured their hunger nor had a digged pit ensnared them: of their own free will they gave themselves up to capture and rejoiced at being seen the prey of so great a goddess. <u>At length the countryside breathes again and the Moorish farmers unbar their now safe huts.</u> Then Latonia collected grey-spotted <u>leopards</u> and other marvels of the south and <u>huge ivory tusks which, carved</u> <u>with iron into plaques and inlaid with gold to form the glistening inscription of the consul's name,</u> should pass in procession among lords and commons. <u>All India</u> stood in speechless amaze to see many an <u>elephant</u> <u>go shorn of the glory of his tusks</u>. Seated upon their black necks despite their cries the goddess shook the fixed ivory and tearing it up from its bloody roots disarmed the monstrous mouths. Nay, she fain would have brought the <u>elephants</u> themselves as a spectacle but feared that their vast weight would retard the ships. **97.** Luxorius, 60: The countryside marvels at the triumphs of the amphitheatre and the forest notices that strange wild beasts are there. The many farmers look at new struggles while ploughing and the sailor sees varied entertainments from the sea. The fertile land loses nothing, <u>the plants grow in greater abun-</u> <u>dance</u> while all the wild beasts fear their fates here. (trans. Rosenblum 1961 n. 60).

Appendix 2

Mass spectacles involving large numbers of people (soldiers, prisoners, gladiators, and *noxii*).
Texts mentioned in Table 2

1. Dion. Hal., *ant.* 6.17.2–4: he [Aulus Postumius] returned to the city with the pomp of a magnificent triumph (ἐκπρεπεῖ θριάμβῳ) … followed by <u>5,500 prisoners</u> taken in the battle. And having set apart the tithes of the spoils, <u>he spent 40 talents in performing games and sacrifices to the gods</u> (τὰς δεκάτας ἀγῶνάς τε καὶ θυσίας τοῖς θεοῖς), and let contracts for the building of temples to <u>Ceres, Liber and Libera</u> in fulfilment of a vow he had made … he made vows to them, when he was on the point of leading out his army, that if there should be the same abundance in the city during the time of his magistracy as before, he would build temples to them and also appoint <u>sacrifices to be performed every year</u>. These gods, hearing his prayer, caused the land to produce rich crops, not only of grain but also of fruits, and all imported provisions to be more plentiful than before; and when Postumius saw this, he himself caused a vote to be passed for the building of these temples. The Romans, therefore, having through the favour of the gods repelled the war brought upon them by the tyrant, were engaged in feasts and sacrifices. (trans. Loeb modified). On this triumph, see ITGENSHORST 2005: 262.

2. Eutr. 2.2: A short time later, Tuscans were overcome by C. Martius, and 8,000 of their prisoners were led in triumph.

Note that Eutropius' account postdates the event by more than 500 years, as underlined by BEARD 2007: 118.

3. Livy 7.27.8–9: The 4,000 who had surrendered were not reckoned a part of the spoils; these the consul [M. Valerius Corvus] sent in chains before his chariot when he triumphed, and they were subsequently sold, and brought in a great sum to the treasury. Some think that this multitude of captives consisted of slaves, and this is more likely than that surrendered men were sold.

On Corvus' triumph over the Antiates, Volsci and Satricani, see ITGENSHORST 2005: 264.

4. Val. Max. 2.7.14: And L. Paullus after defeating king Perseus laid persons of like nationality and guilt in front of elephants to be trampled.

On L. Aemilius Paullus' triumph, see ITGENSHORST 2005: 268.

5. Val. Max. 2.7.13: The younger Africanus after overturning the Punic empire threw deserters of other than Roman nationality to wild beasts in the shows he exhibited for the people.

6. Diod. Sic. 36.10.2: But a 1,000 [of Sicilian rebels] were still left, with Satyrus at their head. Aquillius at first intended to subdue them by force of arms, but when later, after an exchange of envoys, they surrendered, he released them from immediate punishment and took them to Rome to do combat with wild beasts.

This spectacle was a way to dispose of all the prisoners; see VILLE 1981: 97.

7. Sen., *brev.* 13.6–7: see text 25 Table 1 and the 'mimic battle'.

8. Cic., *ad fam.* 7.1.3: What pleasure is there in getting a *Clytemnestra* with 600 mules or a *Trojan Horse* with 3,000 mixing bowls or a variegated display of cavalry and infantry equipment in some battle or other? See also text 26 Table 1.

9. Plut., *Caes.* 5: during his aedileship, he [Julius Caesar] furnished 320 pairs of gladiators, and by lavish provision besides for theatrical performances, processions, and public banquets, he washed away all memory of the ambitious efforts of his predecessors in the office.

10. Suet., *Iul.* 10.2: When aedile, Caesar gave a gladiatorial show besides, but with somewhat fewer pairs of combatants than he had purposed; for the huge band which he assembled from all quarters so terrified his opponents, that the number of gladiators which anyone was to be allowed to keep in the city was capped. (trans. Loeb modified).

11. Plin., *nat. hist.* 8.22: see text 37 Table 1.

12. Suet., *Iul.* 39.1–4: He gave entertainments of divers kinds: a combat of gladiators and also stage-plays in every ward all over the city, performed too by actors of all languages, as well as races in the circus, athletic contests, and a sham seafight (*naumachia*). […] In the gladiatorial contest in the Forum Furius Leptinus, a man of praetorian stock, and Quintus Calpenus, a former senator and pleader at the bar, fought to a finish. A Pyrrhic dance was performed by the sons of the princes of Asia and Bithynia. See also texts 13 and 14.

13. Suet., *Iul.* 39.3: For the races the circus was lengthened at either end and a broad canal was dug all about it; then young men of the highest rank drove four-horse and two-horse chariots and rode pairs of horses, vaulting from one to the other. The game called Troy was performed by two troops, of younger and of older boys. Combats with wild beasts were presented on 5 successive days, and last of all there was a battle between two opposing armies, in which 500 foot-soldiers, 20 elephants, and 30 horsemen engaged on each side. To make room for this, the goals were taken down and in their place two camps were pitched over against each other.

14. Suet., *Iul.* 39.4: In the naval battle, which took place on a lake excavated in the lesser Codeta, there was a contest of ships of 2, 3, and 4 banks of oars from the Tyrian and Egyptian fleets, manned by a huge number of fighters. Drawn by all these spectacles, a vast number of people flooded into Rome from every region, so that many of the visitors had to lodge in tents put up in the streets or along the roads. And the crowds were so great on a number of occasions that many were crushed to death, including two senators. (trans. EDWARDS slightly modified).

This is the earliest naumachy mounted in Rome; see COLEMAN 1993: 50.

15. Cass. Dio 43.23.4–6: As for the men, he [Caesar] not only pitted them one against another singly in the Forum, as was customary, but he also made them fight together in companies in the Circus, horsemen against horsemen, men on foot against others on foot, and sometimes both kinds together in equal numbers. There was even a fight between men seated on elephants, 40 in number […] Finally he produced a <u>naval battle, not on the sea nor on a lake, but on land</u>; for he hollowed out a certain tract on the Campus Martius and after flooding it introduced ships into it. In all the contests the <u>captives</u> and those <u>condemned to death</u> took part; yet some even of the <u>knights,</u> and, not to mention others, the son of one who had been praetor fought in single combat.

16. App., *b. c.* 2.102: He [Caesar] gave also various spectacles with horses and music, a combat of foot-soldiers, <u>1,000 on each side</u>, and a <u>cavalry fight of 200 on each side</u>. There was also another combat of horse and foot together. There was a combat of elephants, 20 against 20, and a naval engagement of <u>4,000 oarsmen</u>, where 1,000 fighting men contended on each side.

17. Cass. Dio 51.22.4: At the consecration of the shrine to Julius there were all kinds of contests, and the boys of the patricians performed the equestrian exercise called 'Troy' and men of the same rank contended with chargers, with pairs, and with four-horse teams; furthermore, one Quintus Vitellius, a senator, fought as a gladiator.

See also text 41 Table 1.

18. Cass. Dio 51.22.6–9: and moreover <u>Dacians and Suebi fought in crowds</u> with one another. The latter are Germans, the former Scythians of a sort […] The whole <u>spectacle lasted many days</u>, as one would expect, and there was no interruption, even though Caesar fell ill, but it was carried on in his absence under the direction of others. On one of the days of this celebration the senators gave banquets in the vestibules of their several homes; but what the occasion was for their doing this, I do not know, since it is not recorded.

19. *RGDA* 22: Three times in my own name I gave a show of gladiators, and five times in the name of my sons or grandsons; in these shows there fought about 10,000 men.

20. *RGDA* 23: I gave the people the spectacle of a naval battle (*navalis proeli spectaclum*) beyond the Tiber, at the place where now stands the grove of the Caesars, the ground having been removed for a length of 1,800 and a breadth of 1,200 feet. In this spectacle 30 beaked ships, triremes or biremes, and even more of smaller size, confronted against each other; in these fleets about <u>3,000 men fought</u>, discounting the rowers. See also text 2.27 and Fron., *de aq.* 11; on this show and on its location: Coleman 1993: 50–4 and Taylor 1997: 475–82.

21. Cass. Dio 55.10.6: see text 52 Table 1.

22. Cass. Dio 59.10.3–5: The same trait of cruelty led him [Caligula] once, when there was a <u>shortage of condemned criminals to be given to the wild beasts</u>, to order that <u>some of the mob</u> standing near the benches should be seized and thrown to them […] He held these contests at first in the *Saepta*, after <u>excavating</u> the whole site and filling it with water, to enable him to bring in a <u>single ship,</u> but later he <u>transferred</u> them to another place, where he had <u>demolished</u> a great many large buildings and <u>erected wooden stands</u>; for he despised the theatre of Taurus.

23. Cass. Dio 60.30.3: In the gladiatorial combats many persons took part, not only of the foreign freedmen but also the <u>British captives</u>. He [Claudius] used up ever <u>so many men</u> in this part of the spectacle and took pride in the fact.

24. Suet., *Claud.* 21.6: He [Claudius] gave representations in the Campus Martius of the <u>storming and sacking of a town</u> in <u>the manner of real warfare</u>, as well as of the surrender of the kings of the Britons, and presided clad in a general's cloak.

25. Suet., *Claud.* 21.6: Even when he was on the point of letting out the water from <u>Lake Fucinus he gave a sham sea-fight first</u>. But when the combatants cried out: 'Hail, emperor, they who are about to die salute thee', he replied, 'Or not', and after that all of them refused to fight, maintaining that they had been pardoned. Upon this he hesitated for some time about destroying them all with fire and sword, but at last leaping from his throne and running along the edge of the lake with his ridiculous tottering gait, he induced them to fight, partly by threats and partly by promises. At this performance <u>a Sicilian and a Rhodian fleet</u> engaged, each numbering <u>12 triremes</u>, and the signal was sounded on a horn by a silver Triton, which was <u>raised from the middle of the lake by a mechanical device</u>.

26. Tac., *ann.* 12.56.1–3: Nearly at this date, the tunnelling of the mountain between Lake Fucinus and the river Liris had been achieved; and, in order that the impressive character of the work might be viewed by a larger number of visitants, a naval battle was arranged upon the lake itself, on the model of an earlier spectacle given by Augustus – though with light vessels and a smaller force – in his artificial pool near the Tiber. Claudius equipped triremes, quadriremes, and 19,000 combatants: the lists he surrounded with rafts, so as to leave no unauthorized points of escape, but reserved space enough in the centre to display the vigour of the rowing, the arts of the helmsmen, the impetus of the galleys, and the usual incidents of an engagement. On the rafts were stationed companies and squadrons of the praetorian cohorts, covered by a breastwork from which to operate their catapults and ballistae: the rest of the lake was occupied by marines with decked vessels. The banks, the hills, and mountain heights were filled, like a theatre *(in modum theatri),* by an uncountable crowd from the nearest municipalities, and some from the City itself, in their desire view or out of duty towards the princeps. He himself in a distinctive military cape – and, not far away, Agrippina in a golden chlamys – presided. The battle, though between convicts, was fought in the spirit of brave men, and after a considerable amount of wounding they were exempted from slaying. (trans. Loeb modified in light of WOODMAN). See also Tac., *ann.* 12.57 (gladiators); on this spectacle and the whole project of drainage of the lake which took 11 years and 30,000 men: Suet., *Claud.* 20; see COLEMAN 1993: 56 and HAMMER 2010.

27. Cass. Dio 61.33.3–4: Claudius conceived the desire to exhibit a naval battle on a certain lake; so, after building a wooden wall around it and erecting stands, he assembled an enormous multitude. Claudius and Nero were arrayed in military garb, while Agrippina wore a beautiful chlamys woven with threads of gold, and the rest of the spectators whatever pleased their fancy. Those who were to take part in the sea-fight were condemned criminals, and each side had 50 ships, one party being styled 'Rhodians' and the other 'Sicilians' … they were ordered to fight just the same, they simply sailed through their opponents' lines, injuring each other as little as possible. This continued until they were forced to destroy one another.

28. Tac., *ann.* 15.44: Therefore, to scotch the rumour [i. e. Rome's fire set by Nero], Nero substituted as culprits, and punished with the utmost refinements of cruelty, a class of men, loathed for their vices, whom the crowd styled Christians. Christus, the founder of the name, had undergone the death penalty in the reign of Tiberius, by sentence of the procurator Pontius Pilatus […] First, then, the confessed members of the sect were arrested; next, on their disclosures, vast numbers were convicted, not so much on the count of arson as for hatred of the human race. And mockeries accompanied their end: covered with wild beasts' skins and torn to death by dogs or fixed to stakes (or crosses), they were set afire in the darkening evening as a form of night lighting. Nero had offered his own gardens for the spectacle. He also gave an exhibition in a circus (*circense ludicrum*), mixing with the crowd in the habit of a charioteer, or mounted on his car. Hence, in spite of a guilt which had earned the most exemplary punishment, there arose a sentiment of pity, due to the impression that they were being liquidated not for the public good but to satisfy a a single man's savagery. (trans. Loeb modified in light of SHAW). On this passage and the 'myth of the Neronian persecution', see SHAW 2015, part. 80–2 et p. 97: 'As far as the available evidence indicates, before the empire-wide assault launched on them by a decree of the emperor Decius in 250 C. E., general persecutions of Christians as defined religious group *never* happened at Rome, the *caput imperii'.*

29. Suet., *Nero* 12.1: see text 76 Table 1.

30. Suet., *Nero* 12.1–2: He also exhibited a naval battle in salt water (*marina aqua*) with sea monsters swimming about in it; besides pyrrhic dances – by some Greek youths, – handing each of them certificates of Roman citizenship at the close of his performance. The pyrrhic dances represented various scenes. In one a bull mounted Pasiphae, who was concealed in a wooden image of a heifer; at least many of the spectators thought so. Icarus at his very first attempt fell close by the imperial couch and bespattered the emperor with his blood; for Nero very seldom presided at the games, but used to view them while reclining on a couch, at first through small openings, and then with the entire balcony uncovered.

31. Jos., *b. j.* 7.118: Of the prisoners, the leaders, Simon and John, together with 700 of the rank and file, whom he had selected as remarkable for their stature and beauty, he ordered to be instantly conveyed to Italy, wishing to produce them at the triumph.

32. Jos., *b. j.* 7.139–48: But nothing in the procession excited so much astonishment as the structure of the moving stages (ἡ τῶν φερομένων πηγμάτων κατασκευή); indeed, their massiveness afforded ground for alarm and misgiving as to their stability, many of them being three or four stories high, while the magnificence of the fabric was a source at once of delight and amazement. For many were enveloped in tapestries interwoven with gold, and all had a framework of gold and wrought ivory. The war was shown by numerous representations, in separate sections, affording a very vivid picture of its episodes. Here was to be seen a prosperous country devastated, there whole battalions of the enemy slaughtered; here a party in flight, there others led into captivity; walls of surpassing compass demolished by engines, strong fortresses overpowered, cities with well-manned defences completely mastered and an army pouring within the ramparts, an area all deluged with blood, the hands of those incapable of resistance raised in supplication, temples set on fire, houses pulled down over their owners' heads, and, after general desolation and woe, rivers flowing, not over a cultivated land, nor supplying drink to man and beast, but across a country still on every side in flames For to such sufferings were the Jews destined when they plunged into the war; and the art and magnificent workmanship of these structures now portrayed the incidents to those who had not witnessed them, as though they were happening before their eyes. On each of the stages was stationed the general of one of the captured cities in the attitude in which he was taken. A number of ships also followed.

33. Cass. Dio 66.25.2: see text 80 Table 1.

34. Suet., *Dom.* 4: see texts 83–84 Table 1.

35. HA, *Gord.* 3.5–8: see text 90 Table 1

36. HA, *Aurel.* 33–34: see text 93 Table 1

See also the texts 81, 82, 86, 92, 94, 95 in Table 1.

Bibliography

ADAMS, J. N. 2016. *An Anthology of Informal Latin 200 BC-AD 900. Fifty Texts with Translations and Linguistic Commentary.* Cambridge.

ALDRETE, G. S. 2007. *Floods of the Tiber in Ancient Rome.* Baltimore.

ALDRETE, G. S. 2014. 'Material Evidence for Roman Spectacle and Sport', in P. CHRISTESEN and D. G. KYLE (eds.), *A Companion to Sport and Spectacle in Greek and Roman Antiquity.* Malden MA, 438–50.

AYMARD, J. 1951. *Essai sur les chasses romaines: des origines à la fin du siècle des Antonins.* Paris.

BAJARD, A. 2008. 'Un décor de navire dans le Grand Cirque sous Septime Sévère', in J. NELIS-CLÉMENT and J.-M. RODDAZ (eds.), *Le cirque romain et son image.* Bordeaux, 335–46.

BAKER, J. and D. BROTHWELL. 1980. *Animal Diseases in Archaeology.* London.

BALL, D. 1984. 'A Bear Hunt Mosaic', *The J. Paul Getty Museum Journal* 12, 123–34.

BALLET, P., P. CORDIER and N. DIEUDONNÉ-GLAD (eds.). 2003. *La ville et ses déchets dans le monde romain: rebuts et recyclages.* Montagnac.

BARBERA, M. 2001. 'Un anfiteatro di corte: il Castrense', in A. LA REGINA (ed.), *Sangue e arena.* Roma, 127–45.

BEARD, M., J. NORTH and S. PRICE. 1998. *Religions of Rome* (2 vols). Cambridge.

BELL, S. 2014. 'Roman Chariot Racing. Charioteers, Factions, Spectators', in P. CHRISTESEN and D. G. KYLE (eds.), *A Companion to Sport and Spectacle in Greek and Roman Antiquity*, Malden MA, 492–504

BERGMANN, B. 1999. 'Introduction: The Art of Ancient Spectacle', in E. BERGMANN and C. KONDOLEON (eds.), *The Art of Ancient Spectacle.* New Haven-London, 9–35.

BERLAN-BAJARD, A. 2006. *Les spectacles aquatiques romains.* Rome.

BERNS, C. and H. A. EKINCI. 2015. 'Gladiatorial Games in the Greek East: a complex of reliefs from Cibyra', *Anatolian Studies* 65, 143–79.

BERTRANDY, F. 1987. 'Remarques sur le commerce des bêtes sauvages entre l'Afrique du Nord et l'Italie', *MEFRA* 99.1, 211–41.

BESCHAOUCH, A. 2006. 'Que savons-vous des sodalités africo-romaines', *CRAI* 150.2, 1401–17.

BESCHAOUCH, A. 2011. '*Invide vide*. La compétition publique entre les sodalités africo-romaine et son écho dans l'espace domestique', in M. CORBIER and J.-P. GUILHEMBET (ed.), *L'écriture dans la maison romaine*. Paris, 315–28.

BEST, H.-J. 2001. 'I sotterranei del Colosseo: impianto, trasformazioni e funzionamento', in A. LA REGINA (ed.), *Sangue e arena*. Rome, 277–99.

BLÁZQUEZ, J. M. (1997): 'El grifo en mosaicos africanos y su significado', *Antiquités Africaines* 33, 155–63.

BODEL, J. 2000. 'Dealing with the Dead: Undertakers, Executioners, and Potter's Fields in Ancient Rome', in V. HOPE and E. MARSHALL (eds.), *Death and Disease in the Ancient City*. London, 128–51.

BOMGARDNER, D. 1992. 'The Trade in Wild Beasts for Roman Spectacles: A Green Perspective', *Anthropozoologica* 16, 161–66.

BOMGARDNER, D. 2009. 'The Magerius Mosaic Revisited', in T. WILMOTT (ed.), *Roman Amphitheatres and Spectacula. A 21st-Century Perspective. Papers from an International Conference held at Chester, 16th–18th February 2007*. Oxford, 165–77.

BOMGARDNER, D. 2013. *The Story of the Roman Amphitheater*. London-New York.

BRADLEY, R. and H. WILLIAMS (eds.). 1998. *The Past in the Past: the Reuse of Ancient Monuments*. London.

BRANDON, C. J., ET ALII (eds.). 2014. *Building for Eternity: The History and Technology of Roman Concrete Engineering in the Sea*. Oxford-Philadelphia.

BRASSOUS, L. 2015. 'Les édifices de spectacles d'Hispanie entre les IIe et IVe s.' in L. BRASSOUS and A. QUEVEDO (eds.), *Urbanisme civique en temps de crise. Les espaces publics de l'Hispanie et de l'Occident romain entre les IIe et IVe s.* Madrid, 272–88.

BRIAND-PONSART, C. 2007. 'Les 'lancers de cadeaux' (*missilia*) en Afrique du Nord Romaine', *Antiquités Africaines* 43, 79–97.

BUONFIGLIO, M. 2008. 'Appunti sui sistemi idraulici del Circo Massimo', in J. NELIS-CLÉMENT and J.-M. RODDAZ (eds.), *Le cirque romain et son image*. Bordeaux, 39–46.

BUTTREY, T. V. 2007. 'Domitian, the Rhinoceros, and the Date of Martial's Liber De Spectaculis', *JRS* 97, 101–12.

CALDELLI, M. L. 1993. *L'agon Capitolinus. Storia e protagonisti dall'istituzione domizianea al IV secolo*. Roma.

CALDELLI, M. L. 1997. 'Gli agoni alla greca nelle regioni occidentali dell'Impero. La Gallia Narbonensis', in *Atti della Accademia Nazionale dei Lincei, Memorie*, ser. IX, IX. 4, 389–481.

CALDELLI, M. L. 2012. 'Associazioni di artisti a Roma: una messa a punto', in K. COLEMAN and J. NELIS-CLÉMENT (eds.), *L'organisation des spectacles dans le monde romain*. Vandœuvres, 131–171 (tables p. 358–363).

CALDELLI, M. L. 2014. 'L'area dello stadio e dell'odeon di Domiziano in età imperiale: condizione della proprietà, funzioni ed uso', in J.-F. BERNARD (ed.), '*Piazza Navona, ou Place Navone, la plus belle & la plus grande*': du stade de Domitien à la place moderne, histoire d'une évolution urbaine. Rome, 39–50.

CAMERON, A. 1970. *Claudian. Poetry and Propaganda at the Court of Honorius*. Oxford.

CAMERON, A. 1976. *Circus Factions, Blues and Greens at Rome and Byzantium*. Oxford.

CARANDINI, A., A. RICCI and M. DE VOS. 1982. *Filosofiana, la villa di Piazza Armerina: immagine di un aristocratico romano al tempo di Costantino*. Palermo.

CARTER, M. 2003. 'Gladiatorial ranking and the *SC de pretiis gladiatorum minuendis* (*CIL* II 6278 = *ILS* 5163)', *Phoenix* 57, 83–114.

CARTER, M. 2004. '*Archiereis* and Asiarchs: A gladiatorial Perspective', *Greek, Roman and Byzantine Studies* 44, 41–68.

CARTER, M. 2014. 'Gladiatorial Ranking and the *SC de Pretiis Gladiatorum Minuendis* (*CIL* II 6278 = *ILS* 5163)', in T. F. SCANLON (ed.), *Sport in the Greek and Roman Worlds. Vol. 2: Greek Athletics Identities and Roman Sports and Spectacle*. Oxford, 229–68.

CARTER, M. 2015. 'Landscaping the Roman arena', *Studies in the History of Gardens & Designed Landscapes: An International Quarterly* 35.2, 115–23.

CARTER, M. J. and J. EDMONSON, 2014. 'Spectacles in Rome, Italy and the Provinces', in C. BRUUN and J. EDMONSON (eds.), *The Oxford Handbook of Roman Epigraphy*. Oxford-New York, 537–58 (cha. 25).

CHAMBERLAND, G. 2001. *The production of shows in the cities of the Roman Empire: A study of the Latin epigraphic evidence*, McMaster (McMasterUniversity Open Access Dissertations and Theses; DigitalCommons@McMaster).

CHAMBERLAND, G. 2012. 'La mémoire des spectacles: L'autoreprésentation des donateurs', in K. COLEMAN and J. NELIS-CLÉMENT (eds.), *L'organisation des spectacles dans le monde romain*. Vandœuvres, 261–303.

CHASTAGNOL, A. 1976. 'Les inscriptions constantiniennes du Cirque de Mérida', *MEFRA* 88.1, 259–76.

CHASTAGNOL, A. 1984. *Histoire Auguste. Les empereurs romains des IIe et IIIe siècles*. Paris.

CIANCIO ROSSETTO, P. 2008. 'La ricostruzione architettonica del Circo Massimo: dagli scavi alla maquette elettronica', in J. NELIS-CLÉMENT and J.-M. RODDAZ (eds.), *Le cirque romain et son image*. Bordeaux, 17–38.

CIANCIO ROSSETTO, P. and G. PISANI SARTORIO. 1994. *Teatri greci e romani alle origini del linguaggio rappresentato*. Roma.

CIANCIO ROSSETTO, P. and G. PISANI SARTORIO. 2002. *Memoria del teatro. Censimento dei teatri antichi greci e romani*. Roma.

COARELLI, F. 1997. *Il Campo Marzio. Dalle Origini alla fine della Repubblica*. Roma.

COARELLI, F. 2001a: 'Gli anfiteatri a Roma prima del Colosseo', in A. LA REGINA (ed.), *Sangue e arena*. Roma, 43–7.

COARELLI, F. 2001b: 'Ludus gladiatorius', in A. La Regina (ed.), *Sangue e arena*. Roma, 147–51.

COLEMAN, K. 1990. 'Fatal Charades: Roman Executions Staged as Mythological Enactments', *JRS* 80, 44–73.

COLEMAN, K. 1993. 'Launching into History: Aquatic displays in the Early Empire', *JRS* 83, 48–74.

COLEMAN, K. 1996. 'Ptolemy Philadelphus and the Roman Amphitheater', in W. J. SLATER (ed.), *Roman Theater and Society*. Ann Arbor, 49–68.

COLEMAN, K. 2000. 'Missio at Halicarnassus', *HSCP* 100, 487–5.

COLEMAN, K. 2003. 'Euergetism in its place: Where was the amphitheatre in Augustan Rome?', in K. LOMAS and T. CORNELL (eds.), *'Bread and circuses': Euergetism and municipal patronage in Roman Italy*. London, 61–88.

COLEMAN, K. 2005. 'Bonds of danger: communal life in the gladiatorial barracks of ancient Rome', in *The fifteenth Todd Memorial Lecture delivered in the University of Sydney 15 August 2002*. Sydney.

COLEMAN, K. 2006. *M. Valerii Martialis Liber spectaculorum*, ed. with introd., transl. and commentary. Oxford.

COLEMAN, K. 2008a. 'Exchanging gladiators for an aqueduct at Aphrodisias (*SEG* 50.1096)', *Acta Classica* 51, 31–46.

COLEMAN, K. 2008b. 'Not Waving but Drowning'. Total Immersion in Aquatic Displays', *JRA* 21, 458–64.

COLEMAN, K. 2010a. 'Spectacle', in A. BARCHIESI and W. SCHEIDEL (eds.), *Oxford Handbook of Roman Studies*. Oxford, 651–70.

COLEMAN, K. 2010b. 'Arena Spectacles', *Oxford Bibliographies Online Research Guide*. Oxford.

COLEMAN, K. and J. NELIS-CLÉMENT 2012. 'Introduction', in K. COLEMAN and J. NELIS-CLÉMENT (eds.), *L'organisation des spectacles dans le monde romain*. Vandœuvres, x–xxvii.

COOLEY, A. E. 2009. *Res Gestae Divi Augusti. Text, Translation and Commentary*. Cambridge.

CRAWFORD, M. H. 1996. *Roman Statutes* (with text, English translation and commentary. 2 vols.). London.

CRUMMY, P. 2008. 'The Roman Circus at Colchester, England' in J. NELIS-CLÉMENT and J.-M. RODDAZ (eds.), *Le cirque romain et son image*. Bordeaux, 213–31.

DAGRON, G. 2000. 'L'organisation et le déroulement des courses d'après le *Livre des Cérémonies*'. Paris.

DAGRON, G. 2011. *L'hippodrome de Constantinople. Jeux, peuple et politique*. Paris.

DAGUET-GAGEY, A. 2015. Splendor aedilitatum. *L'édilité à Rome (1er s. avant J.-C.–IIIe s. après J.-C.)*. Rome.

DAVIES, P. J. E. 2012. 'Pollution, propriety and urbanism in Republican Rome', in M. BRADLEY (ed.), *Rome, Pollution and Propriety. Dirt, Disease and Hygiene in the Eternal City from Antiquity to Modernity*. Cambridge-New York, 67–121.

DECKER, W. 2008. 'Wagenrennen im römischen Ägypt', in J. NELIS-CLÉMENT and J.-M. RODDAZ (eds.), *Le cirque romain et son image*. Bordeaux, 347–58.

DeLaine, J. 2000. 'Building the Eternal City: the construction industry in imperial Rome', in J. Coulston and H. Dodge (eds.), *Ancient Rome. The Archaeology of the Eternal City*. Oxford, 119–41.

Delbrück, R. 1929. *Die Consulardiptychen und verwandte Denkmäler*. Berlin.

Delile, H. 2014. 'Signatures des paléo-pollutions et des paléoenvironnements dans les archives sédimentaires des ports antiques de Rome et d'Éphèse', Thèse présentée et soutenue publiquement le 5 septembre 2014 pour l'obtention d'un Doctorat de géographie / géoarchéologie, Université Lumière Lyon 2, Lyon, published online 2 February 2015, https://tel.archives-ouvertes.fr/tel-01084909.

Deniaux, E. 2000. 'L'importation d'animaux d'Afrique à l'époque républicaine et les relations de clientèle', in M. Khanoussi, P. Ruggeri and C. Vismara (eds.), *L'Africa romana*. Atti del XIII convegno di studio, Djerba, 10–13 dicembre 1998. Roma, 1299–307.

Deschler-Erb, S. and K. Gostenčnik 2008. 'Différences et identités de la vie quotidienne dans les provinces romaines: l'exemple de la tabletterie', in I. Bertrand (ed.), *Le travail de l'os, du bois de cerf et de la corne à l'époque romaine: un artisanat en marge?* Montagnac, 283–309.

Deschler-Erb, S. and Ö. Akeret. 2010. 'Archäobiologische Forschungen zum römischen Legionslager von Vindonissa und seinem Umland: Status quo und Potenzial', *Jahresbericht Gesellschaft Pro Vindonissa*, 13–36.

Di Napoli, V. 2013. *Teatri della Grecia romana: forma, decorazione, funzioni. La provincia d'Acaia*. Athens.

Dodge, H. 2008. 'Circuses in the Roman East: a reappraisal', in J. Nelis-Clément and J.-M. Roddaz (eds.), *Le cirque romain et son image*. Bordeaux, 133–46.

Dodge, H. 2014a. 'Amphitheaters in the Roman Word', in P. Christesen and D. G. Kyle (eds.), *A Companion to Sport and Spectacle in Greek and Roman Antiquity*. Malden MA, 545–60.

Dodge, H 2014b. 'Venues for Spectacle and Sport (other than Amphitheaters) in the Roman Word', in P. Christesen and D. G. Kyle (eds.), *A Companion to Sport and Spectacle in Greek and Roman Antiquity*. Malden MA, 561–77.

Dumasy, F. 2008. 'Les édifices de spectacle dans le paysage urbain de la Gaule tardive', in E. Soler and F. Thelamon (eds.), *Les Jeux et les spectacles dans l'Empire romain tardif et dans les royaumes barbares*. Mont-Saint-Aignan, 69–88.

Dumasy, F. 2011. 'Théâtres et amphithéâtres dans les cités de Gaule romaine: fonctions et répartition', *Études de lettres* 1–2, 193–222 (accessible on-line: http://edl.revues.org/115).

Dunbabin, K. M. D. 1978. *Mosaics of Roman North Africa: Studies in Iconography and Patronage*. Oxford-New York.

Dunbabin, K. M. D. 2015. 'The agonistic mosaic in the Villa of Lucius Verus and the Capitolia in Rome', *JRA* 28, 193–222.

Duncan-Jones, R. 1974. *The Economy of the Roman Empire. Quantitative Studies*. Cambridge.

Dupré Raventós, X. and J.-A. Remolà (eds.). 2000. Sordes Urbis. *La eliminación de residuos en la ciudad Romana*. Rome.

Ebner, C. 2012. 'Die Konzeption der Arenastrafen im römischen Strafrecht', *Zeitschrift der Savigny-Stiftung für Rechtsgeschichte, Rom. Abt.* 129, 245–85.

Edmonson, J. 1996. 'Dynamic Arenas: Gladiatorial Presentations in the City of Rome and the Construction of Roman Society during the Early Empire', in W. J. Slater (ed.), *Roman Theater and Society*. Ann Arbor, 69–112.

Edmonson, J. 1999. 'The Cultural Politics in Rome and the Greek East, 167–166 BCE', in E. Bergmann and C. Kondoleon (eds.), *The Art of Ancient Spectacle*. New Haven-London, 77–95.

Edwards, C. 2015. 'Edward Gibbon and the City of Rome', in T. Fuhrer, F. Mundt, and J. Stenger (eds.), *Constructing and Modelling Images of the City*. Berlin-Boston, 207–26.

Epplet, C. 2001a. *Animal Spectacula of the Roman Empire*. PhD. University of British Columbia Vancouver (accessible on-line : http://www.library.ubc.ca/spcoll/thesauth.html).

Epplet, C. 2001b. 'The Capture of Animals by the Roman Military', *Greece & Rome* 2.48, 210–22.

Epplet, C. 2003. 'The Preparation of Animals for Roman Spectacula. Vivaria and their Administration', *Ludica* 9, 76–92.

EPPLET, C. 2014a. 'Roman Beast Hunts', in P. CHRISTESEN and D. G. KYLE (eds.), *A Companion to Sport and Spectacle in Greek and Roman Antiquity*. Malden MA, 505–19.

EPPLET, C. 2014b. 'Spectacular Executions in the Roman World', in P. CHRISTESEN and D. G. KYLE (eds.), *A Companion to Sport and Spectacle in Greek and Roman Antiquity*. Malden MA, 520–32.

FAGAN, G. G. 2011. *The Lure of the Arena: Social Psychology and the Crowd at the Roman Games*. Cambridge-New York.

FAURE, P. 2013. *L'aigle et le cep: les centurions légionnaires dans l'Empire des Sévères* (2 vols.). Bordeaux.

FAUVINET-RANSON, V. 2006. *Decor civitatis, decor Italiae. Monuments, travaux publics et spectacles au VIe siècle*. Bari.

FAUVINET-RANSON, V. 2008. 'Les spectacles traditionnels dans l'Italie ostrogothique', in E. SOLER and F. THELAMON (eds.), *Les jeux et les spectacles de l'Empire romain tardif et dans les royaumes barbares*. Mont-Saint-Aignan, 143–60.

FEDELI, P. 1990. *La natura violata. Ecologia e mondo romano*. Palermo.

FERRARY, J.-L. 2011. 'La géographie de l'hellénisme sous la domination romaine', *Phoenix* 65.1–2, 1–22.

FERRARY, J.-L. 2014². *Philhellénisme et impérialisme. Aspects idéologiques de la conquête romaine du monde hellénistique*. Rome.

FLEURY, P. 2008. 'Les *sparsiones* liquides dans les spectacles', *REL* 86, 97–112.

FORICHON, S. (forthcoming): *Les spectateurs du cirque à Rome du 1ᵉʳ siècle a. C au VIème siècle p. C.: passion, émotions et politique*, thèse inédite soutenue à l'Université de Bordeaux Montaigne (January 2015).

FREI-STOLBA, R. ET ALII (eds.) 2011. *Das Amphitheater Vindonissa Brugg-Windisch: Kanton Aargau*. Bern.

FRIER, B. W. 2000². 'Demography', in A. BOWMAN, P. GARNSEY, and D. RATHBONE (eds.), *The Cambridge Ancient History: The High Empire, AD 70–192*. Cambridge, 787–817.

GHADDHAB, R. 2008. 'Les édifices de spectacle en Afrique: prospérité et continuité de la cité classique pendant l'Antiquité tardive?', in J. NELIS-CLÉMENT and J.-M. RODDAZ (eds.), *Le cirque romain et son image*. Bordeaux, 109–32.

GIACCHERO, M. 1974. *Edictum Diocletiani et collegarum de pretiis rerum venalium: in integrum fere restitutum e Latinis Graecisque Fragmentis* (2 vols.). Genova.

GOIRAN, J.-P. ET ALII 2011. 'Géoarchéologie des ports de Claude et de Trajan, Portus, delta du Tibre', *MEFRA* 123.1, 157–236.

GOLVIN, J.-C. 1988. *L'Amphithéâtre romain: Essai sur la théorisation de sa forme et de ses fonctions* (2 vols.). Paris.

GOLVIN, J.-C. 2007. 'L'amphithéâtre de Pompéi, un monument de transition?', *Nikephoros* 20, 199–207.

GOLVIN, J.-C. 2008a. 'Réflexion relative aux questions soulevées par l'étude du *pulvinar* et de la *spina* du *Circus Maximus*', in J. NELIS-CLÉMENT and J.-M. RODDAZ (eds.), *Le cirque romain et son image*. Bordeaux, 79–87.

GOLVIN, J.-C. 2008b. 'La restitution architecturale de l'hippodrome de Constantinople. Méthodologie, résultats, état d'avancement de la réflexion', in J. NELIS-CLÉMENT and J.-M. RODDAZ (eds.), *Le cirque romain et son image*. Bordeaux, 147–58.

GOLVIN, J.-C. 2011. 'Comment expliquer la forme non elliptique de l'amphithéâtre de Leptis Magna (Al Khums/Lybie) ?', *Études de lettres* [En ligne], 1–2, published on line 15 May 2014, http://edl.revues. org/123.

GOLVIN, J.-C. 2015. 'Machines et principes de fonctionnement du sous-sol d'un grand amphithéâtre', in P. FLEURY ET ALII (eds.), *La technologie gréco-romaine. Transmission, restitution et médiation*. Caen, 23–41.

GOWERS, E. 1995. 'The Anatomy of Rome from Capitol to Cloaca', *JRS* 85, 23–32.

GRAHAM, S. 2013. 'Counting bricks and stacking wood', in P. ERDKAMP (ed.), *The Cambridge Companion to Ancient Rome*. Cambridge, 278–96.

GUASTI, L. 2007. 'Animali per Roma', in E. PAPI (ed.), *Supplying Rome and the Empire*. Portsmouth RI, 139–52.

GUEYE, M. 2013. *Captifs et Captivité dans le monde romain. Discours littéraire et icinographique (IIIᵉ siècle av. J.-C. – IIᵉ siècle ap- J.-C.)*. Paris.

HAMDOUNE, C. 2012. 'Soldats de l'armée d'Afrique en mission: à propos de CIL VIII, 21567, Agueneb, Djebel Amour', *Aouras, Société d'études et de recherches sur l'Aurès antique* 6, 181–205.

HAMMER, D. 2010. 'Roman Spectacle Entertainments and the Technology of Reality', *Arethusa* 43, 63–86.

HANOUNE, R. 2009. 'La chasse en Afrique. Le cas de Bulla Regia (Tunis)', in J. TRINQUIER and C. VEN-DRIES (eds.), *Chasses antiques: pratiques et représentations dans le monde gréco-romain (IIIe s. av. – IVe s. apr. J.-C.)*. Rennes. 65–72.

HARRIS, W. V. 2013. 'Defining and Detecting Mediterranean Deforestation, 800 BCE to 700 CE', in W. V. HARRIS (ed.), *The Ancient Mediterranean Environment between Science and History*. Leiden-Boston, 173–94.

HASELBERGER, L. ET ALII 2002 [repr. with corr. in 2008]: *Mapping Augustan Rome*. Portsmouth RI.

HAYES, G. and J. HORNE 2011. 'Sustainable Development, Shock and Awe? London 2012 and Civil Society', *Sociology* 45.5, 749–64.

HOPKINS, K. and M. BEARD. 2006 [2005]. *The Colosseum*. London.

HORDEN, P. and N. PURCELL. 2000. *The Corrupting Sea. A Study of Mediterranean History*. Oxford.

HUFSCHMID, T. 2009. *Amphitheatrum in provincia et Italia. Architektur und Nutzung Römischer Amphitheater von Augusta Raurica bis Puteoli* (Forschungen in Augst 43). Augst.

HUGHES, J. D. 2014. *Environmental Problems of the Greeks and Romans: Ecology in the Ancient Mediterranean* (second edition; first published 1994). Baltimore.

HUGONIOT, C. 2008. 'Les spectacles dans le royaume Vandale', in E. SOLER and F. THELAMON (eds.), *Les jeux et les spectacles de l'Empire romain tardif et dans les royaumes barbares*. Mont-Saint-Aignan, 161–204.

HUMPHREY, J. H. 1986. *Roman Circuses. Arenas for Chariot Racing*. Berkeley-Los Angeles.

INGLEBERT, H. (ed.) 2005. *Histoire de la civilisation romaine*. Paris.

ITGENSHORST, T. 2005. *Tota illa pompa. Der Triumph in der römischen Republik*. Göttingen.

IVERSEN, E. 1968–1972. *Obelisks in exile*. Volume one: *The obelisks of Rome*; Volume two: *The obelisks of Istanbul and England*. Copenhagen.

JENNISON, G. 2005² [1937]. *Animals for Show and Pleasure in Ancient Rome*. Manchester.

JIMÉNEZ SÁNCHEZ, J. A. 2008. 'Honorius, un souverain 'ludique'?', in E. SOLER and F. THELAMON (eds.), *Les jeux et les spectacles de l'Empire romain tardif et dans les royaumes barbares*. Mont-Saint-Aignan, 123–42.

JONES, C. 2012. 'The Organization of Spectacle in Late Antiquity', in K. COLEMAN and J. NELIS-CLÉMENT (eds.), *L'organisation des spectacles dans le monde romain*. Vandœuvres, 305–33.

KARAKASIS, E. 2016. *T. Calpurnius Siculus. A Pastoral Poet in Neronian Rome*. Berlin-Boston.

KENAWI, M., E. MACAULAY-LEWIS and J. S. MCKENZIE. 2012. 'A commercial nursery near Abu Hummus (Egypt) and re-use of amphoras for the trade in plants', *JRA* 25, 195–225.

KNOEPFLER, D. 1999. 'Pausanias à Rome en l'an 148 ?', *REG* 112, 485–509.

KOKKINIA, C. 2012. 'Games vs buildings as euergetic choices', in K. COLEMAN and J. NELIS-CLÉMENT (eds.), *L'organisation des spectacles dans le monde romain*. Vandœuvres, 97–130.

KONTOKOSTA, H. 2008. 'Gladiatorial reliefs and élite funerary monuments', in C. RATTÉ and R. R. R. SMITH (eds.), *Aphrodisias Papers 4. New Research on the City and its Monuments*. Portsmouth RI, 190–230.

KYLE, D. G. 1994. 'Animal Spectacles in Ancient Rome: Meat and Meaning', *Nikephoros* 7, 181–205.

KYLE, D. G. 1998. *Sport and Spectacle in the Ancient World*. Oxford.

KYLE, D. G. 2014. 'Animal Spectacles in Ancient Rome: Meat and Meaning', in T. F. SCANLON (ed.), *Sport in the Greek and Roman Worlds*. Vol. 2: *Greek Athletics Identities and Roman Sports and Spectacle*. Oxford, 269–95.

LANCASTER, L. C. 2007. 'The Colosseum for the general public', *JRA* 20, 454–9.

LANDES, C. 2008. 'Le Circus Maximus et ses produits dérivés', in J. NELIS-CLÉMENT and J.-M. RODDAZ (eds.), *Le cirque romain et son image*. Bordeaux, 413–30.

LAVIN, I. 1963. 'The Hunting Mosaics of Antioch and Their Sources. A Study of Compositional Principles in the Development of Early Mediaeval Style', *Dumbarton Oaks Papers* 17, 179–286.

LE ROUX, P. 1990. 'L'amphithéâtre et le soldat sous l'Empire romain', in C. DOMERGUE, C. LANDES and J.-M. PAILLER (eds.), *Spectacula I, Gladiateurs et amphithéâtres*. Lattes, 203–15 [= P. Le Roux, *La toge et les armes. Rome entre Méditerrannée et Océan. Scripta Varia 1*. Rennes 2011, 173–90].

LE ROUX, G., A. VÉRON and C. MORHANGE. 2005: 'Lead pollution in the ancient harbours of Marseilles', *Méditerranée* 104, 2005 (published on-line 28 January 2009).

LETZNER, W. 2009. *Der Römische Circus. Massunterhaltung im römischen Reich*. Wiesbaden.

LEVEAU P. 1993. 'Mentalité économique et grands travaux hydrauliques: le drainage du lac Fucin. Aux origines d'un modèle', *Annales. Économies, Sociétés, Civilisations* 48.1, 3–16.

LOMBARDI, L. 2001. 'The water system of the Colosseum', in A. GABUCCI (ed.), *The Colosseum*. [Translated by MARY BECKER]. Los Angeles, 228–40.

MACKINNON, M. 2006. 'Supplying exotic animals for the Roman amphitheatre games: New reconstructions combining archaeological, ancient textual, historical and ethnographic data', *Mouseion ser.* 3.6, 137–61.

MACKINNON, M. 2014. 'Animals in the urban fabric of Ostia: a comparative zooarchaeological synthesis', *JRA* 27, 175–201.

MADELEINE, S. 2014. *Le théâtre de Pompée à Rome. Restitution de l'architecture et des systèmes mécaniques*. Caen.

MADELEINE, S. 2015. 'La restitution d'un velum sur le théâtre de Pompée', in P. FLEURY, C. JACQUEMARD and S. MADELEINE (eds.), *La technologie gréco-romaine. Transmission, restitution et médiation*. Caen, 43–68.

MANGANARO, G. 1959. 'Aspetti pagani dei Mosaici di Piazza Armerina', *Archeologia Classica* 11, 241–50.

MARCATTILI, F. 2009. *Circo Massimo. Architetture, funzioni, culti, ideologia*. Roma.

MARZANO, A. 2014. 'Roman Gardens and Elite Self-representation', in K. COLEMAN (ed.), *Le jardin dans l'Antiquité*. Vandœuvres, 195–244.

MATTER, M. 2012. 'Des chevaux du cirque: économie et passions à Rome', in S. LAZARIS (ed.), *Le cheval, animal de guerre et de loisir dans l'Antiquité et Moyen-Âge*, Actes des journées d'étude internationales organisées par l'UMR 7044, Strasbourg, 6–7 novembre 2009 (BAT 22). Turnhout, 61–72.

MATTHEWS, M. 1988–1989. 'Some zoological observations an Ancient mosaics', *Bulletin de l'Association internationale pour l'étude de la mosaïque antique (AIEMA)* 12, 334–49.

MAURIN, L. 2008. 'Les édifices de cirque en Afrique: bilan archéologique', in J. NELIS-CLÉMENT and J.-M. RODDAZ (eds.), *Le cirque romain et son image*. Bordeaux, 91–108.

MCLAUGHLIN, R. 2010. *Rome and the distant East: trade routes to the ancient lands of Arabia, India and China*. London.

MEIGGS, R. 1982. *Trees and Timber in the Ancient Mediterranean World*. Oxford-New York.

MIGLIORATI, G. 2003. *Cassio Dione e l'impero romano da Nerva ad Antonino Pio. Alla luce dei nuovi documenti*. Milano.

MOGETTA, M. 2015: 'A new Date for Concrete in Rome', *JRS* 105, 1–40.

MOURITSEN, H. 2015. 'Local Elites in Italy and the Westerns provinces', in C. BRUUN and J. EDMONSON (eds.), *The Oxford Handbook of Roman Epigraphy*. Oxford-New York, 227–49.

NELIS-CLÉMENT J. 2002. 'Les métiers du cirque, de Rome à Byzance: entre texte et images', *Cahiers du Centre Glotz* 13, 265–309.

NELIS-CLÉMENT, J. 2008. 'Le cirque et son paysage sonore', in NELIS-CLÉMENT and J.-M. RODDAZ (eds.), *Le cirque romain et son image*. Bordeaux, 431–57.

NEUBAUER, W. ET ALII 2014. 'The discovery of the school of gladiators at *Carnuntum*, Austria', *Antiquity* 88, 173–90.

NEWBOLD, R. F. 1975. 'Cassius Dio and the Games', *L'Antiquité Classique* 44.2, 589–604.

NG, D. 2015. 'Commemoration and Élite Benefaction of Buildings and Spectacles in the Roman World', *JRS* 105, 101–23.

NOGALES BASSARATE, T. 2008. 'Circos romanos de *Hispania*. Novedades y perspectivas arqueológicas', in J. NELIS-CLÉMENT and J.-M. RODDAZ (eds.), *Le cirque romain et son image*. Bordeaux, 161–202.

ÖSTENBERG, I. 2014. 'Animals and Triumphs', in G. CAMPBELL (ed.), *The Oxford Handbook of Animals in Classical Life and Thought*. Oxford, 491–506.

OLIVER, J. H. and R. E. A. PALMER. 1955. 'Minutes of an Act of the Roman Senate', *The Journal of the American School of Classical Studies at Athens* 24.4, 320–49.

PALME, B. 2006: 'Zivile Aufgaben der Armee im kaiserzeitlichen Ägypten', in A. KOLB (ed.), *Herrschafts-strukturen und Herrschaftspraxis. Konzepte, Prinzipien und Strategie in römischen Kaiserreich*. Berlin, 299–328.

PANCIERA, S. 2000. 'Nettezza urbana a Roma organizzazione e responsabili', in J. A. REMOLÀ and X. DUPRÉ RAVENTÓS (eds.), *Sordes urbis: la eliminación de residuos en la ciudad romana*, Actas de la Reunión de Roma, 15–16 de noviembre de 1996. Roma, 95–105 [= PANCIERA 2006, 478–90].

PANCIERA, S. 2006. *Epigrafi, Epigrafia, epigrafisti. Scritti vari editi e inediti (1956–2005) con note complementari e indici*. Roma.

PAPI, E. (ed.) 2007. *Supplying Rome and the Empire: the Proceedings of an International Seminar Held at Siena-Certosa di Pontignano on May 2–4, 2004, on Rome, the Provinces, Production and Distribution*. Portsmouth RI.

PENTIRICCI, M. 2009. 'Il settore occidentale del Campo Marzio tra l'età antica e l'alto medioevo', in C. L. FROMMEL and M. PENTIRICCI (eds.), *L'antica basilica di S. Lorenzo in Damaso. Indagini archeologiche nel Palazzo della Cancelleria (1988–1993)*. Roma, 1, 15–72.

PISANI SARTORIO, G. 2008. 'Le cirque de Maxence et les cirques de l'Italie antique', in J. NELIS-CLÉMENT and J.-M. RODDAZ (eds.), *Le cirque romain et son image*. Bordeaux, 47–78.

POTTER, D. S. 2001. 'Death as spectacle, and subsequent disposal', *JRA* 14, 478–84.

POTTER, D. S. 2006. 'Spectacle', in D. S. POTTER (ed.), *A Companion to the Roman Empire*. Oxford, 385–408.

PRICE, S. 1984. *Rituals and Power: The Imperial Cult in Asia Minor*. Cambridge.

PURCELL, N. 2013. 'Romans, play on! City of Games', in P. ERDKAMP (ed.), *The Cambridge Companion to Ancient Rome*. Cambridge, 441–58.

REA, R. 2001a. 'Il Colosseo, teatro per gli spettacoli di caccia. Le fonti e i reperti', in A. LA REGINA (ed.), *Sangue e arena*. Roma, 223–43.

REA, R. 2001b. 'Gli animali per la *venatio*: catture, trasporto, custodia', in A. LA REGINA (ed.), *Sangue e arena*. Roma, 245–75.

REA, R., H.-J. BESTE and L. C. LANCASTER 2002. 'Il cantiere del Colosseo', *MDAI/R* 109, 341–75.

REMIJSEN, S. 2014. 'Appendix: Games, Competitors, and Performers in Roman Egypt', in *The Oxyrhynchus Papyri*, 79. London, 190–206.

RICE, E. E. 1983. *The Grand Procession of Ptolemy Philadelphus: Kallixeinos of Rhodes*. Oxford.

ROBERT, L. 1940. *Les gladiateurs dans l'Orient grec*. Paris.

RODDAZ, J.-M. 1984. *Marcus Agrippa*. Rome.

RODDAZ, J.-M. and J.-C. GOLVIN 2014. *Hérode le roi architecte*. Arles-Paris.

ROSEN, B. and E. GALILI 2007. 'Lead Use on Roman Ships and its Environmental Effects', *The International Journal of Nautical Archaeology* 36.2, 300–307 (published on-line 10 April 2007).

ROSENBLUM, M. 1961. Luxorius. *A Latin Poet among the Vandals*. New York.

ROSSITER, J. 1992. '*Stabula equorum*: Evidence for Race-horse Stables in Roman Africa', in: *Spectacles, vie portuaire, religions*. Paris, 41–8.

ROSSITER, J. 2016. '*In ampiζeatru Carthaginis*: the Carthage amphitheatre and its uses', *JRA* 26, 239–58.

ROUECHÉ, C. 1993. *Performers and Partisans at Aphrodisias in the Roman and Late Roman Periods*. London.

ROYO, M. 2008. 'De la *Domus Gelotiana* aux *Horti Spei Veteris*: retour sur la question de l'association entre cirque et palais à Rome', in J. NELIS-CLÉMENT and J.-M. RODDAZ (eds.). *Le cirque romain et son image*. Bordeaux, 481–95.

RUSSELL, B. J. 2014. *The Economics of the Roman Stone Trade*. Oxford

SALOMON, F. ET ALII 2012. 'The Canale di Comunicazione Traverso in Portus: the Roman sea harbour under river influence (Tiber delta, Italy)', *Géomorphologie: relief, processus, environnement* 1, 75–90.

SALZMAN, M. R. 1990. *On Roman time: the Codex-Calendar of 354 and the rhythms of urban life in Late Antiquity*. Berkeley-Los Angeles.

SCANLON, T. F. (ed.) 2014. *Sport in the Greek and Roman Worlds. Vol. 2: Greek Athletics Identities and Roman Sports and Spectacle*. Oxford.

SCHEIDEL, W. 2003. 'Germs for Rome', in C. EDWARDS and G. WOOLF (eds.), *Rome the Cosmopolis*. Cambridge, 158–76.

SCHEIDEL, W. 2013. 'Disease and Death', in P. ERDKAMP (ed.), *The Cambridge Companion to Ancient Rome*. Cambridge, 45–59.

SCHMID, E. 1952/53. 'Der Kamelknochen von Vindonissa', *Jahresbericht Gesellschaft pro Vindonissa*, 23–4.

SCHNEGG-KÖHLER, B. 2002. 'Die augusteischen Säkularspiele', *Archiv für Religionsgeschichte* 4, Leipzig.

SCOBIE, A. 1986. 'Slums, sanitation and mortality in the Roman World', *Klio* 88, 399–433.

SCOBIE, A. 1988. 'Spectator Security and Comfort at Gladiatorial Games', *Nikephoros* 1, 192–243.

SEAR, F. 2006. *Roman Theatres. An Architectural Study*. Oxford.

SETTIS FRUGONI, C. 1975. 'Il grifone e la tigre nella "Grande Caccia" di Piazza Armerina', *Cahiers archéologiques: fin de l'Antiquité et Moyen Âge* 24, 21–32.

SHAW, B. D. 2015. 'The Myth of the Neronian Persecution', *JRS* 105, 73–100.

SHELTON, J.-A. 2014. 'Spectacles of Animal Abuse', in G. CAMPBELL (ed.), *The Oxford Handbook of Animals in Classical Life and Thought*. Oxford, 461–90.

SINTÈS, C. 2008. 'Le cirque d'Arles: l'apport des fouilles depuis 1986', in J. NELIS-CLÉMENT and J.-M. RODDAZ (eds.), *Le cirque romain et son image*. Bordeaux, 203–12.

SOLER, E. 2008. '"*Ludi*" et "*Munera*", le vocabulaire des spectacles dans le Code Théodosien', in E. SOLER and F. THELAMON (eds.), *Les jeux et les spectacles de l'Empire romain tardif et dans les royaumes barbares*. Mont-Saint-Aignan, 37–65.

SORGE, G. 2006. 'Kamele jenseits der Alpens', *Bayerische Vorgeschichtsblätter* 71, 249–59.

SPANNE, A. 2014a. 'Greening the World Cup', *The Daily Climate: A publication of Environmental Heath Science*, (online) http://www.dailyclimate.org/tdc-newsroom/2014/06/green-world-cup-brazil.

SPANNE, A. 2014b. 'Brazil World Cup Fails to Score Environmental Goals', *Scientific American* (online): http://www.scientificamerican.com/article/brazil-world-cup-fails-to-score-environmental-goals.

TAYLOR, R. 1997. 'Torrent or Trickle? The Aqua Alsietina, the Naumachia Augusti, and the Transiberim', *AJA* 101, 465–92.

TAYLOR, R. 2014. 'Movement, vision, and quotation in the gardens of Herod the Great', in K. COLEMAN (ed.), *Le jardin dans l'Antiquité*. Vandœuvres, 145–85.

TEYSSIER, E. 2009. *La Mort en face. Le dossier Gladiateurs*. Arles.

THUILLIER, J.-P. (2012). 'L'organisation des *ludi circenses*', in K. COLEMAN and J. NELIS-CLÉMENT (eds.), *L'organisation des spectacles dans le monde romain*. Vandœuvres, 173–220.

THUILLIER, J.-P. (2015). 'Circensia 2. De toutes les couleurs', *REA* 117.1, 109–29.

THOMMEN, L. 2012. *An Environmental History of Ancient Greece and Rome*, Cambridge.

TONER, J. 2014. *The Day Commodus Killed a Rhino. Understanding the Roman Games*. Baltimore.

TOSI, G. 2003. *Gli edifici per Spettacoli nell'Italia romana*. (2 vol.). Roma.

TOYNBEE, J. M. C. 1973. *Animals in Roman life and art*. Ithaca NY.

TUCK, S. L. 2013. 'The Tiber and River Transport', in P. ERDKAMP (ed.), *The Cambridge Companion to Ancient Rome*. Cambridge, 229–45.

VEASH, N. 2000. ' Brazil's football Mecca corroded by fans' urine', (online) http://www.telegraph.co.uk/news/worldnews/southamerica/brazil/1369447/Brazils-football-Mecca-corroded-by-fans-urine.html.

VELKOV, V. and ALEXANDROV, C. 1988. '*Venatio caesariana*. Eine neue Inschrift aus Montana (Moesia Inferior)', *Chiron* 18, 270–7.

VERGNIEUX, R. 2008. 'Origine de l'usage de la Réalité Virtuelle à l'Institut Ausonius et les premiers travaux sur le Circus Maximus', in J. NELIS-CLÉMENT and J.-M. RODDAZ (eds.), *Le cirque romain et son image*. Bordeaux, 235–42.

VILLE, G. 1981. *La gladiature en Occident des origines à la mort de Domitien*. Rome.

VISMARA, C. 2001. 'La giornata di spettacoli', in A. LA REGINA (ed.), *Sangue e arena*. Roma, 199–221.

VOISIN, P. 2014. *ECOLΩ: écologie et environnement en Grèce et à Rome*, précédé d'un entretien entre Brice Lalonde et Patrick Voisin. Paris.

WEISS, Z. 2014. *Public Spectacles in Roman and Late Antique Palestine*, Cambridge MA.

WELCH, K. E. 2007. *The Roman Amphitheatre. From its origins to the Colosseum*. Cambridge.

WELCH, K. 2014. 'The Roman Arena in Late-Republican Italy: a Re-evaluation', in T. F. SCANLON (ed.), *Sport in the Greek and Roman Worlds*. Vol. 2: *Greek Athletics Identities and Roman Sports and Spectacle*. Oxford, 198–228.

WIEDEMANN, T. 1995. *Emperors and Gladiators*. London-New York.

WITTS, P. 1984. 'Interpreting the Brading "Abraxas" Mosaic', *Britannia* 25, 111–7.

WILSON, A. I. 2013. 'The Mediterranean Environment in Ancient History: Perspectives and Prospects', in W. V. HARRIS (ed.), *The Ancient Mediterranean Environment between Science and History*. Leiden, 259–325.

WILSON, R. J. A. 2004. 'On the Identification of the Figure in the South Apse of the Great Hunt in the Corridor at Piazza Armerina', *Sicilia Antiqua* 1, 153–69.

WILSON, R. J. A. 2014. 'Tile-stamps of Philippianus in Late Roman Sicily: a talking *signum* or evidence for horse-raising?', *JRA* 27, 472–86.

WOODMAN, A. J. 2004. *The Annals* / Tacitus; translated, with introduction and notes. Indianapolis.

WISEMAN, T. P. 2015. *The Roman Audience. Classical Literature as Social History*. Oxford.

ZALESKI, J. 2014. 'Religion and Roman Spectacle', in P. CHRISTESEN, and D. G. KYLE (eds.), *A Companion to Sport and Spectacle in Greek and Roman Antiquity*. Malden MA, 590–602.

ZANKER, P. 2001. 'L'empereur construit pour le peuple', in N. BELAYCHE (ed.), *Rome, les Césars et le Ville aux deux premiers siècles de notre ère*. Rennes, 119–56.

Indices

Ancient sources

Aelianus, *de natura animalium*: 2.11, **225**, **260**

Alcmaeon of Croton, F 24 b 4 DK, **53**

Altbabylonische Briefe: 1.141, **28**; 2.4, 2.5, 2.55, 2.70, 4.19, **30**, **31**; 4.34, **30**; 4.39, 4.80, 4.85 13.5, **31**

Ambrosius, *exameron*: 6.4, **15**

Ammianus Marcellinus, *historiae*: 14.11.25–6, **16**; 22.15.24, **17**, **223**, **257**; 16.10.14–5, **232**

Appianus, *bellum civile*: 2.13, **223**, **257**; 2.102, **224**, **228**, **258**, **270**; *Syriaca* 51, **99**

Aretaeus, *de causis et signis acutorum morborum*: 1.9.1–5, **164**

Aristophanes, *Acharnanes*:1165, **161**; *aves*: 245–9, **159**; *ranae*: 315, **161**; *nubes*: 1, 157–8, **159**; *ploutos*: 537–8, **159**; *vespae*: 143, 147, **181**; 256–7, **158**; 1038, **161**

Aristotle, *Athenaion politeia*: 43.1, 8.844–5, **56**; 50.2, **57**; *politica*: 6.5.4 (1321b), 7.11.4 (1331b), **212**; 7.1330b, 4–14, **54**, **158**; 1330b,9–18, **70**; *problemata*: 19 and 27.2, **158**

Athanasius of Alexandria, *vita Sancti Antoni*: **139**

Athenaeus, *deipnosophistes*: 2.42a, **69**; 2.41f, 2.46d, **123**

Augustinus, *de civitate Dei*: 8.5, **135**; *de doctrina christiana*: 1.24, 2.24, **138**; 2.25, 2.39.58, **139**

Aurelius Victor, *epitome de Caesaribus*: 13.12, **106**; 28.1, **235**, **266**

Bede the Venerable, *historia ecclesiastica Anglorum*: 2.16, 4.2, **141**

Callimachus, *aetia* (F 458 Pfeiffer): **55**

Calpurnius Siculus, *eclogae*: 7.23–73, **226**, **263**

Cassiodorus, *chronica*: 604, **101**

Cassius Dio, *historia romana*: 39.38.1–3, **17**, **224**, **258**; 39.61.1–3, 53.33.7, **99**; 43.22.1–4, 43.23.1–2, **224**, **258**; 43.23.4–6, **228**, **270**; 45.17.7–8, **235**; 48.33.4, **224**, **259**; 49.43, **230**; 50.8.3, **98**; 51.22.4, **228**, **270**; 51.22.5–9, **224**, **228**, **259**, **270**; 53.1.5, **236**; 53.20.1–2, **98**; 53.27.6, **224**, **259**; 54.1, **100**; 54.2.4, **252**; 54.8.5, **224**, **259**; 54.9, **260**; 54.25.2–3, **101**; 54.26.1–2, **225**, **260**; 54.34.1–2, **225**, **260**; 55.8, **101**; 55.10.6–7 and 8, **225**, **228**, **260**, **270**; 55.22.3, **101**; 56.27.4, **102**, **225**, **260**; 56.27.5, **225**, **261**; 57.14.7, **102**; 57.14.8, **101**, **104**, **105**; 58.26.5, **105**, **239**; 59.7.1, **225**, **261**; 59.10.3–5, **228**, **270**; 59.10.6, **246**; 59.11.3, 59.13.8 and 9, 59.20.1–2, **225**, **261**; 59.24.7–8, **261**; 60.5, **241**; 60.7.3, **225**, **261**; 60.13.4, 60.23.4–6, 60.27.2, **226**, **262**; 60.30.3, **229**, **270**; 61.33.3–4, **229**, **271**; 65.1, **120**; 66.25.1–4, **226**, **263**; 66.25.2, **229**, **272**; 68.15.1, **226**, **264**; 72.10.2–3, 72.19.1, **246**, **265**; 76.1.3–5, **227**, **265**; 77.6.2, **246**; 77.9.7–10, **246**; 77.10.4, **246**; 78.19.3–4, **241**, **246**

Cato, *de agricultura*: 1.1–3, **133**

Celsus, *de medicina*: 1.2, **83**; 2.8.30, **167**; 3.3.2, **160**; 5.27.12a, **116**

Cicero, *epistulae ad Atticum*: 13.33.a, **106**; *epistulae ad familiares*: 7.1.3, **17**, **224**, **228**, **258**, **269**; *epistulae ad Quintum fratrem*: 3.7.1, **99**; *de legibus*: 2.5.8, **84**; *de lege agaria*: 1.3, **214**; *de natura deorum*: 2.60, **207**; *topica*: 7.32, **85**; *tusculanae disputationes*: 5.97, **68**, **120**; *in Verrem*: 2.5.27, **66**

Claudianus, *de raptu Proserpinae*: 3.265, **15**; Stilicho: 3.262–355, Theodorus: 291–332, **227**, **267**

Code of Ḫammurapi: 53–6, **32**; 235, **33**; 236–8, 240, **34**

Code of Lipit-Ištar: 8, 9, **33**

Code of Ur-Namma: 30, **31**

Codex Theodosianus: 5.11.1, 6.4.19, 15.7.6, **252**

Columella, *de re rustica*: 1.3.7, **212**

Digesta Iustiniani: 8.5.8.5, **93**, **118**; 8.5.8.5–6, **120**; 8.5.8.5–7, **46**; 21.1.49, **93**; 32.91.5, **92**; 39.3.1*pr.*, **38**; 39.3.1.1–5, **118**; 39.3.3., **45**, **88**; 39.3.30, **118**; 40.7.21, **38**; 41.1.14*pr*, **88**; 43.8.2.8, **118**; 43.8.2.26–8, **118**; 43.8.5, **38**; 43.12.1.17, **88**; 43.12.1.17–22, **118**; 43.20.1.27, **86**; 43.20.3*pr.* **92**; 43.23.1.2, **41**, **119**; 43.23.1.3, **119**; 43.23.2, **41**, **42**; 43.23.7, **41**; 43.24.11, **43**, **86**; 43.24.16.1, **93**; 48.8.1–3, **115**; 48.8.3.3, **116**

Diodorus Siculus, *bibliotheca historica*: 20.36, **41**; 36.10.2, **228**, **269**

Dionysius of Halicarnassus, *antiquitates romanae*: 3.67.5, 4.44.1, **40**; 3.68.1–4, **234**;

Places and Geographical Names

General names

(this list also includes general names of the
Italian papers in English translation)

GEOGRAPHICA HISTORICA

Begründet von Ernst Kirsten, herausgegeben von Eckart Olshausen und Vera Sauer.
Die Bände 1–8 sind in den Verlagen Dr. Rudolf Habelt (Bonn) und Adolf M. Hakkert
(Amsterdam) erschienen.

Franz Steiner Verlag   ISSN 1381–0472

9. Gerhard H. Waldherr
Erdbeben – Das außergewöhnliche Normale
Zur Rezeption seismischer Aktivitäten
in literarischen Quellen vom 4. Jahrhundert v. Chr. bis zum 4. Jahrhundert n. Chr.
1997. 271 S., kt.
ISBN 978-3-515-07070-6

10. Eckart Olshausen / Holger Sonnabend (Hg.)
Naturkatastrophen in der antiken Welt
Stuttgarter Kolloquium zur historischen Geographie des Altertums 6, 1996
1998. 485 S. mit zahlr. Abb., kt.
ISBN 978-3-515-07252-6

11. Bert Freyberger
Südgallien im 1. Jahrhundert v. Chr.
Phasen, Konsequenzen und Grenzen römischer Eroberung (125–27/22 v. Chr.)
1999. 320 S. mit 16 Abb., kt.
ISBN 978-3-515-07330-1

12. Johannes Engels
Augusteische Oikumenegeographie und Universalhistorie im Werk Strabons von Amaseia
1999. 464 S., kt.
ISBN 978-3-515-07459-9

13. Lâtife Summerer
Hellenistische Terrakotten aus Amisos
Ein Beitrag zur Kunstgeschichte des Pontosgebietes
1999. 232 S. und 64 Taf., kt.
ISBN 978-3-515-07409-4

14. Stefan Faller
Taprobane im Wandel der Zeit
Das Śrî-Lankâ-Bild in griechischen und lateinischen Quellen zwischen Alexanderzug und Spätantike
2000. 243 S., kt.
ISBN 978-3-515-07471-1

15. Otar Lordkipanidze
Phasis
The River and City in Colchis
2000. 147 S. und 8 Taf., kt.
ISBN 978-3-515-07271-7

16. Marcus Nenninger
Die Römer und der Wald
2001. 268 S. mit 3 Abb., kt.
ISBN 978-3-515-07398-1

17. Eckart Olshausen /
Holger Sonnabend (Hg.)
Zu Wasser und zu Land – Verkehrswege in der antiken Welt
Stuttgarter Kolloquium zur historischen Geographie des Altertums 7, 1999
2002. 492 S. mit zahlr. Abb., kt.
ISBN 978-3-515-08053-8

18. Maria Francesio
L'idea di città in Libanio
2004. 157 S., kt.
ISBN 978-3-515-08646-2

19. Frauke Lätsch
Insularität und Gesellschaft
Untersuchungen zur Auswirkung der Insellage auf die Gesellschaftsentwicklung
2005. 298 S., kt.
ISBN 978-3-515-08431-4

20. Jochen Werner Mayer
Imus ad villam
Studien zur Villeggiatur im stadtrömischen Suburbium in der späten Republik und frühen Kaiserzeit
2005. 266 S., kt.
ISBN 978-3-515-08787-2

21. Eckart Olshausen /
Holger Sonnabend (Hg.)
„Troianer sind wir gewesen" – Migrationen in der antiken Welt
Stuttgarter Kolloquium zur Historischen Geographie des Altertums 8, 2002
2006. 431 S. mit 58 Abb., kt.
ISBN 978-3-515-08750-6

22. Jochen Haas
Die Umweltkrise des 3. Jahrhundert n. Chr. im Nordwesten des Imperium Romanum
Interdisziplinäre Studien zu einem Aspekt der allgemeinen Reichskrise im Bereich der beiden Germaniae sowie der Belgica und der Raetia